GROUNDWATER GOVERNANCE IN THE INDO-GA
AND YELLOW RIVER BASINS
REALITIES AND CHALLENGES

SELECTED PAPERS ON HYDROGEOLOGY

15

Series Editor: Dr Nick S. Robins
 Editor-in-Chief IAH Book Series
 British Geological Survey
 Wallingford, UK

INTERNATIONAL ASSOCIATION OF HYDROGEOLOGISTS

Groundwater Governance in the Indo-Gangetic and Yellow River Basins

Realities and Challenges

Editors

Aditi Mukherji
International Water Management Institute, Colombo, Sri Lanka

Karen G. Villholth
Geological Survey of Denmark and Greenland (GEUS), Copenhagen, Denmark

Bharat R. Sharma
International Water Management Institute, New Delhi, India

Jinxia Wang
Center for Chinese Agricultural Policy, Chinese Academy of Sciences Beijing, China

CRC Press
Taylor & Francis Group
Boca Raton   London   New York   Leiden

CRC Press is an imprint of the
Taylor & Francis Group, an **informa** business

A BALKEMA BOOK

First issued in paperback 2017

CRC Press/Balkema is an imprint of the Taylor & Francis Group, an informa business

© 2009 Taylor & Francis Group, London, UK

Typeset by Vikatan Publishing Solutions (P) Ltd., Chennai, India

Published by: CRC Press/Balkema
              P.O. Box 447, 2300 AK Leiden, The Netherlands
              e-mail: Pub.NL@taylorandfrancis.com
              www.crcpress.com – www.taylorandfrancis.co.uk – www.balkema.nl

Library of Congress Cataloging-in-Publication Data
Groundwater governance in the Indo-Gangetic and Yellow river basins : realities and challenges / Aditi Mukherji, Karen G. Villholth, Bharat R. Sharma, editors.
        p. cm. -- (Selected papers on hydrogeology, v.15)
    Includes bibliographical references and index.
    ISBN 978-0-415-46580-9 (hardback) -- ISBN 978-0-203-87447-9 (e-book)
    1. Hydrogeology -- Indus River Region. 2. Hydrogeology -- Ganges River Region (India and Bangladesh) 3. Hydrogeology -- China -- Yellow River Region. 4. Groundwater -- Indus River Region -- Management. 5. Groundwater -- Ganges River Region (India and Bangladesh) -- Management. 6. Groundwater -- China -- Yellow River Region -- Management. 7. Crops -- Water requirements -- Indus River Region. 8. Crops -- Water requirements -- Ganges River Region (India and Bangladesh) 9. Crops -- Water requirements -- China -- Yellow River Region. I. Mukherji, Aditi. II. Villholth, Karen G. III. Sharma, Bharat Raj, 1943- IV. Title. V. Series.

GB1140.I53G76 2009
333.91'3095--dc22

                                                                            2009004229

ISBN 13: 978-1-138-11392-3 (pbk)
ISBN 13: 978-0-415-46580-9 (hbk)

Table of contents

Thematic issues on groundwater irrigation

Preface

Groundwater irrigation has emerged as the mainstay of agriculture in the Indo-Gangetic basin of south Asia and the Yellow River Basin in the north China plains. Yet, there is little systematic research and capacity building on sustainable management and governance of groundwater in these regions. Under the Challenge Program for Water and Food, the International Water Management Institute and its partners implemented a highly innovative project that integrated action research with capacity building. It did this by engaging groundwater planners and managers from basin countries in a programme of assisted learning in the class room as well as in the field. Teachers as well as students learned together. Students, who came from NGOs, media, groundwater agencies and academia in India, Pakistan, Nepal, Bangladesh, and China brought distinctive competences and specialized experience to the learning process. An important component was for the students to go back to the field in their respective countries and work on applied research projects guided by experts. This has resulted in a vast corpus of empirical material about how the groundwater socio-ecologies are actually governed in the Indo-Gangetic and Yellow River Basins. This book brings together a selection of these research outputs for a wider global audience. It adds substance to the growing global debate on groundwater use and governance.

If the world's water crisis is "mainly a crisis of governance", groundwater represents the grimmest side of this crisis in Asia. Groundwater governance is also an issue in developed countries like the US and Australia; however, the complexity involved is many times greater in the Indo Gangetic Basin and Yellow River Basin.

This book aims to present a collage of groundwater use and governance experiences from Indo-Gangetic and Yellow River basins. While I cannot claim that this book represents the last word on groundwater governance, I can confidently say that it is a big first step towards understanding the complex socio-ecological issues involved in crafting a groundwater governance regime that might work in the context of south Asia and China. I am confident that the material contained in this book will help to shape the thinking and ideas of several generations of groundwater planners and managers around the world.

Tushaar Shah
Principle Scientist
International Water Management Institute
Anand, India

Foreword

Globally, groundwater resources are facing increasing threats associated with unsustainable extraction and degradation due to contamination. Declining recharge and increasing extraction of groundwater resources is common in many places. In South Asia, groundwater withdrawals, mostly for irrigated farming, have increased from less than 20 cubic kilometers in 1950s to more than 250 cubic kilometers per year at present. In certain parts of India and China, the groundwater resources are being overdrawn and the groundwater table has fallen to levels where extraction is no longer economically feasible. In these instances, the small-scale farmer without access to wells, expensive pumps, secure land and water rights are affected negatively. In some areas, groundwater is available in plenty due to favourable climatic and hydrological conditions, but cannot be taken advantage of by the small farmers due to insufficient or inequitable access to energy and other socio-economic barriers.

Governance constraints are the main contributor to degradation and depletion threats. These constraints relate to: (a) deficiencies in groundwater development, use, and contamination control legislation; (b) uncertainties in land and water rights and lack of institutional enforcement capacities; (c) significant point and non-point pollution, (d) inefficient use of water resources and an emphasis on supply augmentation rather than demand management; (e) inefficient monitoring of groundwater quantity and quality, and (f) a general lack of public and water users' awareness of water quality and overexploitation concerns.

The challenges in managing the significant economies that have evolved around groundwater irrigation are enormous and hence the efforts required to address these challenges and attain sustainable groundwater use will have to be commensurate. The Challenge Program on Water and Food recognized (www.waterandfood.org) this challenge and supported research and capacity building to address this global concern. This publication is one of the contributions that this project has made to complement the global, national and local initiatives aimed at attaining sustainable groundwater use, enhancing groundwater based livelihoods, and supporting the governance of this precious resource.

The book collates studies carried out under the project "Groundwater Governance in Asia: Capacity Building through Action Research in Indo-Gangetic (IGB) and Yellow River (YRB) Basins". The focus of the book is on the challenges that small farmers face when groundwater resources deplete or are out of reach for other reasons, limiting its otherwise great potential for alleviating poverty and securing food. The studies show that:

- Individual farmers have switched from rain-fed farming or surface water irrigation because of the easily accessible, ubiquitous, secure, and reliable source of water provided by groundwater.
- Groundwater problems and their consequences for agriculture and small farmers are heterogeneous across space and not necessarily linked to the resource endowment.
- Energy and access to energy is closely linked to groundwater utilization. Farmers with secure access to energy also benefit more from the groundwater economy.
- In the face of constraints to present day groundwater use, farmers adapt in various ways, such as through livelihood diversification, changing cropping patters, informal

groundwater markets, migration. Water saving, though not widely adopted, is emerging as a means to economise on the resource.

- While market forces and economic incentives can change groundwater use, public initiatives for agricultural groundwater regulation to balance short term economic efficiency with long term resource sustainability are urgently needed.
- Groundwater policies are in some cases not in line with groundwater realities. For example in India, groundwater policies have very little to do with scarcity, depletion and the quality of groundwater and more to do with agrarian politics, manifested among other things, through political power wielded by the farmer's lobby. In China, groundwater management is only implemented in the urban context.

Some generic recommendations include the following:

- Groundwater problems cannot be isolated from broader poverty, food, energy and environmental concerns. Understanding the drivers of groundwater use and the factors governing imbalances in use across geographic as well as social domains is a key to devising policies that may directly or indirectly control the sub-optimal use of groundwater.
- There is scope for significant water and energy optimization for groundwater irrigation through better targeted subsidies (for energy, water saving technology) and charging schemes. With food prices going up globally due to climate change and other factors, farmers will have more of a margin for accepting or implementing such schemes.
- Optimizing present day utilization of groundwater, socially as well as environmentally, also requires looking at options for and support to alternative, less water intensive but lucrative, livelihoods for the farmers, through education and better infrastructure to develop other markets and for transport/migration.

We hope you will enjoy reading this book and that it will inspire new thoughts, criticism and plans to uplift the given conditions of this particular region.

Francis Gichuki
Theme Leader
Integrated Basin Water Management System
Challenge Program on Water and Food

Acknowledgements

This book is an outcome of an inter-disciplinary research and capacity building programme called 'Groundwater Governance in Asia—Capacity Building through Action Research in the Indo-Gangetic and Yellow River basins' (http://www.waterforfood.org/gga/). It was funded by the CGIAR Challenge Programme on Water and Food (CPWF) (http://www.waterandfood.org) and implemented by the International Water Management Institute (IWMI) along with several partner institutions, such as the Indian Institute of Technology, Roorkee (IITR), Indian Institute of Remote Sensing, Dehradun (IIRS), Punjab Agricultural University (PAU), Center for Chinese Agricultural Policy (CCAP), Chinese Academy of Sciences (CAS), China University of Geosciences (CUG), University of South Australia, Adelaide, University of Kansas, the Bangladesh Water Development Board (BWBD), Department of Irrigation (DOI), Nepal and the International Waterlogging and Soil Salinity Research Institute, Pakistan. The editors acknowledge the generous funding given by the CPWF and the collaboration of all partner institutions in implementing this project. Thanks are due to Dr. Tushaar Shah, who designed this project and provided intellectual inputs all throughout. The editors would like to thank Mark Giordano, Niranjan Pant, Adbul Hakeem Khan, Amir Nazeer, MAS Sattar Mandal and Menggui Jin who were involved in supervising the participants of the program and also reviewed various chapters of the book. Thanks are also due to all the participants of the training program who devoted great efforts in collecting primary data and then in analysing and writing their papers which form several chapters of this book. Mala Ranawake of IWMI provided able secretarial assistance to the editors. The editors would like to thank their own respective organisations, namely, International Water Management Institute, Geological Survey of Denmark and Greenland and Center for Chinese Agricultural Policy for providing support.

Editors
Aditi Mukherji
Karen G. Villholth
Bharat R. Sharma
Jinxia Wang

About the editors

Aditi Mukherji is a Researcher (Social Scientist) at the *International Water Management Institute* (IWMI), Colombo. Aditi received her PhD degree from the *University of Cambridge*, United Kingdom in 2007. Her doctoral study at Cambridge was funded by the prestigious *Gates Cambridge Trust*. She has more than 8 years of experience and her area of expertise is institutions and policies of groundwater management in South Asia. She has written around 30 research papers on socio-economic and institutional aspects of groundwater irrigation in South Asia, of which around 15 have been published in refereed international journals. She has participated in a number of national and international conferences as an invited and keynote speaker. In 2006, she co-edited a special issue of the *Hydrogeology Journal* focussing on social and economic aspects of groundwater governance. In 2008, she was awarded the Global Development Network Award for best paper under the category of Natural Resources Management. She has reviewed papers for Hydrogeology Journal, World Development, Economic and Political Weekly and Journal of Environmental Management. Her current research focuses on the impact of electricity reforms in India on the operation of groundwater markets. She has also worked on groundwater issues in Central Asia and on transboundary issues in the Nile Basin in Africa.

Karen Villholth is Senior Researcher at the Geological Survey of Denmark and Greenland (GEUS). She has more than fifteen years experience in soils and groundwater research and water resources management. She has a strong academic background combining chemical engineering, environmental studies and hydrological, hydrodynamic and numerical disciplines. Since 1999, Karen Villholth has been involved in broader issues of water resources management. She was responsible for introducing the first course on "Integrated Water Resources Management" at the Technical University of Denmark. She has been assigned to several international projects concerning water resources management, with long term experience from Denmark (her home country), Sri Lanka, Bolivia and shorter term experience from Thailand, USA, Bangladesh, India, Pakistan, Nepal, Vietnam, Uganda, Burkina Faso, Guatemala, Mexico, and Nicaragua. In addition, Karen has ample experience in teaching and training, with several years of experience of teaching from the Technical University of Denmark and for professionals within environmental and water related courses. Karen joined IWMI (International Water Management Institute), Sri Lanka in April 2004 as a Senior Researcher within groundwater modelling and management. In this position, she has developed work related to the impacts of the tsunami on groundwater in coastal areas of Sri Lanka. In addition, she was the project leader of the large research and capacity building project 'Groundwater Governance in Asia', on which this book is based. Lately, Karen is working for GEUS where she is involved in climate change research related to water resources in Denmark as well as abroad and assumes the role of Senior Research Advisor to the Danish Government. She is the author of more than 25 peer-reviewed journal papers, and is the co-author of the books *'The Agricultural Groundwater Revolution: Opportunities and Threats to Development',* and *'Groundwater Research and Management: Integrating Science into Management Decisions'.*

Bharat Sharma is agricultural water management specialist and has over 30 years research experience in the developing countries. Presently he is Senior Researcher and Head of the New Delhi office of the International Water Management Institute (IWMI). He is also the Project leader of CPWF-IWMI funded research and capacity building project "Groundwater Governance in Asia" and this volume is based on the research and other activities implemented under the Project. He has completed research on the impact of large groundwater inter-basin water transfers, optimisation of groundwater use in large basins in India and safe and productive use of poor quality groundwater. Additionally he is involved in the Strategic Analysis of National River Linking Project of India, assessment and improvement of water productivity and alleviation of poverty in the Indo-Gangetic basin. He has more than 200 scientific publications and a number of books/proceedings to his credit.

Jinxia Wang is the senior researcher of Center for Chinese Agricultural Policy (CCAP), Chinese Academy of Sciences (CAS), and associate professor of Institute of Geographical Sciences and Natural Resources Research. She is also visiting professor of Water Resources Department in the Institute of Water Resources and Hydropower Research (IWHR) of Ministry of Water Resources, member of Scientific Steering Committee of Global Environmental Change and Human Security in the International Human Dimensions Programme on Global Environmental Change (IHDP), board member of the Resources and Environment Statistic Association and member of Water Resources Committee in the China Water Resources Association. She received her BS and MS degree in agricultural economics from Agricultural University of Inner Mongolian separately in 1993 and 1996. In 2000, she obtained her PhD degree in agricultural economics from Chinese Academy of Agricultural Sciences. From 1996 to 1997, she also attended the Leader 21 Winrock International PhD Training Programme on Agricultural Economics. Her research covers management, institution and policy of water resources, climate change and conservation agriculture. She has published more than 70 papers with more than 30 papers in refereed international journals. In addition, she is co-author of three books.

Introduction

Besides being the bread basket for over two billion people, the Indo-Gangetic Basin (IGB), covering parts of India, Pakistan, Bangladesh, and Nepal, and the Yellow River Basin (YRB) in China both have vibrant groundwater irrigation economies. Since the 1970s, growth in tubewell irrigation has created an agrarian boom with massive productivity and benefits to livelihood. In eastern IGB, with proper targeting, groundwater still offers significant opportunities for poverty reduction. Elsewhere, however, the boom will burst due to resource limitations, environmental concerns, and increased cost-benefit ratios, though right now there is evidence to show that groundwater use for agriculture on the whole is still increasing. If the current trends continue, by 2025, India will have nearly 100 million ha of irrigated areas but of these, over 82 million ha will be irrigated by groundwater wells, with the surface water irrigated area falling in relative terms. Pakistan's Punjab and parts of North China plains (NCP) have been experiencing similar exponential growth in groundwater irrigated area. Despite the growing significance of groundwater to agricultural growth, food security and rural livelihoods globally, understanding of the subject has remained limited. Location-specific hydrogeological assessments and numerical groundwater models dominate, while the larger and integrated issues of socio-economic impacts, political economy, groundwater institutions, property rights, and approaches to resource governance and management and specifically integrating science into management decisions have attracted surprisingly little scientific interest.

Therefore, significant challenges are involved in the sustainable, equitable and efficient utilization of groundwater resources in the two major river basins of Asia, the Indus-Ganges River basin (IGB) in South Asia and the Yellow River basin (YRB) in China. Poor understanding of the resource and potential management options and associated weak governance leads to over-exploitation of groundwater in some areas, while under-utilization co-exists in other parts. Both aspects entail a sub-optimal benefit from the resource, which is recognized as a very potent water resource and basis for poverty alleviation in these regions.

Groundwater is used in all sectors in the basins: domestic, industry and agriculture. However, the major battle and challenges to improve overall performance in groundwater utilization in this part of the world lies in the agricultural sector and irrigated farming. This is where the bulk of groundwater is utilized today, and where most livelihoods are at stake. Hence, the scope for optimizing the use and further raising the living standards is greatly linked to addressing agricultural use of groundwater. More than one billion (10^9) people in the basin states of India, China, Pakistan, Bangladesh and Nepal today depend on groundwater for irrigation. This is approximately 40% of their total populations, ranging from 20–25% in China to 60–65% in Pakistan.

Proper and efficient groundwater management in these regions is an immense task and cannot be achieved in the short term. There are no blue print solutions to transfer from the developed part of the world where groundwater is also under stress, and most national authorities are struggling with keeping up with new challenges and responding to previous and new impacts of excessive use and degradation of the resource. In the Asian context, the problem is complex. The fact that a great part of the population is traditionally reliant on income from farming, the constant population pressure and poverty implications means that hard decisions are even harder to make and there needs to be alternative approaches to managing the resource. Registering and regulating the groundwater use of these farmers and potentially compensating them for refraining from using groundwater, as in countries such as USA and Australia, may not be a viable approach and alternatives need to be sought. The approach of yesterday, i.e. doing nothing, may also prove critical and could risk the life-saving basis of millions of poor families in rural Asia. To compound the problem, there is not much research on groundwater and that which is carried out often excludes the people and institutions which use and manage the resource. If the institutional perspective is brought into the analysis, it still tends to be borrowed from western models of direct and legal management, which are quite untenable in the Asian context.

It was against this backdrop of groundwater governance vacuum that a three-year research and capacity building programme was designed. The aim of the programme was to engage key public groundwater agencies in the Indo-Gangetic basin and Yellow River basin in a collaborative enterprise to increase the integrated understanding of contemporary groundwater use and develop a toolkit and network for proactive groundwater governance, and in the process catalyze incentives to encourage these agencies to assume a wider mandate than they have so far. The project 'Groundwater Governance in Asia—Capacity Building through Action Research in the Indo-Gangetic and Yellow River basins' (http://www.waterforfood.org/gga/) was funded by the Challenge Programme on Water and Food (http://www.waterandfood.org) and implemented by the International Water Management Institute along with several partner institutions, such as the Indian Institute of Technology, Roorkee (IITR), Indian Institute of Remote Sensing (IIRS), Punjab Agricultural University (PAU), Center for Chinese Agricultural Policy (CCAP), Chinese Academy of Sciences (CAS), China University of Geosciences (CUG), the Bangladesh Water Development Board (BWBD) and the Department of Irrigation (DOI), Nepal. The training part of the project, 'The International Training and Research Programme on Groundwater Governance in Asia—Theory and Practice', was designed to develop cross-disciplinary learning and understanding of the realities and challenges of groundwater use and management in the Asian context. It had two components, namely a five-week classroom training course followed by five weeks of fieldwork in various locations within the IGB and YRB.

This book is an outcome of this inter-disciplinary research and capacity building programme. It aims, at least partially, to address the research and knowledge gaps mentioned earlier, and provide a coherent, consolidated and cross-disciplinary source of information and contemporary thinking on agricultural groundwater use and management in poverty-prone areas of Asia. This volume also provides a vehicle for this thinking to be broadcast and disseminated to other workers in different regions of the world who are faced with similar issues. An overall goal of the book is to serve as a tool and input to improved, better informed, and integrated management of groundwater within the broader development aspirations of these countries. The contributors are drawn from a wide range of academic disciplines from both the physical and social sciences. Most of the authors are groundwater professionals working in groundwater or related departments in their respective countries and they bring in a wealth of location-specific experience and expertise. Practically all the chapters in this book rely on primary data collected by the authors in the course of innovative, cross-cutting, and inter-disciplinary fieldwork of the training programme, covering areas across the regions that significantly depend on groundwater for agricultural livelihoods.

The first set of chapters (Chapters 1–3) introduces and summarizes the groundwater governance challenges in the IGB and the YRB. The second part (Chapters 4–11), which is the main section of the book, presents an inter-disciplinary situation analysis of the physical conditions of the groundwater resources, the socio-economic impacts of groundwater use, the policy-institutional mechanisms, and potentials for its management based on data collected through primary fieldwork in eight representative areas within the IGB and the YRB. The intention of this section is to present a cross-regional, comparable and cross-disciplinary perspective on groundwater use in the two basins. The third part (Chapters 12–17) of the book is devoted to thematic and comparative issues of groundwater governance. Unlike the inter-disciplinary and geographic focus of the previous section, the authors in this section use specific disciplinary tools to unravel some specific aspects of groundwater use and governance.

Karen G. Villholth
Senior Researcher
Geological Survey of Denmark and Greenland
Copenhagen, Denmark
Member of IAH, International Association of Hydrogeologists

Aditi Mukherji
Researcher
International Water Management Institute
Colombo, Sri Lanka

Bharat R. Sharma
Senior Researcher
International Water Management Institute
New Delhi, India

Jinxia Wang
Senior Researcher
Center for Chinese Agricultural Policy
Beijing, China

Glossary of non-English words

Adharmi	A caste categorized as Schedule caste as per Constitution of India
Aman	Monsoon (June-July to November) rice paddy crop grown in West Bengal, India and Bangladesh
Aus	Pre-monsoon (May to August) rice paddy crop grown in West Bengal, India and Bangladesh
Balu	Sand
Boro	Summer (January to April-May) rice paddy crop grown in West Bengal, India and Bangladesh
Chapakal	Domestic hand tubewells
Chaur	Depressions where rain water accumulates
Chira	Traditionally water carrier caste
Choes	Torrents
Crore	10 million
Dhekul	Counterpoise lift
Ghar	A household that eats from a common kitchen
Gots	Kins of clans men of same caste and descent. *Gots* are divided into *khandans* and *khandans* into *tabbars* in rural Punjab
Gram Panchayat	Elected village council
Haq	Right
Jat	An agricultural caste inhabiting north western India and Pakistan Punjab
Kandi	Foothills of the Himalayas
Kankar	Gravel
Khandan	Kins men of maximum lineage
Khara	Saline
Kharif	Wet monsoon season from June-October in south Asia
Khet	An unit of land in Punjab, India equivalent to 0.404 ha
Kuccha	Unlined
Kuh halt	Wells with a Persian wheel
Lakhs	100,000
Mallah	Castes or communities whose livelihood is well drilling or boat ferrying
Mandals	An administrative unit in India
Mitti	Clay
Nadi	River
Noong	Green gram (pulses)
Pathar	Stones
Patwaris	Land record officials at village level
Pucca	Permanent structures made of brick and mortar
Rabi	Winter dry season (October-November to February-March) in south Asia
Sanjha khuh	A commonly owned well

Sarpanch	Elected head of the village
Semi-pucca	Semi-permanent structures partially made of brick and mortar
Surangams	Underground channels
Tabbar	Kins men of minimal lineage who can trace patrilineal descent from a common grandfather or great grandfather
Tarkhan	Caste whose traditional occupation is Carpentry
Terai	The narrow strip of land between the Himalayas in the north and the Gangetic flood plains in the south in Nepal and India
Wari	Water turns for irrigation

Introduction

CHAPTER 1

The role of groundwater in agriculture, livelihoods, and rural poverty alleviation in the Indo-Gangetic and Yellow River basins: A review

K.G. Villholth
Geological Survey of Denmark and Greenland (GEUS), Copenhagen, Denmark

A. Mukherji
International Water Management Institute (IWMI), Colombo, Sri Lanka

B.R. Sharma
International Water Management Institute (IWMI), New Delhi, India

J. Wang
Centre for Chinese Agricultural Policies (CCAP), Chinese Academy of Sciences, Beijing, China

ABSTRACT: Groundwater and rural livelihoods are intricately linked in many parts of south Asia and China where millions of farmers depend on this resource for irrigated farming. This chapter summarizes and synthesizes the results of field-based investigations related to groundwater conditions, development, use, and present constraints for small scale farmers in rural parts of the Indo-Gangetic (IGB) and Yellow River basins (YRB) and in the North China Plains (NCP). Evidence from primary surveys and supplementary studies shows that intensive use of groundwater takes place under various constraints such as threats of resource depletion and lack of energy supply and other necessary agriculture-facilitating measures. Such constraints afflict mostly the poorest farmers, and potentially jeopardize food security in a wider and longer-term perspective. Influencing and optimizing these groundwater-based economies has proven to be extremely difficult and this paper tries, through a comprehensive, integrated and multidisciplinary approach, to point to various means and focus areas for research and policy that may contribute to the maintenance and enhancement of accrued benefits from groundwater irrigation in these regions.

1 INTRODUCTION

Groundwater is gradually being recognized as a major source of water in irrigated agriculture, food production, food security, and support to livelihoods of significant number of people in south Asia and northern China. It is estimated that approximately one billion people in the countries of India, China, Pakistan, Bangladesh and Nepal rely on groundwater, often as their only source, for irrigated agriculture (Villholth & Sharma, 2006). These countries account for the bulk of the world's groundwater use for agriculture (Shah et al., 2003), probably about 60%. At the same time, official statistics underestimate the extent and importance of groundwater irrigation (FAO, 2008). Though, or maybe because, this scale of groundwater development is relatively recent, with significant increases in use since the 1970s, this recognition has lagged behind for several reasons.

Firstly, the green revolution was spurred in the 1960's primarily by large surface water irrigation schemes and support to farming, based on public investments, subsidies, and interventions. Hence, the focus and continued attention from governments, donors, development assistance organizations, and researchers remained with surface water development and how to deal with the problems related to management of these public surface water irrigation schemes. Groundwater development has been called the 'silent revolution' (Llamas & Martínez-Santos, 2005), because the exploitation of groundwater to a large extent evolved in the hands of the private sphere. Farmers saw the possibility of increasing their crop yields and incomes through the secure, reliable and self-managed supply of water that groundwater provides. This was further driven and facilitated by new developments and spreading of improved and affordable drilling and pumping technologies and the electrification of rural areas providing energy for labour-free lifting of the water. In many places, groundwater development has been assisted by governments in the form of subsidies and loans for the development of wells and for the energy used by the farmers. So, while public support may have played a facilitating role, the groundwater revolution would not have taken off if the farmers did not perceive this recourse as beneficial to them.

Another reason for the less conspicuous role of groundwater in irrigated farming is the fact that groundwater, derived from underground in individual units, i.e. wells, extracting water in relatively small incremental volumes does not call the same degree of attention as large-scale networks of surface irrigation canals. Yet, groundwater irrigation can turn previously unproductive or less productive rain-fed land into a large oasis of highly productive, in fact often more productive, land than when irrigated with surface water (Deb Roy & Shah, 2003), primarily because of the reliability of the resource and its access.

It is generally accepted that access to groundwater and groundwater irrigation is a major contributing factor to poverty alleviation in Asian countries, through the more wide-spread and equal use opportunities provided by groundwater (Deb Roy & Shah, 2003), but it is also increasingly realized that several factors may limit this poverty reduction potential now and into the future. One is the fact that groundwater resources in many places today are exploited to a level which for most practical purposes must be considered unsustainable, because of intensive pumping, degradation of the quality, or both. Such degradation relates both to the groundwater resource itself but also to the environment, the land and soils providing the media for farming (Alauddin & Quiggin, 2008; Qureshi et al., 2008; Fang et al., 2006; Ju et al., 2006; Liu et al., 2005; Agrawal et al., 1999). Another important factor that limits the poverty alleviating potential of groundwater in these regions is the poor targeting of support to groundwater development and the inherent low capabilities of various population groups to take advantage of the benefits that groundwater may offer. Since the groundwater development for irrigation is relatively recent and despite its scale of importance, our understanding of both the positive and negative implications remains limited.

The objective of this introductory chapter is to give an integrated description of the present workings of groundwater-based irrigated farming and associated livelihoods in IGB and YRB based on a summary and synthesis of comprehensive surveys and primary data from interviews with farmers, village level leaders, and public servants involved in groundwater use, irrigated farming and water management in representative groundwater dependent areas. Because groundwater development is more prevalent and significant in the downstream reaches of the Yellow River and in fact in the major plains area (the North China

Plains) northeast of the Yellow River basin, home to more than 200 million people, focus in China is on these areas. Together, these areas represent a large fraction of farmers in these regions relying on groundwater for their livelihood and subsistence. From a comparison and analysis of issues and challenges across the regions, some pointers as to how to ensure a sustained benefit from this resource while adapting the societies to other drivers of global change, like population growth, climate change, and urbanization, are given.

Various existing papers describe the historic and present development and significance of groundwater use in South Asia and China (Shah, 2007; Wang et al., 2007) as well as specific elements of importance to this, such as the link to poverty alleviation (Narayanamoorthy, 2007; Moench, 2003), the link to the energy sector (Shah et al., 2007; Scott & Shah, 2004), groundwater markets (Mukherji, 2008; Zhang et al., 2008), and environmental and health aspects of groundwater use in the region (van Geen et al., 2006; Datta & de Jong, 2002; Zheng & Routray, 2002). While building on the existing understanding, this chapter intends to paint an integrated and updated picture of the situation and derive some general and revised views on how to address possible imbalances in groundwater use and associated livelihood impacts in these countries that to a large extent depend on farming as the major productive sector, the major sector for employment of poor people, the major sector influencing the natural environment, and the major sector providing stable food to a large proportion of the world's population in these countries.

2   METHODOLOGY

The present synthesis is based on a set of field studies conducted simultaneously during 2006–2007 in eight study areas across the IGB and YRB/NCP (Fig. 1). The studies were developed as part of a major capacity building programme and benefited from additional parallel research and literature reviews. For further description of the programme, see Villholth (Chapter 2, this Volume). The individual studies, covering specific representative geographic areas within the two major basins are documented as separate chapters of this Volume (Chapters 4 to 11). In Pakistan, Bangladesh and Nepal, one site was chosen, while in China two sites and in India three sites were selected to represent the variability across the river basins (Fig. 1). Most authors of the chapters were participants of the capacity building programme. They had previously worked professionally with water issues in the respective regions and had a good understanding of the groundwater conditions in their area. Analogous approaches and analysis methodologies were applied across the regions to facilitate comparative analysis. Issues compared across the regions include physical groundwater conditions, drivers and significance of groundwater use in a historical perspective, agronomic and economic conditions related to groundwater use, present constraints for groundwater use and implications for the small and marginal farmers[1], their coping mechanisms, and the scope for adapting to or alleviating constraints to enhance groundwater-derived benefits to the poor. To broaden the analysis and obtain a more comprehensive understanding of the situations, the approach adopted was very inter-disciplinary

---

[1] 'Small' and 'marginal' farmers (though no strict definition exists because it may vary with local physical and socio-economic conditions) signify the poorest farmers, generally with small landholding size and a significant level of subsistence farming. Figures for landholding size distribution for the studied areas are given in Appendix 1.

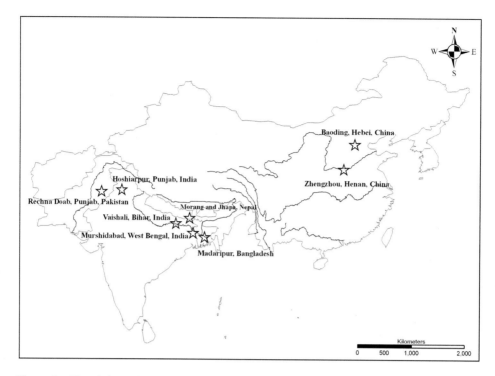

Figure 1.   The eight study sites across the Indo-Gangetic and Yellow River basins/North China Plains.

attempting to integrate issues of more technical disciplines related to the resource as well as more socio-economic and policy oriented issues.

The primary sources of data were derived from village level surveys among village leaders as well as surveys based on structured interviews and more informal talks with households and owners of tube-wells (common term used to describe groundwater-accessing wells). Finally, interviews with local level officials involved in water management and secondary data from public sources were used to inform the studies. On average, 73 farmers were interviewed in each of the eight case studies, covering a number of villages of between 2 and 22 for each study area (Appendix 1).

3   RESULTS

3.1   *Groundwater development and relation to physical conditions*

Groundwater development for agriculture has been possible in these regions because of favourable hydrogeological conditions, fertile soils, and level terrain (Jain et al., Chapter 3, this Volume). Groundwater resources are primarily available in alluvial sedimentary uncon-solidated deposits, providing generous aquifers, in plains area (NCP) and along tributaries or main rivers of YRB and IGB. Only the study site in Nepal was underlain by primarily hard rock formations, however, still productive enough for development at present day levels. Groundwater quality has mostly been of adequate quality to meet irrigation water standards, though limitations are encountered in Pakistan due to salinity in many areas. Arsenic is widely present in the lower reaches of the Ganges River basin aquifers, though it presents more of a hazard to drinking water supply than to irrigation and crop production (van Geen et al., 2006). Annual rainfall ranges from 375 mm in the Punjab Province of Pakistan to approximately 2000 mm in Bangladesh (Appendix 1).

Groundwater constitutes the most significant source of irrigation water in all the case study areas, with groundwater being practically the sole source in the study areas of Punjab (India), Bihar (India), West Bengal (India), Bangladesh, and Hebei (China). In Punjab (Pakistan), Nepal, and Henan (China), some surface water is also used (up to 30%) (Appendix 1).

There has been a huge increase in numbers of wells in these regions, starting back as early as the 1950s, but for most part accelerating from the 1970s (Fig. 2) though Nepal has had a slower and later development (Kansakar et al., Chapter 9, this Volume). High densities of wells in the irrigated areas are encountered with densities up to 40 per km^2 (Appendix 1). Well discharge rates are also favourable, ranging from 10 to 300 m^3/hr or 28 to 83 l/s. Many wells are used intensively, with most productive wells being pumped about one third of the time (Appendix 1).

There is a quite clear relationship between the areas under increasing stress from groundwater over-exploitation and rainfall. This is pronounced in the two sites in China where rainfall is relatively low (550–640 mm/yr) and groundwater levels during recent years have dropped by 1 to 2 m per year (Appendix 1). Similar declines are observed in Punjab (India) where rainfall is 768 mm/yr. Lowest overall rainfall is observed in Punjab, Pakistan, but here groundwater levels are dropping less (up to 0.5 m/yr) due to the near proximity to major rivers and canals in many irrigation areas. As a consequence, groundwater levels in the semi-arid areas are now down to 25 to 300 m below the ground surface while in the areas with higher rainfall or close to major surface water streams they are less than 15 m (Appendix 1). This indicates that groundwater utilization today is governed more by demand and inherent availability of groundwater from present storage and not by rates of replenishment and sustainability considerations.

3.2 *Groundwater development and relation to agricultural productivity and socio-economic parameters*

Except for Punjab (India) and the two sites in China, the productivity of these areas, in terms of agricultural output, are still relatively low compared to productivity potentials estimated by FAO. As an example, wheat yields on average for the five sites in Punjab (Pakistan), Bihar and West Bengal (India), Nepal, and Bangladesh, were 2,158 kg/ha, as compared to an average of 6,484 kg/ha for China and Punjab (India), and to 3,600 kg/ha reported as an attainable goal in 'developing countries' by 2030, and a goal of 7,600 kg/ha in the 'developed' countries, as reported by FAO (2002). A similar trend is observed for rice. Hence for most of these areas, and especially the poorest ones, productivity levels remain relatively low, despite the use of groundwater which normally is considered to favour efficiency gains due to its reliable nature when it comes to access and availability (Deb Roy and Shah, 2003). The low outputs, and contrasting outputs between the areas, may be explained by sub-optimal use and access to other agricultural inputs in the low-productivity areas, like high quality seeds, fertilizers, and difficult working conditions in general (Mandal, 1989). For example, field operations in many areas still use animal

Figure 2.   The old meets the new. Within a generation, groundwater use in Punjab, India has changed from shallow wells and simple extraction mechanisms to mechanic pumps and deep (>20 m) wells. In the foreground, a derelict and defunct water wheel is seen and in the background, a deep tube-well with a modern diesel-driven submersible pump.

or human labour. Access to markets and other physical and institutional infrastructure is another factor limiting farmers' realistic opportunity to make profits from cropping and hence their incentives to increase production (Kishore, 2004).

Also, despite the observed increases in cropping intensity in many of the areas with the advent and intensification of groundwater use (current values ranging from 146 to 250), crop diversification to higher value crops, normally observed as a natural development in groundwater irrigated areas, does not constitute a major fraction of the crop production. The agricultural production of these areas is still dominated by the cultivation of traditional stable crops, like rice, wheat, and maize.

It is clear from the above, that constraints are associated with the unlocking of productivity potentials in these areas. From Appendix 1, it is apparent that the continued population growth and associated small landholding sizes and land fragmentation and poor socio-economic conditions in general are significant and general contributing factors to limited productivity in the poorest areas. Only in China is the annual population growth rate stagnating, today at approximately 0.5% as compared to sustained high levels in all the other sites (2.2% on average).

3.3 *Drivers for groundwater development*

An overall driver of groundwater development has been the aspiration of the poor, and less poor, farmers to increase their income and secure their livelihoods, through better income and/or stable access to basic food supply. However, other factors have worked as drivers, pushers or enabling circumstances for groundwater development in the various areas. In Pakistan and China, lack of, or unreliable poor quality surface water, or poor-functioning surface water irrigation systems have prompted farmers to turn to groundwater. In some areas, where rain-fed agriculture was previously marginally sustainable, like in some areas of Punjab (India) and Bihar, decrease in rainfall over recent years has driven farmers to develop groundwater. Lack of proper water storage and surface water irrigation systems despite plenty of surface water, and in fact a risk of flooding, particularly in Bihar and Bangladesh, create a need for alternative water sources to cropping during the dry season, when flood risk is reduced. New cropping practices, like switching to paddy rice due to land reclamation (Punjab, India), and introduction of high yield variety seeds (HYV) which decreased the risk of production failure and increased the profitability of cropping (Bangladesh) also has meant that more farmers made the transition and necessary investments to use groundwater. The technological development of well drilling (all over the areas) and the diffusion of cheap (mostly Chinese) pumps (Bangladesh) make the entry into the market easier. Finally, direct donor or governmental support to well drilling and groundwater irrigation (China, Nepal, Bihar, and West Bengal), or indirect support, e.g. via subsidies to electricity (Punjab (India) and China) have also been clear incentives for farmers to use groundwater.

3.4 *Present workings of groundwater economies*

A multitude of mechanisms have been developed or emerged in these areas to enable farmers to benefit from groundwater. Though groundwater development for irrigation is very feasible in these areas hydrogeologically and environmentally, there is often a great economic barrier for the predominantly small and marginal farmers to enter into this irrigation technology and associated economy. Reported capital cost of installation of a single well ranges from US$15–100 for the relatively shallow wells (<75 m) in Pakistan, West Bengal, and Bangladesh to US$2,500–5,700 for the deep (60–600 m) wells in Punjab (India) and China (Appendix 1). This can be compared to farmers' income from agriculture, which may be in the range of US$500 per year for the poorest farmers across the whole region, signifying that especially in the deep aquifer areas, the cost of well installation may be prohibitive for the individual farmer. In these areas, the drilling costs are also increasing due to declining levels of the water table while in other places, like West Bengal, the costs have reportedly come down, due to the development of the techniques and competition between well drillers.

As mentioned, external financing, through favourable loaning schemes or subsidies to drilling and well installation have made the entry easier for many farmers. In Nepal, however, a concern was raised over the equitable access to such government-implemented schemes, and it was found that the smallest farmers did not benefit from these despite a lax requirement for entry in terms of ownership of land. Another, or complementary, means of optimizing on capital investments for well installation has been for farmers to pool resources in order to get shared access to a well. Except for the Pakistan and Bangladesh cases where all wells were reported to be private, individual wells, this approach has been implemented in all the other cases, to various degree with various level of success. Despite several benefits of joint ownership and management, like shared costs of maintenance, there is a general tendency for farmers to prefer to have their own well, for the simple convenience of self-governance. This is seen in the case of Nepal where group formation tended to dissolve as farmers became wealthy enough to exit on their own, or as monopoly or other internal imbalances developed due to unequal power relations. Similar situations were reported from Bihar and West Bengal. In China, the collectively owned (privately-owned) wells or community-wells (government-owned) seem to work satisfactorily because of the more even landholding size distribution and because access of one farmer to more shared wells helps alleviate the problem of land fragmentation. But even in these regions, there seems to be a drive towards individual well ownership due to water scarcity and a possibility for entering into lucrative water markets, despite high initial costs.

In Punjab, India, a unique arrangement of well-sharing was observed in a single village (Tiwary & Sabatier, Chapter 12, this Volume). Here, well-sharing was based on kinship, with clear rules on water sharing, sharing of costs of maintenance, crops to be grown, and inheriting of rights. Productivity (both in terms of *land productivity*, i.e. yield per area, as well as *water productivity*, i.e. crop yield per input of groundwater for irrigation) in these areas for fully irrigated crops were reported to be higher than for areas with only private wells and without these sharing mechanisms (Selvi et al., Chapter 5, this volume) indicating that such systems, when working, are quite efficient. However, solidarity between farmers beyond family members, or except as part of traditional collectivist systems, may not be easy to foster. Hence, the success of driving such a group-sharing processes by donors, public entities, or NGOs in villages where they are not already in place, is key for the success of groundwater-related projects that take this approach (Kansakar et al., Chapter 9; Rama Mohan, Chapter 13, this Volume).

Another very common means of entering into groundwater irrigation in these areas, particularly for the least wealthy farmers left without direct access to wells through individual or joint ownership, is to buy access to other farmer's wells. This system of water 'purchasing', and 'selling', has popularly been termed 'groundwater markets' though the term may be somewhat misleading. Basically, the farmers without access buy the service, or the access, to using a well of another farmer with a well. In principle, the water is not sold *per se*, as the well owners do not have strict ownership of the water or particular user rights except the right that their access convey to them. The markets are informal, spontaneous, local (as farmers sell to nearby farmers), private, and unregulated as there are no formal rules in place to govern the transactions. Nevertheless, multiple ways of interacting have developed into customary rules in different areas, responding to local conditions. For example, in China, the price is calculated in monetary terms (Zhang et al., 2008) while in India and Bangladesh, payment in terms of part of the harvested crop is widely practiced (also termed share-cropping) (Chapters 6, 7, and 8, this Volume). The system of share-cropping (like one fourth or one third of harvested crop paid to the well owner) may ease the burden of payment on part of the poor farmers as they pay only when the harvest is secure.

In some areas, like Bihar (India) and Punjab (Pakistan), water well installation did not constitute the major obstacle or economic barrier for farmers to enter into groundwater irrigation but rather the cost of or access to pumps. Here, a market related to pump rental has developed to service the poorest farmers without pumps. A thriving economy related to pump manufacturing and well drilling has emerged in all case sites, showing that the groundwater economy has generated new livelihoods for the peasant communities.

In general, the farmers buying access to groundwater are the smaller farmers compared to the sellers. Though the fraction of farmers participating in the markets (as either water seller or buyer) and the rate of number of water sellers to water buyers gives an indication of the significance of the market and the degree of dependence of smaller farmers on the larger farmers, respectively, a certain levelling out of these disparities were noticed from the fact that many well owners simultaneously acted as water buyers and purchased water from other well owners as their land was fragmented. The land fragmentation in combination with private wells in fact may be a major driver for the markets as it enables access to groundwater and irrigation for most farmers by either buying or selling water.

Even though no detailed comparative analysis of the financial aspects of groundwater markets and relation to poverty aspects were performed as part of this study, it was found that the cost of irrigation (involving primarily the cost for the energy to extract groundwater) relative to the total cost of crop production was not necessarily higher for the water buyers as compared to the water sellers. The share of irrigation cost to total cost of cultivation was found to vary from 4.4% for the most water saving crops to 58.6% for paddy rice (i.e. flooded rice) (Appendix 1). This lack of significant difference between buyers and sellers may indicate that the price charged to the water buyer for the energy consumption in connection with pumping for water extraction may not be significantly higher than the well/pump owner has to pay himself. This means the market is relatively competitive, partly because many well owners also act as buyers, which may tend to equalize the price across all farmers. However, the water sellers always had an additional income from their water selling, automatically favouring them economically relative to the water buyers.

In summary, the groundwater markets seem to be a win-win situation for most farmers as they provide a means to enter into groundwater irrigation for small farmers that may not otherwise be able and they provide an income for the larger farmers. In general, one can rank the farmers in terms of prosperity according to their access to water in the following order: well (or pump) owners > water buyers > purely rain-fed farmers. This is often related to their landholding size or quality of land. Below these groups, and particularly in South Asia, as the poorest, are the landless farmers which for a large part subsist as hired (maybe seasonal or migratory) workers for the farmers with land entitlements.

The groundwater markets were found to be prevalent in most of the study areas while only emerging in the Nepalese and Chinese sites. For the Nepalese case, this could be explained by the fact that groundwater development only now is reaching a level and intensity where a demand from non-well owners can be satisfied by existing well owners. In China, the explanation given for the almost complete lack of groundwater markets in the investigated areas was that wells were owned collectively (by the government) or by groups of farmers (shareholders) rather than by individuals who would normally be more inclined to make a business out of selling water services. Also because of a relatively equal access to wells, and land, there was no demand for water from a market. This is somewhat different from what has been observed in other parts of north China where groundwater markets are more developed and where well privatization of wells to individual farmers is more prevalent, a process partly driven by increasing water scarcity (Zhang et al., 2008).

Another interesting issue related to water markets is the question of who are more productive, in terms of their water use, the water buyers or the water sellers. There is no clear answer to that and variable results appear from the studies. In Pakistan and Bangladesh, the water buyers were found to obtain higher groundwater productivities (crop yield per input of groundwater for irrigation) than the water sellers. This was explained by the fact that the water buyers were more scrupulous with their water use, optimizing it as their parcels were smaller and their financial assets generally less. In Pakistan, the water buyers also relied more on conjunctive use of cheaper surface water. In Punjab (India), the water buyers likewise obtained higher water productivity in the complementarily irrigated wheat crop, presumably again because of more rational use of groundwater. In contrast, in Bihar and West Bengal (India), the water sellers were found to get better yields and outcomes for their input of water. Here, the poorer performance of the water buyers was attributed to the fact that the water buyers are 'second in line' in terms of getting access to water and a consequent failure in timing of irrigation could imply lower productivities. Other factors, such as the well owners' higher attention to soil fertility may also be important. In summary, the secured and timely access to groundwater along with access to other production inputs is essential for the poor farmers to reach profitable limits of their production.

3.5   *Present constraints for groundwater use and agricultural productivity*

Besides the initial constraint for poor farmers to enter into groundwater irrigation which is related to the access to or ownership/sharing of a well, a pre-condition for taking advantage of groundwater is the access to energy for its extraction. Practically everywhere, a source of mechanical energy is now used, and only in the very shallow aquifer regions can groundwater be acquired through manual power, e.g. the treadle pumps in Bihar (India), used only by the very poorest farmers because the time and effort needed to extract sufficient water for irrigation are demanding. Generally, electricity is the preferred option, because it is cheaper compared to diesel, which is the other major source of energy. This is because the electricity supply for irrigation is widely (except in Pakistan (Shah et al. (2006)) subsidized. It is also cleaner and less labour intensive. However, diesel is still widely used for a number of reasons, but mainly because of the lack of electrification in the rural areas or because of its irregular and unreliable supply. As such, diesel and diesel-driven pumps are also used as a 'back-up' for the electric mechanisms. So, diesel is used commonly by small-scale farmers in eastern India, Pakistan and Bangladesh due to insufficient availability of electricity or due to a high initial cost or other obstacles related to obtaining new power connections. Even in western India, here represented by Punjab, where groundwater irrigation has thrived on free power to the farmers, complaints about poor quality of electricity supply were expressed.

In summary, across the south Asian region, farmers are under a general energy squeeze, either due to inefficient supply or due to high, and increasing, prices (primarily diesel)[2] (Shah et al., 2006). In contrast, in China, where only electricity is used, the supply is partially subsidized, and energy does not appear to be a constraint for irrigation, not even under the present conditions of increasing depth of extraction. Relatively speaking and across the study regions, the farmers using electricity are better off. On the other hand, and particularly pronounced in India, the regions with good electricity network coincide, maybe not surprisingly, with the areas where the more affluent rural societies reside. These are also the areas where the agricultural productivity traditionally has been comparatively high, as particularly in Punjab, which is considered the 'bread basket' of India. In the north eastern parts of India, where farmers resort to expensive diesel for lack of electricity, farmers find it increasingly difficult to sustain their groundwater-based livelihoods. In China, the relative cost of irrigation to the total cost of crop production is in the range of 10–15% for wheat, while it is up to 31.5% in Bihar (Appendix 1). In Punjab (India), the water buyers spend in the same range as in China for their growing of wheat, showing that they are much better off than their counterparts in the eastern parts of the country, though the pump owners still pay nothing. This is ironical and counter-intuitive as the water resources are much scarcer in China and Punjab relative to eastern India, and the cost of using groundwater in these areas does not reflect its scarcity value. A similar trend as in eastern India is seen in Bangladesh where small-scale farmers pay up to one third of the crop output value for irrigation.

In summary, while groundwater is abundant in the lower parts of the Ganges basin in eastern India, Nepal Terai, and Bangladesh, its utilization is hampered by economic constraints faced by the farmers due to relatively high cost of diesel or lack of access to free or cheaper electricity. In contrast, in China and Punjab (India), where electricity is relatively widely available and cheap, groundwater irrigation is reaching its physical limits due to extractions exceeding replenishment rates and continuously declining groundwater levels. For the farmers, this constraint is mostly felt through the need for drilling deeper wells and buying higher capacity pumps rather than due to excessive energy costs and environmental degradation. As an example, many rivers have run dry in these areas due to intensive exploitation, and reverting them to pre-development conditions where groundwater and surface water systems were directly connected hydraulically, may now be difficult. The brake, though not efficient, on further groundwater development is the need for continuous investments to 'chase' the groundwater.

3.6   *Adaptation and coping strategies of the poor farmers in face of constraints for groundwater use and agricultural productivity*

In the face of perceived and real constraints for further development of groundwater or for the assuring of benefits from investments already made and efforts to increase

---

[2]   A part of this 'energy-squeeze' however has eased recently as a result of the increase in food prices, whereby farmers get higher returns on their crops.

agricultural productivity and prosperity in these areas, farmers are diligent in trying to cope with conditions on the ground and optimizing available assets and opportunities. The groundwater markets as informal institutions have developed spontaneously as a response to unequal access. Though external support, from governments, NGOs and donors, may have facilitated technically and financially the access to the groundwater, the markets have developed without any external effort, hence demonstrating how free market mechanisms can regulate access to water and to some extent equalize the access among farmers. This is also becoming apparent in China though the prevailing principles of equity in terms of landholding size and community-sharing of wells are still shaping the conditions for groundwater access and land productivity.

With respect to the energy squeeze, the farmers find various ways to optimize resources. Diesel is used primarily because of lack of (good) electricity sources. However, with diesel prices going constantly up (Mukherji, 2007), the farmers try to find alternatives. One is to use kerosene which is a cooking oil widely subsidized in eastern India. Mixing it with diesel or using special kerosene-driven pumps, the farmers save on energy costs (Shah et al., Chapter 15, this Volume). Substituting traditional pumps for more energy-efficient pumps is another means, and here cheap Chinese pumps have become very popular in many parts of India. Farmers with pumps lease them out to even smaller farmers to make their investment more attractive.

Crop diversification is a common strategy for farmers to save on costs for water, either by switching to less water-intensive crops or moving to crops with higher market value. Higher value crops, or cash crops, such as horticulture crops, fruits and vegetables, generally require secure inputs in terms of fertilizers and water, but the additional investment and vigilance are usually justified by a much higher profit margin. For farmers with groundwater available on demand, it is a viable alternative. Often, farmers have a dual strategy when substituting part of their traditional stable grain crops (maize, wheat and rice), increasing the share of high value crops and at the same time increasing the share of rain-fed or less water-demanding crops. This way, the valuable groundwater is prioritized for the more profitable crops, for the right season and potentially for the best soil while less valuable crops are grown with somewhat higher risk of failure but less water input. Such risk spreading trends were observed in all the study sites.

Besides these indirect measures for water saving and optimization implied in crop diversification, the move to more direct water saving irrigation technologies was not pronounced in the study areas. The most common method for groundwater irrigation is still flooding (also called surface) irrigation, often combined with border and furrow irrigation. Some trends were apparent, especially in the Chinese sites where various simple community or household based water-saving irrigation techniques are adopted, for example using impervious distribution canals or pipes, plastic covers to limit evaporation in the fields, land levelling to optimize water distribution, and minimum tillage. The more advanced water-conservation technologies like drip irrigation and sprinkler irrigation were not applied, though some demonstration projects have been implemented in China by the government in an attempt to encourage farmers to introduce them. Simple rainwater harvesting schemes were also advocated but generally the reaction from the farmers was that these technologies were not economically feasible, considering the high cost of these technologies in relation to the low economic return from their crops (Sun et al., Chapter 10, this Volume).

In West Bengal, use of plastic pipelines was reported as an adapted measure by the farmers to maximize the beneficial use of their groundwater. It was observed that these methods were adopted more in the fields dominated by diesel-driven pumps compared to fields irrigated with electric driven pumps. This shows that, when the costs associated with groundwater extraction is high enough, even the poorest farmers will invest in water saving technologies. However, it must be concluded that there is a huge potential for reducing groundwater use in the majority of these areas through irrigation improvements and water conservation technologies.

In the Chinese cases, it is argued that the cost of water extraction is too low for farmers to adopt water saving irrigation. Basically, in all areas (except in Punjab, India where electricity is free for irrigation), the farmers only pay the nominal cost of the energy used to lift the water, and this apparently is insufficient to constrain farmers from over-utilizing the aquifers. It is argued that the price of water extraction is an instrument to control the groundwater extractions. By increasing the price, the farmers will be discouraged from maintaining their present abstraction rates. This may be feasible in China where wells are still in relatively central control through community wells and there is a relatively efficient electricity fee collection (Shah et al., 2004). However, in South Asia, such control does not exist. Furthermore, where the control may be most required, in the western part of India, the farming lobby is too strong to allow the implementation of such measures (Mukherji, 2006). And in the eastern part, the major problem is not over-use of groundwater, but rather inefficient agricultural production, partly due to inefficient access to groundwater. Here, groundwater extraction is already very costly to the farmers and a better energy supply is needed along with access to and better use of other crop inputs, extension services on optimal cropping technologies, better markets for crop products, and other livelihood options for the farmers. In Punjab (India) and China, subsidies for (and supply of and extension services for) water saving technologies may be appropriate (Shen, 2006; Feng, 1999). Such subsidies could be raised from a premium on the electricity tariff.

Another, maybe often overlooked, coping strategy by the farmers is migration. Though not studied in detail as part of this work, it was mentioned specifically in several case studies (Punjab, India; Bihar; Bangladesh, and China) that migration was a means to adjust and diversify livelihoods. In China, the major pattern is for the younger farmers to move to the cities where job opportunities are better. Similar trends were observed for the western part of India, while for the poorest farmers in eastern India and Bangladesh it may involve seasonal migration as contracted farmers. Both the rich and the poor seem to migrate, but to different areas and for different jobs (Deshingkar et al., 2006). In Punjab (India), immigration to other countries (even in Europe) was quite common for the richer segments, and remittances to the remaining family become a significant income source. Ironically, there is a history of worker migration from the poor areas in eastern India, e.g. Bihar, to the richer areas of Punjab (Singh, 1997). This is interesting in the light of how the groundwater resource is distributed, as this trend may exacerbate the intensification of groundwater use in already over-exploited areas while areas in the east remain underutilized. This again shows how groundwater use is driven by the socio-economic conditions and fabric rather than considerations of its physical endowment. Equally apparent from the interviews is the deliberate choice of the Punjabi farmers to maximize the benefit of their groundwater-based farming for financing the education of their children enabling the next generation to leave the traditional rural livelihoods.

3.7   *Formal groundwater management initiatives*

In most of these areas, it was found that very limited attention to pro-active groundwater management was in place, and even less so institutionalized and enforced regulations. India has a model bill on groundwater management (Romani et al., 2006) and it has been adapted and adopted in certain states, such as Andra Pradesh, Goa, Tamil Nadu, Kerala, West Bengal, Karnataka, and Himachal Pradesh (India Water Portal), but in no place has the law had any significant effect on practical groundwater use in agriculture (see e.g. Rama Mohan, Chapter 13, this Volume). In China, certain groundwater management initiatives have been implemented with respect to urban use, basically trying to curb use in these areas (Sun et al., Chapter 10; Cao et al., Chapter 11, this Volume). However, groundwater use in agriculture remains practically uncontrolled.

4   DISCUSSION

This study has shown that groundwater continues to provide a key input and resource to irrigated farming across many parts of South Asia and northern China and as such to livelihoods, food security and relative prosperity. But it also points to crucial constraints, in terms of the resource itself or other socio-economic factors, limiting the further or sustained benefits from groundwater irrigation. Generally, groundwater use is not governed or controlled by its availability, implying that with the high levels of intensity of use of today in the semi-arid areas, North China and western India face secular declining stores of groundwater, while areas of intensive use in Pakistan, Nepal, eastern India and Bangladesh, where recharge from rainfall or rivers are much higher, face constraints related more to inadequate access to energy for pumping of the groundwater, and in Pakistan due to salinity. Overall, energy is a key factor in groundwater use, and price, access and availability of energy to a large extent determines where and who is benefitting from groundwater irrigation. So, where energy is free and widely available to the farmers (Punjab, India) or is cheap and broadly available such as in North China, intensive groundwater use implies declining groundwater levels whereas in areas where energy availability and access is poor (Pakistan, Nepal, Eastern India, Bangladesh), groundwater may not be used optimally despite being widely available. Despite having improved the livelihoods of huge population groups in these areas, the full development potential of groundwater development has not been unlocked, and poverty remains chronic and widespread in these areas.

In terms of food productivity and security, groundwater has supported overall increases in all the areas investigated (via increased cropping intensities, crop yields, and diversification to other crops), but in the low-efficient areas (north eastern parts of South Asia), yields of stable crops remain low indicating that further gains are possible, through improved groundwater access, but possibly also other factors, like access to other farming inputs, technologies, extension services, markets, and infrastructure, that collectively have to come together to raise food production from a mere subsistence type occupation to more lucrative enterprise. In summary, the potential for groundwater to alleviate poverty has not materialized in those areas where there is an even greater need and physical possibility.

Furthermore, it was found that the cost of groundwater extraction is excessive where groundwater is shallow, while this is not the case in areas where groundwater is deep. In the latter areas, the problem of accessing groundwater was more related to primary entry via obtaining electricity connections and the increased *drilling* costs as groundwater levels decline. This is contrary to what is typically reported, namely that increasing costs of *pumping* due to declining groundwater levels presents a major constraints for the farmers. This lack of significant relationship between increasing groundwater levels and increasing pumping costs relate directly to the skewed influence of the energy access and price.

It is useful to understand the views of the farmers in the face of resource limits. In the resource-depleted areas, it appears that the farmers may be aware of restrictions in resource availability or feasibility of use but they do not seem to be concerned about the resource in an ecological sustainability sense. This could be because the resource depletion has occurred over many years practically from before their generation and they may not feel any environmental impacts from increased abstraction of groundwater, e.g. from drying out of rivers, because this has occurred already many years back. Generally, they view the problem from an economic point of view, raising their drilling costs to access groundwater. Also, they incorporate estimates of risks and opportunities into strategies for exiting irrigated farming, by maximizing their incomes from groundwater and thereby providing opportunities for their children to get an education and leaving agriculture.

It has proved extremely difficult to actively control the trends of continued groundwater exploitation and resource exhaustion through policies and regulations. Self-regulation through economic disincentives to stay in the business may very well be what determines the future of groundwater irrigation in these areas. Such trends are already seen in China, and western India where urbanization may be partly driven by groundwater exhaustion and related decreases in profitability from irrigated farming. Migration in general was found to be linked to restricted groundwater access, even in resource-rich but economically depressed areas.

Water-saving irrigation technologies beyond more traditional methods of land levelling, furrow irrigation and impervious canal or pipe conveyance systems were not found to be widely adopted in the study region, for various economic, institutional, and market-deficient reasons. Despite documented saving potentials (Yang et al., 2006; Zhang et al., 2003), research also indicates that with present cropping intensities and expected yields, water-saving technologies may not arrest present groundwater level declines, and harder choices, like cuts in production, are required (Kendy et al., 2007; Ambast et al., 2006). More than reverting declining trends of groundwater levels, advanced water saving technologies may reduce impacts of water logging, salinization, and N leaching (Fang et al., 2006). Water logging and salinization in western India and Pakistan remain a problem in certain areas. Sustaining crop productivity increases in these areas depend on the control and reversal of land degradation (FAO, 2002). Where water conservation technologies were encountered they were found to be driven by either constraint in access to water or energy, again demonstrating the intricate link between groundwater and energy.

The physical resource conditions as well as the socio-economic situation and its variability across the study region have shaped the different groundwater irrigation situations. The findings of this study agree with the findings from South Asia of Shah et al. (2006) while also providing comparable insights into the context of groundwater irrigation in China. Though similar trajectories are discernible from a comparison with the Punjab case, China also provides distinct features in terms of its groundwater development. Groundwater markets seem to have developed slower in China and seem to correlate with the privatization of wells (Zhang et al., 2008) which has occurred over the last two to three decades in China (Wang et al., 2006). This indicates that a significant drive towards marketing of the water is the private ownership, maybe not of the water *per se*, but rather the private access (whether formally stipulated or not) to the groundwater. This finding confirms theoretical considerations of the prerequisites for resource marketing, i.e. that property rights are needed for a market to develop (Thobani, 1995).

Land distribution in China is more equal than in South Asia. Though equally fragmented as in South Asia, this seems to promote equity in groundwater access and the benefits derived from it. There also seem to be a better recovery of the cost of energy. While farmers in China still only pay the nominal cost of energy, and hence no cost of the water resource or any externality costs associated with its development and exploitation, there seem to be a better fee collection efficiency and energy is not free like in Punjab, where there are consequent strains on the energy companies (Malik, 2002). As groundwater development has been shaped by these variable boundary conditions, so will future prospects of groundwater irrigation be determined by the societal and political context.

As far as counter measures and possible means of controlling the groundwater economies of these regions are concerned, it is emphasized that current experiences derive from the development of groundwater and not from direct and pro-active formal and institutionalized control of the resource users. The stakes are high and contradictory goals are at hand, one of maximizing poverty alleviation by the use of groundwater and the other, the concern for sustainability of the endeavour. The impending and increasing dilemma as the resource gets scarcer in many areas is how to secure the access to the resource of the poorest and generally deprived farmers. In water-rich, but poor regions a still unclear strategy needs to be formulated of how to better increase access to energy, as well as other basic inputs and requirements for production increases, for the millions of small scale farmers with inadequate livelihood opportunities to escape poverty.

The obvious divide between the resource-poor, but richer parts of western India and the poor, but groundwater-rich eastern parts of India needs further attention in a broader sense, not just from the groundwater perspective. Punjab, like other western Indian states will find it increasingly difficult to sustain its agricultural productivity levels and associated prosperity. Consideration as to how a greater proportion of India´s food production can be produced in the eastern parts in the future may well be a key to sustained groundwater use and food security in India. It will be a matter of devising less water-intensive livelihood options for people in the west while securing food access, and at the same time catering for a greater productivity in the east.

In China, groundwater level declines could be partly controlled through the increased price of electricity for groundwater pumping combined with subsidies or cheap loans for advanced water-saving technologies. With increased food prices as seen today, farmers are likely to remain engaged in irrigated farming and the necessary investments. Whether such approaches may work in Punjab, is not clear, as the political resistance against electricity charging is so intense. Here, it may be a matter for the farmers of exiting safely to other enterprises. Such strategies may be warranted, from an overall food security point of view, provided compensatory food production take place in the east.

In Nepal, we are still to see the further development of groundwater irrigation, alongside, hopefully, improved energy supply. Groundwater quality problems, in terms of arsenic, may pose a potential future constraint to groundwater development, but more in terms of securing safe domestic water supply than in supplying water for irrigation.

Bangladesh is in a similar energy squeeze, and struggles with arsenic for public water supply as groundwater is developed. However, in this country, the perpetual challenges to water management, stemming from recurrent extreme flooding and lack of water storage in dry seasons, continue to make groundwater irrigation a somewhat risky business and the immediate prospects of agricultural productivity increasing beyond subsistence levels are questionable. Many other factors seem to limit the poverty alleviating potential of groundwater, like proper infrastructure and hence access to markets and other farmer inputs. Further complicating the matter, climate change is likely to have an unsettling influence on overall development.

Pakistan may be compared to eastern India in that further development of groundwater is restricted by unequal access to land and poor energy supply, but compounded by increasing issues of land and water quality degradation (salinity). In this case, focus on energy supply will need to go hand in hand with measures to optimize the conjunctive and equitable use of surface water and groundwater.

In general, the views of policy makers with regard to groundwater are still very traditional, focused on price regulations, laws, etc., but also lately through resource augmentation through artificial recharge (Roy et al., 2006; Naik et al., 2006). However, a higher degree of awareness of indirect methods of controlling groundwater use, including access to energy, is necessary.

5   CONCLUSIONS

Groundwater irrigation in South Asia and Northern China faces severe constraints. Maintaining the benefits already accrued and ensuring future food security and livelihoods in these regions constitute a major challenge deserving continued interest from researchers and policymakers alike. Climate change and other factors of global uncertainty, like population growth and urbanization, pose further complicating dimensions to the issue.

What can be learned from this study and direct further research and policy focus is interesting. Firstly, continued groundwater irrigation requires huge, increased, and secured energy supply. Cost recovery of energy production and efficiency gains will become increasingly important (Nelson & Robertson, 2008).

Secondly, making policy makers aware of the close link between energy and groundwater irrigation is critical for the development of properly targeted policies, investments and subsidies. Such policies need to be viewed at a macro-level ensuring a more balanced, resource-based and equity-enhancing development and support across states and traditional political influence. Research and development needs to focus on the use of renewable energy sources, such as solar energy. Though groundwater (or rather access to it) appears to be a significant constraining factor for many poor farmers, it has to be recognized that in many cases, general poverty and poor infrastructure systems and poorly developed markets for other inputs limit the farmers in fully optimizing their production. Hence, other poverty alleviating initiatives are required in these regions, especially the northeastern parts of South Asia (Eastern India, Nepal, Bangladesh), to release the development potential inherent in groundwater irrigation.

Farmers are ingenious in their adaptation and coping strategies when it comes to optimizing benefits of groundwater irrigation, at the personal as well as community levels. Such strategies should be investigated in more detail and used to further support farmers' escape from poverty. Migration, to other regions or abroad, whether for contract farming or urban livelihoods is a common trend. The drivers, conditions and obstacles for this trend need to be better understood in order to improve these adaptation measures and facilitate the conditions under which they occur. Such measures may enhance other policies and direct economic activities towards more sustainable development (Deshingkar et al., 2006).

There may be scope for local groundwater management through support to capacity building, awareness raising and funding of specific activities and investments. However, experience shows that local groundwater initiatives require intensive external effort and need to be well documented, disseminated and replicated widely, to make investments viable and ensure impacts on broader scale.

Better understanding and information about the resource base is also crucial. Often, lack of groundwater-related data is a critical constraint for developing such understanding and policy recommendations based on informed knowledge. However, a comprehensive, integrated and multi-disciplinary understanding of the resource base along with knowledge of the functioning of the socio-economic, developmental, and political environment in which the groundwater irrigation occurs is fundamental. As the problems are multi-dimensional, the solutions need to take all these aspects into account.

REFERENCES

Agrawal, G.D., S.K. Lunkad, and T. Malkhed, 1999. Diffuse agricultural pollution of groundwaters in India. Water Sci. and Technol., 39, 3, 67–75.

Alauddin, M. and J. Quiggin, 2008. Agricultural intensification, irrigation and the environment in South Asia: Issues and policy options. Ecol. Econ., 65, 1, 111–124.

Ambast, S.K., N.K. Tyagi, and S.K. Raul, 2006. Management of declining groundwater in the Trans Indo-Gangetic Plain (India): Some options. Agric. Water Mgt., 82, 279–296.

Datta, K.K. and C. de Jong, 2002. Adverse effect of waterlogging and soil salinity on crop and land productivity in northwest region of Haryana, India. Agric. Water Mgt., 57, 223–238.

Deb Roy, A. and T. Shah, 2003. Socio-ecology of groundwater irrigation in India. In Llamas, R. and E. Custodio (Eds.): Intensive Use of Groundwater: Challenges and Opportunities. A.A. Balkema Publishers, the Netherlands, pp. 307–335.

Deshingkar, P.S. Kumar, H.K. Chobey, and D. Kumar, 2006. The Role of Migration and Remittances in Promoting Livelihoods in Bihar. Overseas Development Institute, London.

Fang, Q., Q. Yu, E. Wang, Y. Chen, G. Zhang, J. Wang, and L. Li, 2006. Soil nitrate accumulation, leaching and crop nitrogen use as influenced by fertilization and irrigation in an intensive wheat-maize double cropping system in the North China Plain. Plant Soil, 284, 335–350.

FAO, 2008. Climate change, Water, and Food Security: A Discussion Paper for an Expert Meeting 26–28 February 2008, Rome. Zero Draft NRLW, 20 February 2008.

FAO, 2002. World Agriculture: Towards 2015/2030. Summary Report. Rome, 2002. ISBN 92-5-104761-8.

Feng, Y., 1999. On water price determination and fee collection in irrigation districts, J. Econ. Water Res., 4, 46–49.

India Water Portal, http://www.indiawaterportal.org/Network/askq/kb/?View=entry&EntryID=80 Accessed Nov. 14, 2008.

Ju, X.T., C.L. Kou, F.S. Zhang, and P. Christie, 2006. Nitrogen balance and groundwater nitrate contamination: Comparison among three intensive cropping systems on the North China Plain. Env. Poll, 143, 117–125.

Kendy, E., J. Wang, D.J. Molden, C. Zheng, C. Liu, and T Steenhuis, 2007. Can urbanization solve inter-sector water conflicts? Insight from a case study in Hebei Province, North China Plain. Water Policy 9 Supplement 1, 75–93.

Kishore, A., 2004. Understanding agrarian impasse in Bihar. Econ. And Polit. Weekly. July 31, 2004.

Liu, G.D., W.L. Wu, and J. Zhang, 2005. Regional differentiation of non-point source pollution of agriculture-derived nitrate nitrogen in groundwater in northern China. Agric., Ecosyst. and Environ., 107, 211–220.

Llamas, M.R. and P. Martínez-Santos, 2005. Intensive groundwater use: Silent revolution and potential source of social conflicts, J. Water Resour. Plng. and Mgmt. 131, 5, 337–341.

Malik, R.P.S., 2002. Water-energy nexus in resource-poor economies: The Indian experience. Water Resour. Dev. 18, 1, 47–58.

Mandal, M.A.S., 1989. Declining returns from groundwater irrigation in Bangladesh. Bangladesh. J. Agric. Econ., 12, 2, 43–61.

Moench, M., 2003. Groundwater and poverty: exploring the connections. In Llamas, R and E. Custodio (Eds.): Intensive Use of Groundwater: Challenges and Opportunities. A.A. Balkema Publishers, the Netherlands, pp. 441–455.

Mukherji, A., 2006. Political ecology of groundwater: the contrasting case of water-abundant West Bengal and water-scarce Gujarat, India. Hydrogeol. J., 14, 392–406.

Mukherji, A., 2008. Spatio-temporal analysis of markets for groundwater irrigation services in India: 1976–1977 to 1997–1998. Hydrogeol. J., 16, 1077–1087.

Naik, P.K., G.C. Pati, P.K. Mohapatra, and A. Choudhury, 2006. Watershed development approach in groundwater augmentation in Nawapada District of Orissa. In: Romani, S., K.D. Sharma, N.C. Ghosh, and Y.B. Kaushik (Eds.): Groundwater Governance—Ownership of Groundwater and its Pricing. Proceedings of the 12th National Symposium on Hydrology. Nov. 14–15, 2006, New Delhi, India. Chaman Enterprises, New Delhi, India. pp. 497–507.

Narayanamoorthy, A., 2007. Does groundwater irrigation reduce rural poverty? Evidence from Indian states. Irrig. and Drain. 56, 349–362.

Nelson, G.C. and R. Robertson, 2008. Estimating the contribution of groundwater irrigation pumping to CO_2 emissions in India. Draft technical note. International Food Policy Research Institute (IFPRI), Washington D.C. 16 p.

Qureshi, A.S., P.G. McCornick, M. Qadir, and Z. Aslam, 2008. Managing salinity and water logging in the Indus Basin of Pakistan. Agric. Water Mgt., 95, 1–10.

Romani, S., K.D. Sharma, N.C. Ghosh, and Y.B. Kaushik, 2006. Groundwater Governance—Ownership of Groundwater and its Pricing. Proceedings of the 12th National Symposium on Hydrology. Nov. 14–15, 2006, New Delhi, India. Chaman Enterprises, New Delhi, India. 514 p.

Roy, M., H.C. Sharm, and A. Kumar, 2006. Artificial groundwater recharge planning in northern Utar Pradesh using remote sensing and GIS. In: Romani, S., K.D. Sharma, N.C. Ghosh, and Y.B. Kaushik (Eds.): Groundwater Governance—Ownership of Groundwater and its Pricing. Proceedings of the 12th National Symposium on Hydrology. Nov. 14–15, 2006, New Delhi, India. Chaman Enterprises, New Delhi, India. pp. 105–113.

Scott, C.A. and T. Shah, 2004. Groundwater overdraft reduction through agricultural energy policy: Insight from India and Mexico. Water Resour. Dev., 20, 2, 149–164.

Shah, T., A. Deb Roy, A.S. Qureshi, and J. Wang, 2003. Sustaining Asia's groundwater boom: An overview of issues and evidence. Natural Resources Forum, 27, 130–141.

Shah, T., M. Giordano, and J. Wang, 2004. Irrigation institutions in a dynamic economy—What is China doing differently from India? Econ. and Polit. Weekly. July 31, 2004.

Shah, T., O.P. Singh, and A. Mukherji, 2006. Some aspects of South Asia's groundwater irrigation economy: Analyses from a survey in India, Pakistan, Nepal Terai, and Bangladesh. Hyd. J., 14, 286–309.

Shah, T., 2007. The groundwater economy of South Asia: An assessment of size, significance and socio-ecological impacts. In: Giordano, M. and K.G. Villholth (Eds.): The Agricultural Groundwater Revolution: Opportunities and Threats to Development. CABI, in ass. w. IWMI. ISBN-13: 978 1 84593 172 8. pp. 7–36.

Shah, T., C. Scott, A. Kishore, and A. Sharma, 2007. Energy-irrigation nexus in South Asia: Improving groundwater conservation and power sector viability. In: Giordano, M. and K.G. Villholth (Eds.): The Agricultural Groundwater Revolution: Opportunities and Threats to Development. CABI, in ass. w. IWMI. ISBN-13: 978 1 84593 172 8. pp. 211–242.

Shen D.J., 2006. Theoretical base of water resources free and its pricing method. Shuili Xuebao, 1(37), 120–125.

Singh, M., 1997. Bonded migrant labour in Punjab agriculture. Econ. and Polit.Weekly, March 15, 1997.

Thobani, M., 1995. Tradable Property Rights to Water—How to improve water use and resolve water conflicts. World Bank. FPD Note No. 34. Feb. 1995.

Van Geen, A., Y. Zheng, Z. Cheng, Y. He, R.K. Dhar, J.M. Garnier, J. Rose, A. Seddique, M.A. Hoque, and K.M. Ahmed, 2006. Impact of irrigation rice paddies with groundwater containing arsenic in Bangladesh. Sci. of the Total Environ., 367, 769–777.

Villholth, K.G. and B.R. Sharma, 2006. Creating synergy between groundwater research and management in south and south east Asia. In: Groundwater Research and Management: Integrating Science into Management Decisions (Ed. by B.S. Sharma and K.G. Villholth). Proceedings of IWMI-ITP-NIH International Workshop on: "Creating Synergy between Groundwater Research and Management in South and Southeast Asia", Feb. 8–9, 2005, Roorkee, India. Groundwater Governance in Asia Series. International Water Management Institute, South Asia Regional Office, New Delhi, India. ISBN: 92-9090- 647-2. 270 p.

Wang, J., J. Huang, and Q. Huang, 2006. Privatization of tubewells in North China: Determinants and impacts on irrigated area, productivity and the water table. Hydrogeol. J., 14, 275–285.

Wang, J., J. Hunag, A. Blanke, Q. Hunag, and S. Rozelle, 2007. The development, challenges and management of groundwater in rural China. In: Giordano, M. and K.G. Villholth (Eds.): The Agricultural Groundwater Revolution: Opportunities and Threats to Development. CABI, in ass. w. IWMI. ISBN-13: 978 1 84593 172 8. pp. 37–62.

Yang, Y., M. Watanabe, X. Zhang, J. Zhang, Q. Wang, and S. Hayashi, 2006. Optimizing irrigation management for wheat to reduce groundwater depletion in the piedmont region of the Taihang Mountains in the North China Plain. Agric. Water Mgt., 82, 25–44.

Zhang, L., J. Wang, J. Huang, and S. Rozelle, 2008. Development of groundwater markets in China: A glimpse into progress to date. World Dev., 36(4), 706–726.

Zhang, X., D. Pei, and C. Hu, 2003. Conserving groundwater for irrigation in the North China Plain. Irrig. Sci., 21, 159–166.

Zheng, L. and J.K. Routray, 2002. Groundwater resource use practices and implementations for sustainable agricultural development in the North China Plains: A case study in Ningjin County of Shandong Province, PR China. Water Resour. Dev., 18, 4, 581–593.

Appendix 1. Key figures from the eight research sites across five south and south east Asian countries.

	1	2	3	4	5	6	7	8
	Rechna Doab, Punjab	Hoshiarpur, Punjab	Vaishali, Bihar	Murshidabad, West Bengal	Madaripur	Morang and Jhapa	Zhengzhou, Henan	Baoding Hebei
	Pakistan	India	India	India	Bangladesh	Nepal	China	China
Coordinates (latitude, longitude)	31° 10.80"N 72° 39'33.28"E	31° 16'7.695"N 76° 8'12.895"E	25° 44'0.00"N 85° 20'30.00"E	24° 10'3.18"N 88° 16'18.11"E	23° 27'33.57"N 90° 108.02"E	26° 36'51.64"N 87° 40'21.39"E	34° 44'55.79"N 113° 38'13.39"E	38° 50'56.07"N 115° 28'36.87"E
Chapter	4	5	6	7	8	9	10	11
Representativity of areas	Study villages in fresh GW area	Study villages in relatively water well-endowed region	Representative of the state	Representative of the state	Representative of the low-lying flood plains in Bangladesh	Representative of the southern low-lying Terai of Nepal	Representative of lower Yellow River setting	Representative of the NCP
No. of villages investigated	4	3	3	4	5	2	22	20
No. of farmers per village	20	25	19–25	20	9–26	32–40	3	3
Total number of farmers interviewed	80	75	65	80	83	72	65	60
Rainfall (mm/year)	375–650	768	1232	1389	2000	1500	641	550
Annual population growth rate for state, %	2.6	1.6	2.5	1.8	2.2	2.3	0.42	0.60
Landholding size distribution (percentage of farmers)	<1 ha: 38.0% 1–2 ha: 24.0% >2 ha: 38.0%	<0.2 ha: 20.0% 0.2–0.5 ha: 33.3% 0.5–2 ha: 26.7% >2 ha: 20.0%	<1 ha: 92.5% 1–2 ha: 5.1% >2 ha: 2.4%	Landless: 47% <1 ha: 41% 1–2 ha: 10% 2–4 ha: 2% >4 ha: 0.3%	<0.2 ha: 64% 0.2–1 ha: 24% 1–2 ha: 7% >2 ha: 5%	<0.2 ha: 13.1% 0.2–0.5 ha: 27.8% 0.5–2 ha: 55.7% >2 ha: 3.4%	0.08 ha per person	0.3 ha per household, on average 0.08 ha per person

(Continued)

Appendix 1. (*Continued*)

	1	2	3	4	5	6	7	8
	Rechna Doab, Punjab	Hoshiarpur, Punjab	Vaishali, Bihar	Murshidabad, West Bengal	Madaripur	Morang and Jhapa	Zhengzhou, Henan	Baoding Hebei
	Pakistan	India	India	India	Bangladesh	Nepal	China	China
Major crops	Rice, wheat, sugarcane, and cotton	Wheat, rice, and maize	Rice, maize, and wheat	*Aman* (monsoon) paddy, *Boro* (summer) paddy, jute, mustard, wheat, and vegetables	Summer *Boro* paddy, transplanted *Aman* paddy, wheat, mustard, sugarcane, onions, and jute	Paddy, wheat, maize, mustard, banana, and vegetables	Wheat and maize	Wheat and maize
Cropping intensity	160	179	222	197 for water buyers 174 for pump owners	146	250	156	171
Crop yields for major crops (kg/ha)	Wheat: 2600 Rice: 2750 Sugarcane: 49,539	Wheat: 7500 Rice: 7000 Maize: 5700	Wheat: 2116 Rice: 1271 Maize: 3752	Monsoon paddy: 3703 Summer paddy: 5054 Jute: 2148 Wheat: 2098 Mustard: 1010	HYV *Boro*: 5000 Transplanted *Aman*: 900 Wheat: 1976 Mustard: 380 Sugarcane: 30,035 Onion: 5691 Jute: 4804	Wheat: 2000 Rice: 2500	Wheat: 6198 Maize: 6874	Wheat: 5753 Maize: 6709
Percentage of irrigation water from GW	74–79	90	91	100	100	72	68	100
Water productivity for major crops (kg/m³)	–	Wheat: 1.34 Rice: 0.242 Maize: 3.37	–	Monsoon paddy: 0.4 Summer paddy: 0.3 Jute: 1.2 Wheat: 0.9 Mustard: 0.6	Summer paddy: 0.38 for water sellers 0.43 for water buyers	–	Wheat: 2.85 Maize: 4.45	Wheat: 1.3 Maize: 3.5
Groundwater level (mbgl)	4–15	75–300	<10	<6 m	<6 m	<6 m	40	23–50

	Punjab Province	Punjab State	Study area	West Bengal State	Whole of Bangladesh	Terai	(tube-wells)	In study area
Groundwater level decline rate (m/year)	Since 1980: 0.11–0.49	Since 1980: 1.3	0	Less that 10 cm in a year	0	0	1–2 m/year	0.9
Increase in well numbers	Punjab Province: 890,000 over 1989–2005	Punjab State: 162,199 over 1999–2006	Study area (5.3 km²): From 5 wells in 1970 to 112 in 2005	West Bengal State: By 40% from 1991 to 2001	Whole of Bangladesh: From 133,800 in 1986 to 1,182,500 in 2006	Terai: from 0 in 1970 to — 800.000 in 2005	Number of tube-wells remained steady at 6000 from 1993 to 2003, but deep tube-wells replaced shallow tube-wells	In study area (42 km²): From 902 to 1214 over 1995–2205, primarily deep tube-wells
Depth of wells (m)	24–76	95–650	6–67	18–45	15–40	15–40	60–100	20–158
Density of wells (no. per km²)	11–54	4.2–23.6	22.2–40.0	23	—	3.2	5.7	14
Well discharge (m³/hr)	78	13–301	10–25	30	25–30 for shallow tube-well 150 for deep tube-well	46–75	38	41
Hours of extraction per well/year	403–3087	1123	250	1800 for electric 380 for diesel	1900	–	680	
Predominant energy source	Diesel	Electricity	Diesel	Diesel for private pumps, electricity for government	Diesel	Electricity and diesel	Electricity	Electricity
Drivers of GW development	Lack of/unreliability of surface water	Affordable well installation, free electricity since 2004, reclamation of land to paddy cultivation, declining rainfall	Flood-proneness in monsoon season that restrict cultivation to only dry season, low cost of borings, plentiful availability of GW, declining rainfall	Introduction of summer *Boro* paddy in late 1970s, cheap drilling, plentiful availability of GW	Flood-proneness, introduction of HYV seeds, import liberalization in 1987 making it easier to import cheap Chinese pumps, population pressure	Push from government and prospect of farmer income enhancement	Cheap electricity, collective and government funds	Scarcity of surface water, cheap electricity fee, and encouragement of government to drill tube-wells

(Continued)

Appendix 1. (*Continued*)

	1	2	3	4	5	6	7	8
	Rechna Doab, Punjab	Hoshiarpur, Punjab	Vaishali, Bihar	Murshidabad, West Bengal	Madaripur	Morang and Jhapa	Zhengzhou, Henan	Baoding Hebei
	Pakistan	India	India	India	Bangladesh	Nepal	China	China
Groundwater extraction mechanisms and ownership	Private wells with diesel centrifugal pumps, few electric pumps	Private wells with electric centrifugal pumps and kinship-shared wells	Private wells with portable diesel pumps. Few community wells.	Mix of private wells mostly with diesel pumps, community wells, and government deep tube-wells	Private wells with diesel pumps	½ Group-owned and individual wells	Collective (governmental) tube-wells (96%) and others (4%)	Collective and private (shareholding) tube-wells
Capital cost of installation of typical well at present depth and technology	40–100 USD	2460 USD	–	50–100 USD (<10 m deep)	15 USD	360–570 USD	5700 USD	–
Groundwater access/allocation institutions	Informal GW market	Informal GW markets, family/cast based sharing system	Informal pump rental markets, community managed group wells	Informal GW markets and government tube-wells	Informal GW markets	Sharing of group-owned wells and emerging informal GW markets	Sharing of collective tube-wells	Sharing of collective and shareholding tube-wells, little informal GW markets

Subsidies provided	None	Electricity is free for the farmers	Previous government support to well drilling. Diesel is subsidized, but still the prices are very high given the crops grown, electricity is subsidized but is not available in the area	Diesel is subsidized, but still the prices are very high given the crops grown, electricity is subsidized but is not available in the area	Government provides subsidies for tube-wells owned by groups of farmers	Diesel is subsidized, but still irrigation costs account for 20–40% of total cost of cultivation	The local government funds installation of new tube-wells in some villages	Government provides some subsidies for digging tube-wells in some counties
Share of GW irrigation cost in total crop production cost (percent)	—	Maize: 7.8 % Wheat: 12.4 % Rice: 28.3 %	—	—	—	—	Wheat: 12.6 % Maize: 10.3 %	Maize: 15.0 %
Share of GW irrigation cost in total crop production cost for *water sellers* (percent)	Wheat: 32.9% Rice: 58.6% Sugarcane: 18.7%	0	Rice: 12.9 % Wheat: 22.4 %	Rice (*Aman*): 13.8% Rice (*Boro*): 22.6% Jute: 4.4% Mustard: 7.1% Wheat: 7.2%	—	Rice: 41.1% (diesel motive power) Rice: 27.5% (electric motive power)	—	—
Share of GW irrigation cost in total crop production cost *for water buyers* (percent)	Wheat: 21.8% Rice: 40.9% Sugarcane: 47.3%	Maize: 7.8% Wheat: 12.4%	Rice: 8.9 % Wheat: 31.5 %	Rice (*Aman*): 14.2% Rice (*Boro*): 14.5% Jute: 9.4% Mustard: 11.9% Wheat: 14.0%	—	Rice: 33.0% (1/3 crop share)	—	—
Perceived problem(s) related to groundwater use	Increasing diesel cost, difficulty in getting new electricity connections, and falling water table	Declining GW levels, poor quality of electricity supply	Increasing diesel cost, difficulty in getting new electricity connections, high cost for pumps	High cost of well installation, increasing diesel cost, difficulty in getting new electricity connections	High capital costs of tube-well installation, requirement for land ownership and group formation to obtain loan	Increasing diesel costs, lack of electricity, lack of markets and infrastructure	Groundwater table decline, high cost of new tube-well installation	Groundwater table is declining, the cost of digging tube-wells is increasing

(Continued)

Appendix 1. (*Continued*)

	1	2	3	4	5	6	7	8
	Rechna Doab, Punjab	Hoshiarpur, Punjab	Vaishali, Bihar	Murshidabad, West Bengal	Madaripur	Morang and Jhapa	Zhengzhou, Henan	Baoding Hebei
	Pakistan	India	India	India	Bangladesh	Nepal	China	China
Responses or adaptation strategies of the poorest farmers	Conjunctive use of GW and surface water, water buyers crop intensively and grow more water saving crops or cash crops	Crop diversification, income diversification, non-farming enterprises, hired labor	Rain-fed farming, leasing out land to tube-well owners, use of kerosene to replace diesel, rental market for pumps	Shift away from summer paddy to low water consuming crops, use of fuel-efficient Honda pumps, use of plastic pipelines for conveyance, and mixing of diesel with kerosene	The poor continue to cultivate rain-fed crops only or *Boro* paddy on crop sharing basis, diversify livelihoods and work for larger farmers	Group formation to obtain loan, water buying from well owners, switch to high-value crops	Install deeper tube-wells, change to more efficient pumps, grow more high value crops, diversify livelihoods, some move to the cities	Change to private ownership of wells, change the cropping pattern, drilling deeper tube-wells, some simple water saving techniques

CHAPTER 2

Towards better management of groundwater resources—lessons from an integrated capacity building project in the Indo-Gangetic and Yellow River basins

K.G. Villholth

Geological Survey of Denmark and Greenland (GEUS), Copenhagen, Denmark

ABSTRACT: It is increasingly recognized that human capacity is a major shortfall when it comes to addressing pressing challenges of groundwater management in Asia. Here, the bulk of this subsurface resource is extracted for irrigation purposes with extensive, but increasingly threatened, poverty alleviation impacts. A half-year training and research programme was carried out for the capacity building of professionals involved in groundwater management in two major river basins under water stress, the Indo-Gangetic and Yellow River basins in south Asia and China, respectively. The programme specifically addresses the five countries of India, China, Bangladesh, Pakistan and Nepal. The objective of the programme was to enhance the capacity of groundwater managers in the region through interactive and inter-disciplinary training and research activities. The aim was to contribute to a better understanding of the groundwater challenges and devising ways and means of confronting those challenges.

1 INTRODUCTION

Significant challenges are involved in the sustainable, equitable and efficient utilization of groundwater resources in the two major river basins of Asia, the Indus-Ganges River basin (IGB) in South Asia and the Yellow River basin (YRB) in China. Poor understanding of the resource and potential management options and associated weak governance leads to over-exploitation of groundwater in some areas, while under-utilization co-exists in other parts. Both aspects entail a sub-optimal benefit from the resource, which is being increasingly recognized as a very potent water resource and basis for poverty alleviation in these regions.

It is estimated that the groundwater extraction in India has more than quadrupled (from 50 to 210 km^3/year) between 1970 and 2000 (Shah, 2005). Similar figures for China are from 20 to 75 km^3/year over the same period. These numbers can be compared to the USA where groundwater over-abstraction is also causing concern today, but where the levels have only risen from 90 to 107 km^3/year during the same time (Shah, 2005).

Proper and efficient groundwater management in these regions is an immense task and cannot be achieved in the short term. There are no blue print solutions to transfer from the developed part of the world where groundwater is also under stress, and most national authorities are struggling with keeping up with new challenges and responding to previous and new impacts of excessive use and degradation of the resource. In the Asian context, the problem is complex. The fact that a great part of the population is traditionally reliant on income from farming, the constant population pressure and poverty implications means that hard decisions are harder to make and there needs to be alternative approaches to managing the resource. Registering and regulating the groundwater use of these farmers and potentially compensating them for refraining from using groundwater, as in countries such as USA and Australia, may not be a viable approach and alternatives need to be sought. The approach of yesterday, i.e. doing nothing, may also prove critical and could risk the life-saving basis of millions of poor families in rural Asia.

The human capacity available for addressing these challenges is limited in most parts of Asia, including the two basins in question. Overall, human capacity is required in three key areas: a) equipping authorities engaged in water management to undertake groundwater management as an integral part of the whole; b) enhancing and directing ongoing research towards applicable and cross-disciplinary research relevant for the region; and c) raising awareness of the general public on the importance of safeguarding the resource for the benefit of all. None of these aspects are covered sufficiently at the moment, although some initiatives are beginning to develop and discussions and attempts across the region and within countries are shaping possible approaches. Examples are a national level meeting on groundwater governance in India, discussing the present and future role of legislation, water rights and pricing (Romani et al., 2006), research into the relationship between energy and irrigation, and how irrigation performance and groundwater use is influenced by energy policies, prices and availability (Shah et al., 2003b, Mukherji 2007), and NGO initiatives to encourage groundwater management and protection at the community level (van Steenbergen, 2006; Rama Mohan, Chapter 13 this Volume).

Against this backdrop, a large-scale project within the Challenge Programme on Water and Food (CPWF)[1] was designed to address deficiencies in human capacity of managing groundwater in the two river basins. The premise of the project was that proper groundwater management needs to be built on informed knowledge of professionals from the region, with emphasis on inter-disciplinary knowledge and understanding of the actual groundwater situation in the rural areas.

2   THE GGA TRAINING AND RESEARCH PROGRAM

The CPWF project on Groundwater Governance in Asia (GGA) was designed and directed by the International Water Management Institute (IWMI). The objective of the capacity building (CB) component of the project was to contribute to the enhancement of the capacity of existing institutions in the basin states involved in groundwater research and management so that they could undertake more integrated, multi-disciplinary and sustainable approaches to groundwater governance. By involving practitioners directly in training and applied groundwater research, the project intended to seed a process of enduring change in the groundwater management sectors of the basin states.

The programme consisted of two annual cycles of a comprehensive training and research programme called 'The International Training and Research Program on Groundwater Governance in Asia: Theory and Practice'. The target group for the CB programme was

---

[1]   http://www.waterandfood.org/

professionals actively involved in water from the five basin countries, preferably with groundwater research, management and media coverage. The project was quite ambitious in the sense that it tried to integrate three major aspects of the groundwater management challenge:

1. It intended to *work across two major basins in Asia*, where conditions are variable but at the same time subjected to similar constraints and problems.
2. It intended to *integrate various disciplines*, both by inviting people with different professional backgrounds and present roles, but also by addressing a multitude of topics related to groundwater use and management in the curriculum.
3. It intended to *address theoretical as well as practical issues* of groundwater exploitation and management, by having both class work teaching as well as hands-on field work incorporated in the programme.

2.1   *Focus on major river basins in Asia*

The project area is large, covering two major river basins in the world, the Indus Ganges basin (2,251,500 km^2) and the Yellow River basin (795,125 km^2)[2] (Figs. 1 and 2). The key issues related to groundwater management varied across the region. The Indus River basin, principally covering Pakistan, is faced with over-use of groundwater particularly in the arid areas while the wetter areas or areas supplied by the river are mostly plagued with salinity problems either of geological origin or due to mismanaged irrigation with insufficient drainage. This is particularly the case in the lower reaches, where groundwater use today is more or less abandoned.

The upper reaches of the Ganges River, in Nepal and north-eastern India, have abundant groundwater resources, although arsenic could prove to be a problem requiring more attention. Groundwater development in Nepal holds great promise for small-scale tube-well development and livelihood improvement. The middle reaches, located in the north-western part of India face similar problems to those in Pakistan, namely, over-utilization and salinization. Moving down the Ganges, the areas become increasingly wet and well supplied by the Ganges and by precipitation. However, the promise of groundwater development has not developed accordingly, due to structural and institutional hindrances.

For the Yellow River, groundwater is increasingly being developed to supplement surface water irrigation in upper and middle reaches, while groundwater irrigation is more common in the lower reaches as well as in the NCP, raising concerns over constantly declining groundwater levels. Though groundwater use in agriculture was a main concern in this project, it is clear that groundwater conditions (quality and quantity) are severely impacted by urban and industrial use in many parts of the basins, especially in the middle reaches of Indus, middle reaches of Ganges and the whole of NCP. The quality aspects are not documented very well at present but it is recognised that quality could cause major long-term impacts. By addressing these diverse river basins and making comparisons across the regions, a comprehensive and integrated understanding of the situations was obtained.

---

[2]   It covers more basins if Indus-Ganges is considered as two separate river basins, and the Yellow River basin is extended to include the North China Plains (NCP) (320,000 km^2), which are the flat alluvial plains situated in the northeast of China, in as much as groundwater problems prevail in these regions.

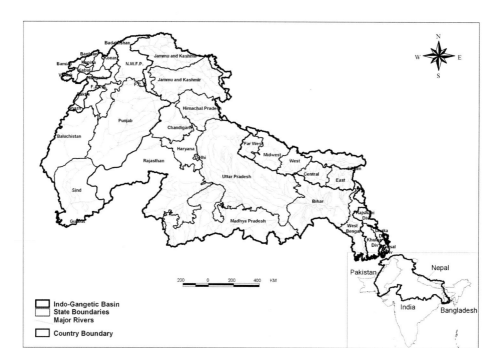

Figure 1.   Indo-Gangetic basin.

2.2   *The inter-disciplinary approach*

Recognizing the challenge that groundwater management has to achieve through integrated approaches and knowledge beyond a single discipline, the whole programme was designed to incorporate knowledge from various fields of learning. Three thematic areas or perspectives were reflected in the curriculum: physical science of groundwater, agronomic aspects of groundwater irrigation, socio-economics and institutional aspects related to groundwater use. The lectures covering these topics were a mix of general theory, explaining concepts, tools and terminology, applied theory, and cases illustrating these concepts and ideas.

The list of participants, categorized per country, background and profession is given in Table 1. In total, 33 participants, or fellows, enrolled and completed the programme. The target number of participants per country was based on the estimated significance of the groundwater economy of that particular country (Shah et al., 2003a), i.e. the number of nationals decreased in the order: India, China, Bangladesh, Pakistan, Nepal. The candidates were characterized into three groups: junior fellows (22), senior fellows (9) and media fellows (2). The distinction between junior and senior fellows was based on the seniority and level of experience and competence of the applicants.

Media people were also invited and two journalists, from Nepal and China, participated. The idea behind this was to investigate the existing and potential role of the news media in disseminating groundwater policy discussions and increasing public awareness in the region. All participants were made more aware of the role that communication at various

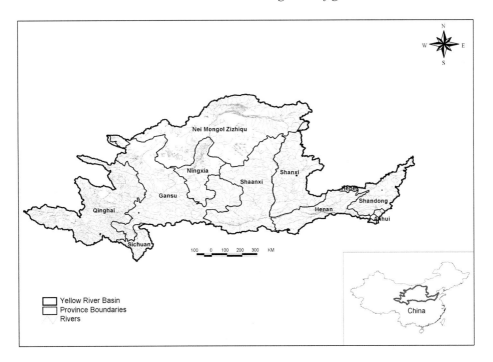

Figure 2.   Yellow River basin.

Table 1.   Description of participants of the GGA training and research program.

Total number of participants	33
Average age (for Junior fellows)	32.5
Average age (for Senior Fellows)	46.2
Percentage of women participants	12%
Average number of years of experience (Junior fellows)	5 years
Average number of years of experience (Senior fellows)	15 years
No. of participants from:	
India	11
China	9
Bangladesh	5
Pakistan	4
Nepal	2
Background	
Natural sciences and engineering	25
Economics and social sciences	6

[a] Non-Governmental Organization.

levels and across disciplines plays in making ground water issues better known to the general public. The media/communication angle was incorporated in various ways throughout the programme: special sessions in the curriculum were devoted to the importance of communication, knowledge sharing and awareness raising, the media fellows were given assignments to develop videos from the field that could be used in explaining to the general public of that country what groundwater meant to the farmers today and what constraints were associated with its (further) development. The gender concern was addressed by ensuring the entry of qualified women to the programme. In total, four women fellows entered the programme.

2.3   *Theory and practice*

The overall schedule for the programme is shown in Table 2. Basically, the programme was split into two major parts, a theoretical training course (up to five weeks) and an action research phase (up to fourteen weeks). Participants had to complete both parts successfully to fulfil the programme requirements and obtain the programme certificate. The programme concluded with a joint summary workshop of one week where all participants attended to present their research findings, discuss cross-cutting issues, and evaluate the programme. The senior fellows' programme was more condensed than the junior and media fellows' in an attempt to accommodate their normally busy working schedules. The common language throughout the programme was English. The advertising for the course was done through internet, though national newspapers and through email and regular postage to relevant institutions.

2.3.1   *The training course*

The more detailed programme for the 5-week training course is given in Table 3. The split between the three thematic areas mentioned earlier are indicated as different modules. In addition, a part was dedicated to scientific methods, research preparation and field tours. This part constituted about 30% of the time and included field trips, hands-on to software and modelling tools, personal work and group discussions, films and interactive games. The last week was dedicated to the integrated aspects of groundwater governance (Table 4). The seniors participated only in this week and gained a compressed version of the discussions that had taken place throughout the previous four weeks and at the same time contributed to a lively and practical view on the debated topics. The detailed curriculum was developed or adapted specifically for this programme and dedicated papers and presentations were prepared by the lecturers and trainers. In total, 40 trainers, including six from outside the basin countries, were engaged in the 5-week course.

Table 2.   Schedule for the GGA training and research program.

Component	No of weeks	Duration for the various fellows	
		Junior and media fellows	Senior fellows
1. Training course	5	5 weeks	1 week (the last of the five weeks)
2. Research phase	14	5 weeks in the field, 9 weeks for analysis and reporting	3 weeks in the field, 1 week for report writing
3. Writing workshop	1	1	0
4. Summary workshop	1	1	1

Table 3.   Programme for theoretical training component.

Modules and sessions: / Day	Resource characterization and mobilization, Environment — Week 1	Agricultural water use — Week 2	Social sciences, Economics, Institutions — Week 3	Policies, Governance — Week 4	Scientific methods, Research preparation, Field tours	Senior Fellows — Week 5
(Senior Fellows, Weeks 1–4)						Holiday
Monday	Arrival of fellows	Geophysics Data analysis and RS/GIS GW modelling	Game: Naranpur Express-Simulation of a South Asian Village	Crop water requirements Water productivity Irrigation water savings Drainage and salinity		Introduction Paradigms for GW govt., local to global scale Awareness raising
Tuesday	Introduction GW in a global context Intro to CPWF basins	Intro. to agronomy	Water harvesting techniques	GW/Irrigation institutions GW markets		GW govt. in Pakistan Adoptive GW mgt.
Wednesday	Basic hydrology Hydro-geology	Physical landscape field trip	Social, cultural and economic anthropology	Irrigation management Farming systems in the IGB and YRB		Water rights Community mgt. of GW Legal aspects of GW mgt. in Australia and India
Thursday	Intro. to social sciences	Public holiday, Diwali	Rural sociology Sociology of technology	Local and global experience on GW institutions		GW-energy nexus GW politics in north and south India
Friday	GW flow processes Soil-water processes		Political anthropology	GW irrigation field trip Intro. to socio-economic survey in Punjab		GW policies and political economy of Northern China
Saturday	GW quality, health, and environmental impacts GW lifting devices		State and community			Mgt. of arsenic GW GW in Nepal Evaluation and closing

Table 4.   Topics for the groundwater governance theme during week 5 of the training course.

Theme
Understanding water institutions: structure, environment and change process
Integrated approach to groundwater law
The energy-irrigation nexus—regulation in North India
The electricity-groundwater conundrum in India
Groundwater as Cinderella of water laws, policies and institutions in Australia
Groundwater legislation in Tamil Nadu
Groundwater use and abuse in India: some socio political, economic and institutional characteristics
Managing arsenic contamination of groundwater
Arsenic in the groundwater of Santiago de Estero Province, North Western Argentina
Water saving technology adoption in China
Groundwater markets in China: evolving trends and determining factors
Demand for groundwater and water pricing in China
Groundwater quality protection and management
Bhakra Canals and groundwater in Punjab and Haryana

The course was conducted in collaboration with four different institutions in north India, all with locations within the Indo-Gangetic basin. The main host institution was the Department of Water Resources Development and Management (DWRDM), located at the Indian Institute of Technology (IIT), Roorkee, Uttaranchal. The Central Soil and Water Conservation Research and Training Institute (CSWCRTI), Dehradun, the Indian Institute of Remote Sensing (IIRS), Dehradun, and Punjab Agricultural University (PAU), Ludhiana also hosted parts of the program. Finally, the Punjabi University, Patiala provided significant input to some of the field visits and subsequent research in Punjab, India.

2.3.2   The research component
2.3.2.1   Senior fellow research
The senior fellows' research phase consisted of a three-week study tour to a developed country which uses significant amounts of groundwater for irrigation and are faced with considerable management challenges. The objective of the study tours was to expose the participants to contemporary groundwater issues, management challenges and approaches pertinent to a developed country. By doing so, it was expected that the senior fellows would be equipped with tools and knowledge as well as a useful network with which to expand their ability to address groundwater governance challenges in their own management environment.

An initial survey among several potential host institutions in relevant countries showed that there was great interest in participating in such a programme. Nevertheless, it was apparent that only the USA and Australia were preferred as study countries. Hence, agreements with the University of Kansas-School of Law and the School of Commerce, University of South Australia, with Prof. John Peck and Prof Jennifer McKay, respectively as the main coordinators and senior fellow resource persons were made. Individual programmes for each group of fellows (6 fellows to the USA and 3 fellows to Australia) addressing groundwater relevant problem areas were drawn up. These included lectures,

Table 5. Programme components and topics covered in the senior study tours.

USA	Australia
Kansas socio-economics	State-level administration of laws and policies
Kansas water rights administration	Environmental court, conflict resolution
Interstate issues on Missouri River	Aquifer storage and recovery work
Interstate issues with Colorado and Nebraska	Water recycling
Municipal water reuse, reservoir sedimentation, aquifer storage and recovery project, water pollution	Groundwater irrigation techniques and regulation
Land use change—from ranch to game reserve	
Kansas irrigation economy	
Ogallala aquifer and sustainable agriculture	
Retracting groundwater rights and farmer compensation	

institutional visits, and field trips to areas with significant groundwater use and various innovative technologies or approaches (Table 5). Based on their learning, personal interaction with experts, and literature searches the senior fellows were requested to write an individual scientific paper on a preferred topic related to groundwater management in the study country but viewed from the perspective of the context of their own country. For this purpose, one week was set aside for them to work from their home base upon returning from the study tour.

2.3.2.2   Junior and media fellow research

The junior and media fellows undertook research in their own respective countries. This research was designed to address contemporary issues of groundwater use in a representative range of locations (Figure 3 and Table 6) spanning the IGB and YRB. This research component was called the Cross-Cutting Research (CCR) because a main goal was to synthesize realities of groundwater use and management across the geographical region as well as across the traditional disciplinary boundaries. The other primary objective was to train, equip and enable present and future groundwater managers to grasp complex challenges of groundwater governance through first hand field experience and to facilitate interaction, discussion and cross learning among the participants with diverse backgrounds. The media fellows were given a role to focus on media/awareness aspects of the groundwater use in their particular study area.

Groups of two to four fellows worked on a particular site and conducted surveys in between 3 to 20 villages each. Fieldwork was conducted for a period of five to six weeks based on research instruments, primarily questionnaires, drafted by IWMI but discussed and improved during a three-day pre-CCR training session prior to the field work. All groups were provided with a site co-ordinator to take care of logistic and two resource persons to provide intellectual guidance and support on the technical and contextual part of the research and to review the reports. A mid-term field work review visit was undertaken by IWMI scientists.

Research questions and instruments for the CCR (Table 7) were framed in order to understand groundwater issues from each of the three overarching perspectives of the

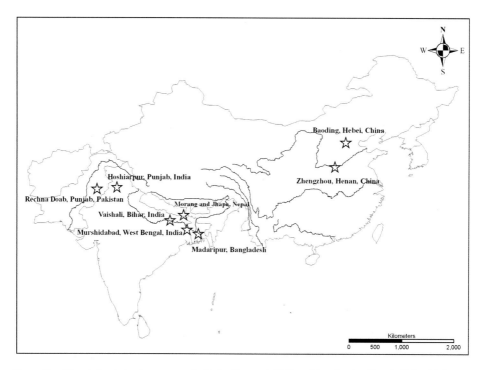

Figure 3.   The eight study sites across the Indo-Gangetic Yellow River basins/North China Plains.

Table 6.   Locations of study sites of the cross-cutting research.

Country	Location
Pakistan	Punjab province, Gujranwal and Jhang district
India	Punjab State, Hoshiarpur district
	Bihar State, Vaishali district
	West Bengal State, Murshidabad district
Bangladesh	Madaripur district
Nepal	Eastern Nepal *Terai* districts of Morang and Jhapa
China	Hebei province, Qingyuan and Mancheng counties of Baoding municipality
	Henan province, Xinmi and Xingyang counties of Zhengzhou municipality

GGA Training Programme: the physical resource, socio-economic, and policy and institutional perspectives (Table 8). Each group wrote a comprehensive paper on their findings, integrating all three perspectives and including some key parameters to enable comparisons across the sites. Following the field research phase, a period of nine weeks was devoted to the analysis and synthesis of results and for initial report writing within the groups. A report writing template was made available to the fellows to facilitate paper preparation around a common structural and content framework. Fellows were also encouraged to write individual theme papers on specific topics of relevance and personal interest.

Table 7. Research instruments of the CCR.

RI 1: Village level interview schedule
RI 2: Individual (pump owners and non-pump owners) interiew schedule
RI 3: History of irrigation checklist
RI 4: Well driller's information schedule
RI 5: Groundwater officials interview checklist
RI 6: Media news content analysis checklilst

Table 8. The research perspectives of the CCR.

The physical resource perspective	• Groundwater resource inventory
	  – Nature and characteristic of aquifer
	  – Groundwater quality
	  – Discharge from wells and tube-wells
	• Groundwater extraction technologies
	  – Historical evolution of groundwater extraction technologies
	• Technical options for groundwater supply augmentation

The social and economic perspective	• Economics of groundwater use
	  – Estimation of land and water productivity
	  – Cost and benefits of groundwater irrigation
	  – Comparisons with rainfed and canal irrigation
	• Groundwater irrigation and equity
	  – Who has access to groundwater and at what cost
	• Positive and negative externalities of groundwater use
	  – Coping strategies in case of negative externalities

The policy and institutional perspective	• Groundwater institutions
	  – Formal institutions such as groundwater law
	  – Informal institutions such as water markets and water sharing arrangements
	• Groundwater authorities and their role
	  – Vision, mission and strategies
	  – Legal aspects of groundwater governance
	• Politics of groundwater
	  – Peasant mobilisation around GW issues?
	• Managing the energy-irrigation nexus

2.3.3 *The writing and summary workshop*

The training and research programme was concluded with a two-week session comprising a one week writing workshop for the junior and media fellows and a one-week summary workshop for all the participants, including the senior fellows, project team members, site coordinators, and resource persons. The purpose of the writing workshop was to bring together all the groups and help them with their group papers and presentations in the summary workshop. Short lectures on 'good' scientific writing and presentation skills were provided, the last also addressing the senior fellows.

3   RESULTS AND EXPERIENCES

3.1   *Outcomes*

The tangible outputs of the first cycle of the GGA Training and Research Programme include:

- This volume. It contains eight comprehensive chapters written by junior and media fellows as leading authors and covering key results from the junior fellow research, the CCR. It also contains six additional chapters (focussing on special thematic issues) that resulted from research carried out either from a grant made under the GGA project, or through synthesis of data compiled during the CCR fieldwork.
- Three journal papers written by the senior fellows as first authors, published international scientific journals and describing how to transpose lessons learned in the study country in the context of groundwater management in the Indo-Gangetic and Yellow River basins (Dutta and McKay, 2007; Lashari et al., 2007; Sudan and McKay, 2007).
- A project homepage[3] describing the project and the capacity building approach and specific program. Furthermore, here all training material, including notes, presentations and software packages, are available for downloading
- A platform and forum for continuous debate among the involved participants on topics relevant to groundwater governance in the region.

3.2   *Programme evaluation and feedback*

The programme underwent a systematic evaluation and feedback process in order to assess the strong and weak points and derive suggestions and recommendations for improving it for the second round, scheduled for 2007–2008. This evaluation was based on questionnaires administered to the participants at various times and after various milestones in the programme in addition to a plenum discussion held during the summary workshop. The questionnaires covered each lecture or course session, the various modules, the programme components (the classroom course and the research component), and finally the programme in its entirety.

The main positive points derived from the evaluation and the aspects appreciated most by the fellows were the inter-disciplinary approach and the interaction with a vast group of individuals with diverse background and experience. The combination of theory and practice was also perceived as an asset of the programme. The fellows also highlighted the programme as a valuable forum for learning and sharing experience. The overall score for the whole program was 8.4 on a scale of 10. This may be difficult to judge absolutely, but gives a good indication of the overall satisfaction with the programme.

On the critical side, some general points were brought forward:

- The course was too India-centric. There was relatively little focus on the groundwater aspects of the other four project countries.
- The linkage between the course and the research phase was not strong enough.
- The programme was very intense and ambitious which meant that the fellows felt overwhelmed by the programme schedule and the requirements in terms of outputs.

---

[3] http://www.waterandfood.org/gga/

On this basis, some changes for the second cycle of the programme were implemented, e.g. shortening the length of the course from 5 to 4 weeks, conducting it in 2 locations instead of 5 different locations, devotion of greater proportion of time to hands-on-learning exercises and finally better integration of class room sessions with the research phase.

4   DISCUSSION AND CONCLUSIONS

The GGA Training and Research Programme clearly responded to a demand and a need—the need for strengthening human resources and capacity building within groundwater management. This need is increasingly recognized, e.g. at the World Water Week in Stockholm, 2007. Similarly, it is increasingly recognized that the Millennium Developments Goals will not be achieved without increased attention to groundwater and capacity building within groundwater and that IWRM necessitates more focus on groundwater aspects (BGR et al., 2007). The number of applicants for the GGA Training and Research Programme was significant[4], considering that it was conducted for the first time, indicating a sincere demand.

The programme was unique and innovative in its approach, striving to face various challenges, of which the inter-disciplinary and the inter-regional were the major ones. Both aspects were appreciated by the participants and this assessment is taken as an indication that the programme succeeded in meeting its objectives. The programme was successful in bringing out a comprehensive understanding of the contemporary issues related to groundwater use across the different parts of the focus river basins and increasing the awareness and understanding of the participants with respect to the inter-linkages of groundwater use with socio-economics, legal, policy and institutional aspects, and the physics of the resource.

The programme was ambitious and adjustments to address this point for the second phase have been suggested. In terms of costs, the project was expensive. A quick estimate of total costs associated with the programme shows that approximately US$630,000 were spent on the first cycle. It translates into a cost of between US$15,000 to US$19,000 per fellow, and approximately US$150 per day of training per fellow.

To optimize the investment made and to ensure a larger impact and more sustainable effort, several considerations need to be made. First of all, the programme will be repeated in its current approximate form on order to gain further experience and consolidate the experiences and relations with collaborating partners. In the longer term, various initiatives will be taken to anchor the programme with an existing institution in the region ensuring ownership and at the same time building up the capacity and funding options for conducting this type of programme on a recurrent basis. Secondly, more active involvement is needed to build a recognized platform for debate, consisting of a network of individuals and institutions involved in the management of the groundwater resources in the region. This could be called the Groundwater Management Research Alliance, as proposed in Villholth and Sharma (2006).

---

[4] The number of applicants for the programme was approximately four times the number of possible candidates.

ACKNOWLEDGEMENTS

The Challenge Programme on Water and Food is acknowledged for its funding of the project. All fellows, lecturers, resource persons, project team members are gratefully appreciated for their personal involvement and contributions to the programme. The institutions taking part in devolving this endeavour are also thanked for hosting, supporting and promoting the programme. It would not have been possible without the input of all these institutions and individuals.

REFERENCES

BGR, Cap-Net, WA-Net, WaterNET, (2007). Capacity Building for Groundwater Management in West and Southern Africa. http://www.bgr.bund.de/cln_011/nn_459796/DE/Themen/TZ/Politikberatung_GW/Downloads/groundwater_capacitybuilding,templateId=raw,property=publicationFile.pdf/groundwater_capacitybuilding.pdf

Dutta, D. and J. McKay, (2007). Development of an integrated capacity building framework for building spatial data infrastructure for integrated groundwater governance and management in India: Lessons from Australia. Int. J. Environ. Dev. 4 (1): 32–44.

Lashari, B., J. McKay, and K. Villholth, (2007). Institutional and legal groundwater management framework: Lessons learnt from South Australia for Pakistan. Int. J. Environ. Dev., 4 (1): 45–59.

Mukherji, A. (2007). The energy irrigation nexus and its impact on groundwater markets in eastern Indo-Gangetic basin: Evidence from West Bengal, India, *Energy Policy*, 35 (12): 6413–6430.

Romani, S., K.D. Sharma, N.C. Ghosh, and Y.B. Kaushik, (2006). Goundwater Governance—Ownership of Groundwater and its Pricing. Proceedings of the 12th National Symposium on Hydrology. Nov. 14–15, 2006, New Delhi, India, 514 p. Capital Publishing Company, New Delhi, Kolkata, Bangalore. ISBN-81-85589-56-9.

Shah, T., (2005). Groundwater and human development: challenges and opportunities in livelihoods and environment. *Water Science & Technology*, 51 (8): 27–37.

Shah, T., A. Deb Roy, A.S. Qureshi, and J. Wang, (2003a). Sustaining Asia's groundwater boom: An overview of issues and evidence. *Natural Resources Forum*, 27: 130–141.

Shah, T., C. Scott, A. Kishore, and A. Sharma, (2003b). Energy-Irrigation nexus in South Asia—Improving groundwater conservation and power sector viability. *IWMI Research Report 70*. Colombo, Sri Lanka. IWMI, International Water Management Institute.

Sudan, F.K.,and J. Mc Kay, (2007). Institutional aspects of groundwater governance: Experiences from south Australia and lessons for India. *Int. J. Environ. Dev.* 4 (1): 1–32.

van Steenbergen, F. (2006). Promoting local management in groundwater. *Hydrogeology Journal*, 14 (3): 380–391.

Villholth, K.G. and B.R. Sharma, (2006). Creating Synergy between Groundwater Research and Management in South and South East Asia. In: Sharma, B.S. and K.G. Villholth (Eds.). Groundwater Research and Management: Integrating Science into Management Decisions. Proceedings of IWMI-ITP-NIH International Workshop on: "Creating Synergy between Groundwater Research and Management in South and Southeast Asia", Feb. 8–9, 2005, Roorkee, India. Groundwater Governance in Asia Series. International Water Management Institute, South Asia Regional Office, New Delhi, India.

CHAPTER 3

A comparative analysis of the hydrogeology
of the Indus-Gangetic and Yellow River basins

S.K. Jain[1], B.R. Sharma[2], A. Zahid[3], M. Jin[4], J.L. Shreshtha[5],
V. Kumar[1], S.P. Rai[1], J. Hu[4], Y. Luo[4] & D. Sharma[2]

[1] *National Institute of Hydrology, Roorkee, India*

[2] *International Water Management Institute, New Delhi office, India*

[3] *Bangladesh Water Development Board, Dhaka, Bangladesh*

[4] *China University of Geosciences, Wuhan, China*

[5] *Department of Irrigation, Kathmandu, Nepal*

ABSTRACT: The Indus, Gangetic and Yellow River basins have supported thriving agriculture economies from time immemorial. Of late, unplanned over-exploitation of their resources, especially groundwater, has raised concerns about the long-term hydrological sustainability of the irrigated agriculture in the region. This calls for a fresh look at the hydrogeological resources and the management options for these basins. The Indus basin with a drainage area of 1.1 M km^2 supports the world's largest contiguous irrigation system called the Indus Basin Irrigation System. In both the Indus Basin and western parts of the Gangetic basin, there is intensive use of groundwater and this has led to a decline in the groundwater table. In the eastern part of the Gangetic basin groundwater is relatively under utilized. The Huang-Huai-Hai plain of the Yellow River Basin is one of the largest groundwater irrigated regions in China and produces the bulk of China's food grains.

1 INTRODUCTION

The Indus-Gangetic basin and the Yellow River basin have large surface and groundwater resources and support some of the oldest civilizations in the world. Both the basins have thriving irrigated economies and the surplus food produced in the basins ensures food security for the region (India, Pakistan, Bangladesh, Nepal and China). The Indus basin has the world's largest continuous surface irrigation system and the Ganges and Yellow Rivers also support vast canal networks. However, since the 1980s, the runaway growth in groundwater tubewells has outpaced surface irrigation. This has created an agrarian boom with massive productivity and livelihood benefits (Shah et al., 2003). However, such a phenomenal growth over a short period has also created several problems including a decline in water tables over large areas, increase in energy requirements for extracting groundwater, emergence of groundwater quality issues such as salinity, nitrate, fluoride and arsenic in groundwater and intrusion of sea water into coastal areas. At the same time there are regional imbalances in groundwater exploitation with vast untapped groundwater potential in the eastern Ganges basin and southern parts of the Yellow river basin. All these existing and emerging problems of groundwater governance and management require a better understanding of the hydrogeology.

2   THE INDUS AND GANGES RIVER BASIN

The Indus basin is one of the largest river basins in Asia extending over four countries, China, Afghanistan, Pakistan and India. The Indus River Basin is 3,199 km long with a basin area of about 1.1 million km² of which 63% of the area is in Pakistan, 29% in India and the rest 8% is shared between China and Afghanistan (IUCN, 2005; Fig. 1). The drainage area within Pakistan is 692,700 km². According to a 2001 estimate (UNESCO, 2001) the population of the basin is 150 million. However, Fahlbusch et al. (2004) placed the population at 196 million.

The annual flow of the entire Indus river system ranges from 120 billion m³ to 230 billion m³ (41 year average, 1957–1997). The mean annual system inflows are 175 billion m³ with a coefficient of variability of 13%. About 83% of the flow occurs during the monsoon months (April to September). There is not much yearly variation in the system inflow during the winter *rabi* season (Khan, 1999). The Indus Basin Irrigation System (IBIS) of Pakistan is the largest integrated irrigation network in the world, serving 17 million hectares of land. In the Indian part of the Indus basin, an average annual surface water potential of 73.3 km³ has been assessed. Out of this, only about 46.0 km³ is utilizable.

The Ganga River is one of most important rivers of India and is considered sacred by its majority Hindu population. The Ganga basin extends over an area of 1,089,370 km². It flows

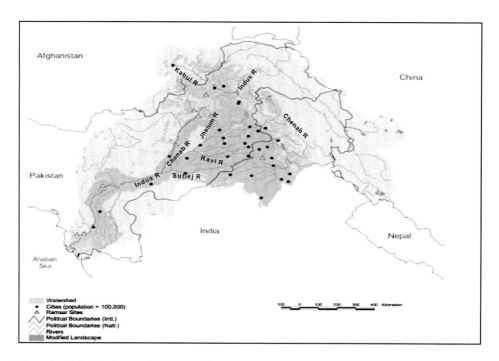

Figure 1.   Indus Basin map (Revenga et al., 1998) (See colour plate section).

through four countries, namely India (79.2%), Nepal (12.85%), Tibet (China) (3.67%), and Bangladesh (4.28%). The total length of the Ganga River is 2,525 km (Fig. 2). The Ganga basin is one of the most densely populated regions of the world. The average population density is 550 persons/km^2 and about 42% of India's population lives in this basin.

The average annual discharge of the Ganga River is 16,650 m^3/s. The surface water resource potential of the Ganga and its tributaries in India has been assessed at 525 billion m^3 out of which 250 billion m^3 is considered to be useable. There are large temporal variations in the flow. The Ganga River carries one of the world's highest sediment loads, equal to nearly 1,451 million metric tons per annum. Figure 3 shows

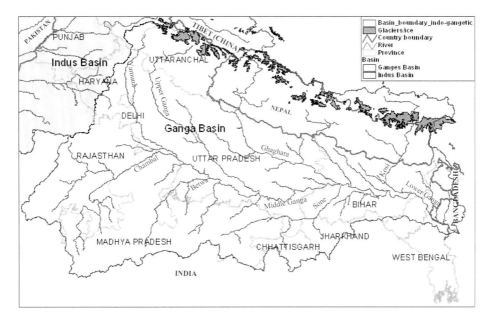

Figure 2.   Ganga Basin map (IWMI 2008, available at http://dw.iwmi.org/) (See colour plate section).

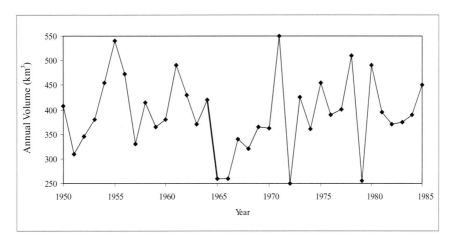

Figure 3.   Annual discharge of Ganga at Farakka (Rogers et al., 1989).

the annual discharge of Ganga at Farakka for the period 1950 to 1985 (Jain et al., 2007).

2.1   Hydrogeology of Indus and Ganga basins

2.1.1   Hydrogeology of Indus in Pakistan

The geology of the Indus Basin is dominated by Quaternary sediments which are often hundreds of metres thick. The Indus sediments are mainly alluvial and deltaic deposits, consisting of fine-to medium-grained sand, silt and clay. Coarser sands and gravels occur on the margins of the plain abutting upland areas (WAPDA/EUAD, 1989). Wind-blown sands occur to the east of the Indus Plain (Thar and Cholistan desert areas). Mesozoic and Cenozoic sedimentary rocks occur in a north-south tract to the west of the Indus Plain stretching from Peshawar to the coast. Older (palaeozoic) sediments and crystalline basement rocks (granites, metamorphic rocks) are mainly restricted to the north, including North West Frontier Province, Gilgit and Jammu and Kashmir.

Groundwater yields from these sediments are typically around 50–300 m^3/hour down to 150 m depth. Lower yields (up to 50 m^3/hour) are obtained from the sediments of the Thar and Cholistan desert regions on the eastern fringes of the Plain (south-eastern border with India)—see Figure 4. Yields are also restricted in the fine-grained Quaternary tidal and deltaic deposits in the southern coastal area (Shamsi, 1989). A district level study across the country by NESPAK (1991) reported a specific yield of 0.09 to 0.2. These estimates broadly show a high yielding aquifer with a substantial storage capacity. Before the introduction of surface irrigation systems, groundwater tables in the Indus Plain were typically 20–30 m deep. Over the years, water levels in the aquifers have risen significantly as a result of increased recharge from earthen canals and irrigated fields. Waterlogging with accompanying salinisation of groundwater has become a major problem in parts of the Indus Plain.

Groundwater is important for irrigated agriculture, which allows for an extensive exploitation and recycling over the major portion of the Indus Basin. Irrigated land provides 90% by value of Pakistan's agricultural production, and accounts for 25% of its gross domestic product (GDP) and employs 60% of the labour force.

2.1.2   Hydrogeology of Indus-Ganges Basin in India

Hydrogeologically, the Indo-Gangetic basin from north to south is divisible into six regions, namely, the Himalayan region, sub-Himalayan region, Bhabar zone, Tarai zone, Central Ganga Plains (CGP) and marginal alluvial plains.

The Himalayan Region confined between the Siwalik range in the south and the Zanskar range in the north represents hilly and rugged terrain consisting of a variety of rock formations, which are continuously undergoing disintegration through glacio-fluvial action. Larger part of this region remains under snow cover throughout the year. The Great Himalayas are the gathering grounds which feed a multitude of glaciers, some of which are among the largest in the world outside the Polar circles (Wadia, 1990). The Indus basin has the largest number of glaciers (3,538), followed by the Ganga basin (1,020) and the Brahmaputra (662) (WWF, 2005). The inter-granular pore spaces, openings, fissures, fractures, joints and bedding planes developed promote the infiltration of rainwater which reappears down slope as spring and seepage (Valdiya and Bartarya, 1989). Here groundwater occurs in the secondary porosity of the formation and is unconfined.

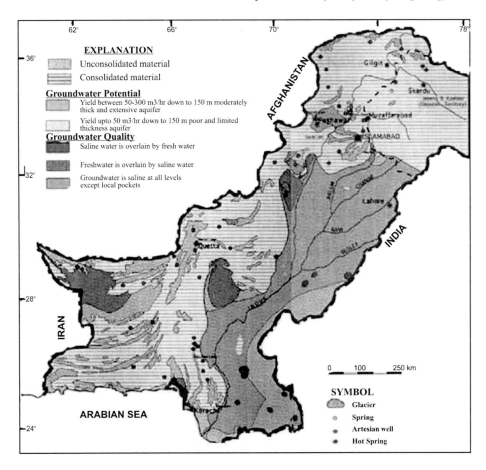

Figure 4.   Hydrogeology map of Pakistan (Geological Survey of Pakistan, 2000) (See colour plate section).

The sub-Himalayan region lies to the south of the Himalayan zone and is occupied by the Siwalik ranges. They form a system of low foothills with an average height of 900–1,500 m (Wadia, 1990). The groundwater here mostly is unconfined and sometimes semi-confined. The depth to the water table ranges between 10 and 20 m below ground level (bgl). The tubewells are reported to be capable of yielding 50 to 80 m^3/hr in Siwaliks of Uttarakhand and 100 to 120 m^3/hr in the inter-montane valleys in Himachal Pradesh (CGWB, 2008). The coefficient of permeability varies between 15 and 250 m/day.

The northern belt of the Indo-Gangetic alluvial tract (near the Himalayan foothills) is characterized by coarse materials (principally boulder-gravel) forming the piedmont terrain. It is referred to as *Bhabar* in Uttar Pradesh and *Kandi* in Jammu & Kashmir and Punjab. This formation has been formed by lateral coalescence of alluvial cone and fan deposits brought down by innumerable streams (Wadia, 1990). The rivers crossing Bhabar lose large quantity of flow to the gravels. Groundwater is unconfined and the water table is deep (30 m or more). The groundwater has a hydraulic gradient of around 3 m/km. The hydraulic

Table 1. Aquifer variables in Central Ganga Plain (Karanth, 1987).

S.No.	Parameter	Right bank aquifer	Left bank aquifer
1	Transmissivity (m²/day)	520	3530
2	Storage coefficient	1.13×10^{-3}	1.5×10^{-3}
3	Hydraulic conductivity (m/day)	37	55

conductivity ranges between 25 to 250 m/day. The Bhabars are capable of yielding about 100–300 m³/hr of water (CGWB, 2008).

The Bhabar belt is overlain by the Terai belt of stratified bands of dominantly coarse sediments with clay. This occupies a narrow belt and its contact with the Bhabar is well marked by a spring line. The presence of highly porous and permeable fan deposits ensures large supplies of groundwater from the *Terai* belt for agricultural and industrial use. The *Terai* has an upper unconfined aquifer and a lower interconnected system of confined aquifers. Between the *Bhabar* and the confined aquifers of the *Terai* belt is a vast zone where recharge is encouraged by high rainfall and hilly streams. The piezometric head of the aquifer ranges between 6 to 9 m above ground level. The coefficient of permeability ranges between 17 and 108 m/day. The yield of tubewells tapping the *Terai* zones ranges between 50–200 m³/hr (CGWB, 2008).

The Central Ganga plain forms one of the richest aquifers of the world. The typical channel deposit of the Ganga River, from the bottom upward, comprises coarse sand mixed with gravel, medium-to fine—grained sand to silt and a capping of thin clay. This clay cap and some fine sand layers are washed away during the flood period, and a fresh body of sand with a fining upward sequence is deposited again each year during the flood, thus building up a thick terrigenous clastic deposit until the river next changes its course. In the Central Ganga Plains, extensive exploratory studies have indicated the presence of four aquifer groups within a depth of 700 m bgl. The individual aquifers vary in thickness from a few meters up to 300 m. Although locally separated, aquifers are hydraulically connected at a regional scale (Karanth, 1987). The range of aquifer parameters in the Central Ganga Plains is shown in Table 1. Groundwater occurs under water table conditions in the shallow aquifers, and is semi-confined to confined in the deeper aquifers. The yield of tubewells in this area ranges between 90–200 m³/hr (CGWB, 2008).

CGWB (1984) mapped the Upper Yamuna Basin (UYB) showing different potential aquifer groups up to a depth of 450 m. The regional picture of sub-surface geology revealed the existence of three aquifer systems. Table 2 presents the profile and aquifer characteristics of these three systems.

2.1.3   *Hydrogeology of Ganges in Nepal Terai*

The Nepal *Terai* plain is the northern extension of Gangetic plain and is about 30 km in width. It comprises a significant thickness of alluvial clays and silts with important but subordinate sand and gravel layers. In the foothills, the basement hard rocks are believed to be at depth of 4–6 km. The overlying Eocene-Pliocene deposits are an upward coarsening gently northward dipping sequence of sedimentary rocks of the Siwaliks (shale, sandstone conglomerate) of about 2 km thickness. The unconsolidated strata of alluvium overlie the Siwalik sequence. The thickness exceeds the maximum depth of the investigatory drilling at around 500 m. It is a leaky aquifer, with vertical exchange between shallow and deep

Table 2. Aquifer parameters of UYB area (CGWB, 1984).

Parameter	Range	Average values
Aquifer group—I		
Profile: Extend up to 167 m bgl over the basin and underlain by a clayey horizon, 10–15 m thick, except in the foothill region. The group is unconfined and semi-confined		
Transmissivity (m^2/day)	800–5,210	2,200
Lateral hydraulic conductivity 'K' (m/day)	14–47	24
Specific yield (S$_y$)%	6–24	12
Aquifer group—II		
Profile: Numerous sand and clay lenses occurring at variable depth (65–283 m bgl). Sediments are less coarse and are occasionally mixed with gravel (*kankar*). Ground water occurs under confined to semi-confined conditions. Aquifer is underlain by another clayey horizon of considerably thick at places and appears to be regionally extensive.		
Transmissivity (m^2/day)	750–1,050	700
Lateral hydraulic conductivity 'K' (m/day)	4–11	7.2
Storativity	5.6×10^{-4} to 1.7×10^{-3}	1.0×10^{-3}
Specific yield (S$_y$)%	3.35×10^{-4} to 2.7×10^{-3}	1.9×10^{-3}
Aquifer group—III		
Profile: Thin sand layers alternating with thicker clay layers occurring at variable depths (197–346 m bgl). The granular material is generally finer in texture. Kankar occurs in the southern parts. Groundwater normally occurs under confined conditions.		
Transmissivity (m^2/day)	345–830	525
Lateral hydraulic conductivity 'K' (m/day)	3.5–10.7	7.1
Storativity	6.6×10^{-4} to 2.4×10^{-4}	4.5×10^{-4}

confined aquifers, dependent upon thickness and lateral persistence of the low permeability beds (Sharma, 1984, 1995). In the south the aquifers material is predominantly sands, while the aquifers in the north are predominantly gravels. Further, south the aquifer material is composed of finer sediments. The static water level in the south is at around 3 to 5 m, while in the north it is between 6 m to over 10 m depth. The drawdown in the south is comparatively small compared to the northern part (Figure 5).

2.1.4 *Hydrogeology of Ganges Basin in Bangladesh*

Bangladesh constitutes a part of the Ganges Basin, and lies at the head of the Bay of Bengal. The basin is about 200 km wide in the northeast and broadens to about 500 km in the vicinity of the Bay of Bengal (Aggarwal et al., 2000). The basin comprises geosynclinal deposits of late Cretaceous age and a thick sequence of Tertiary marine continental deposits. Immense sedimentation took place in the Ganges-Brahmaputra-Meghna delta complex during the Cenozoic time and more than half of it was deposited during the Plio-Pleistocene and Holocene time leading to the southward growth and development of the Bengal Delta. Much of the sediment is deposited by the Ganga-Brahmaputra- Meghna (GBM) river systems during Miocene to Holocene time. Holocene floodplain deposits cover most of the surface area of present-day Bangladesh. Arsenic contamination mainly occurs at shallow depths beneath the Holocene floodplains. The area is underlain by poorly consolidated or unconsolidated sediments of Tertiary and Quaternary ages.

The tropical monsoon climate with favorable geological and hydrogeologic conditions ensures very high potential storage of groundwater in Bangladesh. Major aquifer systems belong to the Late Pleistocene Holocene sediments. In the floodplains of the major rivers

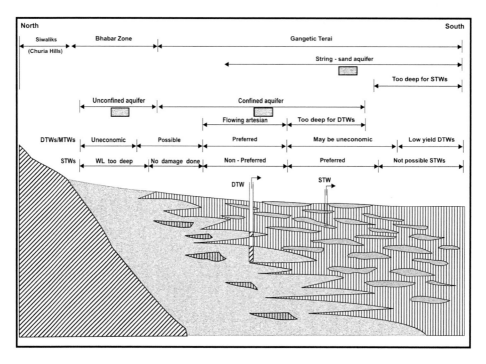

Figure 5.   Schematic cross section through the aquifer system of main Terai plains (Sharma, 1995).

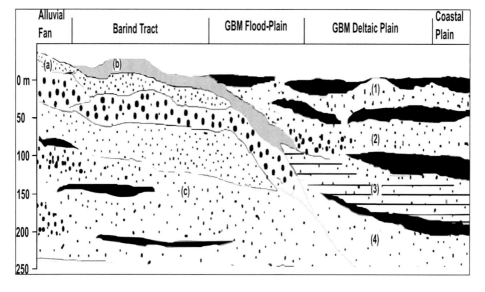

Figure 6.   Distribution of BWDB-UNDP aquifers (BWDB-UNDP, 1982).

Table 3. Aquifer properties in the shallow and main aquifers in Bangladesh (BWDB, 1994, 1989).

Regions	Transmissivity (m^2/day)		Storage co-efficient		Permeability(m/day)	
	Minimum	Maximum	Minimum	Maximum	Minimum	Maximum
North-East	200	3000	0.002	0.10	3	90
North-West	300	4000	0.003	0.23	12	114
South-West	900	3200	0.010	0.15	11	6.5
South-East	140	1900	0.0007	0.07	5	23

and the delta plain of the GBM Delta Complex regions, major aquifer systems occur in the Late Pleistocene to Holocene sediments. In part of the southeast region multi-layered aquifer conditions exist. On a regional basis BWDB-UNDP (1982) described three aquifers between Holocene and Plio-Pleistocene formations. These are the 1) Upper Shallower or the Composite Aquifer, 2) Main Aquifer, and 3) Deep Aquifer (Figure 6). The aquifer sediments are composed of very fine to fine sand, in places inter-bedded or mixed with medium sand in very thin layers. Discontinuous thin clay layers often separate these sand layers. In the coastal region water in this aquifer zone is saline with occasional fresh water pockets. Water in this layer is severely contaminated by arsenic. Beneath the composite aquifer, the main aquifer occurs at depths ranging from less than 5 m in the northwest to 140 m in the south east and is generally underlain and overlain by a silty clay bed. In the Meghna floodplain areas it up to 300 m deep (Zahid and Hassan, 2007) and in large areas of the Pleistocene terraces, the main aquifer is encountered below 75–300 m. This aquifer is comprised of medium-and coarse-grained sediments, in places inter bedded with gravel. It is either semi-confined/leaky or consists of stratified interconnected, unconfined water-bearing zones. Irrigation water is drawn predominantly from these strata.

The deep aquifer is separated from the overlying main aquifer by one or more clay layers of varied thickness. Geologically and hydro-stratigraphically the deeper aquifer (Holocene/Late Pleistocene) is separated by impermeable or leaky clay/silty clay layers from the upper alluvial aquifer (GWTF, 2002). The depth of the deep aquifers varies from region to region depending on the geology and depositional environment of the sediment. This aquifer comprises mainly grey to dark grey fine- to medium-grained sand that in places alternates with thin sandy shale/clay lenses. The aquifer properties for the shallow and deep aquifers are given in Table 3, and for the different physiographic units existing in the Ganges dependent areas in Table 4.

2.2 *Groundwater resources of Indus and Ganga basins*

Groundwater resources can be classified as static and dynamic. The static resource is the amount of groundwater available in the permeable portion of the aquifer below the zone of water level fluctuation. The dynamic resource is the amount of groundwater available in the zone of water level fluctuation. Sustainable groundwater development requires that only the dynamic resources are tapped. Exploitation of static groundwater resources could be considered during extreme conditions, but only for essential purposes. The static fresh groundwater resource of the Indus and Ganga basin are listed in Table 5.

Table 4. Physiographic units and acquifer types of the Ganges dependent district of Bangladesh.

Physiographic unit	District	Lithology	Aquifer (Type/Thickness/Depth)	Aquifer properties
Northwest Region **Ganges** *(Ganges flood plain, ridges and basins)*	Rajshahi, Pabna	Silty clay bands occur to over 30 m in places thin sand lenses common with small aquifer potential. Below 30 m, coarser sediments with 70 m or more thickness	• Semi-confined • Thickness Range: 10–70 m • Average Depth Range: 30–35 m	Transmissivity (m²/day) *Regional:* 300–4000 Rajshahi: 418–2399 Pabna: 1200–4000 Storativity *Regional:* 0.003–0.23 Rajshahi: 0003–0.10 Pabna: 0.04–0.10 Permeability (m/day) *Regional:* 11–114 Rajshahi: 14–40 Pabna: 26–82
Southwest Region **Old Gangetic Floodplain** *(Part of Gangetic moribund delta with linear ridges comprising of level to very gently undulating levees, inter ridge depressions and stream beds of dominant topographic elements)*	Kushtia, Jessore, Faridpur, Khulna	Predominantly medium-to coarse-grained, well sorted sand grading upward to fine sand to silty sand and clay with distinct fining up sequences. Capped by 10 to 60 m silty/sandy surface clay	• Extremely unconfined, locally semi-confined and confined • Thickness: around 100 m; • Depth: 10 to 60 m with an average of 30 m	Transmissivity (m²/day) *Regional:* 900–3200 Jessore: 161–2300 Khustia: 710–2500 Faridpur: 900–2300
Young Gangetic Floodplain *(Meander floodplains landscape of ridges, young channels and widespread development of past basins)*	Faridpur, Khulna, Barisal	Medium to fine grained sand constitutes the suffer materials. 30–60 m of surface clay with abundant peat layers covers the aquifer	• Semi-confined to unconfined, locally multiple • Thickness: average 55 m • Depth: averages 40 m	Storativity *Regional:* 0.01–0.15 Jessore: 0.004–0.07 Khustia: 0.007–0.252 Faridpur: 0.01–0.15
Coastal Plain *(Coastal deltaic plain with mangrove forest)*	Khulna, Barisal Patuakhali	Extensive thick surface clay up to 90 m, locally up to 150 m. Well defined aquifer of coarse-to medium-grained sand locally occurs at depths between 225 to 325 m with extensive clay cover	• Confined and multiple	Permeability (m/day) *Regional:* 11–65 Jessore: 24–31 Khustia: 15–35 Faridpur: 11–36

Source: GWTF, 2002.

Table 5.   Static fresh ground water fesource (km^3) of IG Basin (CGWB, 1999).

River basin	Alluvium/ Unconsolidated rocks	Hard rocks	Total
Indus	1,334.9	3.3	1,338.2
Ganga	7,769.1	65	7,834.1

Groundwater resources in the Ganga basin are nearly six times that of the Indus basin. The dynamic groundwater resources of different Indian states which fall in the Indo-Gangetic basin are listed in Table 6. Table 7 shows the comparative analysis of groundwater availability and consumption in three basin countries. The Ganga basin falls under 'safe' category on (GW development <70%) compared to the Indus basin (over-exploited category) based their status of groundwater development. This implies that annual groundwater consumption is more than the annual groundwater available in the Indus basin.

3   YELLOW RIVER BASIN (YRB)

The Yellow River, the sixth longest river in the world and second longest river in China, is the "Cradle of Ancient Chinese civilization", because human inhabitants have existed in this region since prehistoric times (Wang et al., 2000). The Yellow River is 5,464 km long with a basin area of 794,712 km^2. It is a home to a population of over 171.9 million (NBS, 1999). It flows from the high mountain areas as high as 4,500 m in Qinghai, eastward from Gansu, Ningxia Hui, Inner Mongolian, Shanxi, Shaanxi, Henan, Hebei and Shandong, and into the Bohai Sea (Yellow sea). The Yellow River water contains sediment concentration as high as 35 kg/m^3 on average after it flows across the loess plateau. The lower reaches of the Yellow River flow through the North China Plain formed by the flooding deposits from the Huang (Yellow) River, the Hai River and the Huai River. The lower channel is about 700 km long through the Huang-Huai-Hai Plain . Flooding occurred 1500 times during the past 2000 years, because of the instability of the river. Fortunately, the flooding problem has been almost controlled in the past five decades with the construction of levees and flow regulation.

  The average flow rate of the river is 58 billion m^3/yr, of which 56% is from the upstream of Lanzhou. Long term annual precipitation in the Yellow River Basin is 452 mm which is unevenly distributed in time and space. Rainfall ranging from 60 to 80% of the annual total occurs between July and October, while only 10 to 20% precipitation takes places during the growing season of winter wheat (March to May). The mean annual evaporation ranges from 1,000 to 3,000 mm. The total water resource is 70.7 billion m^3/yr. Water availability per capita is in the order of 428 to 647 m^3 and irrigation water availability per ha is in the order of 3,765–4,350 m^3/ha. Table 8 shows water resources in the Yellow River and Huang-Huai-Hai plain. Per capita water resource availability is low, varying from 121 to 428 m^3/year.

3.1   *Hydrogeological characteristics of the Yellow River Basin*

Geologically, the Yellow River Basin covers three large geotectonic units: the north China platform, Qinling-Qilian-Kunlun geosyncline and Yunnan-Tibet geosyncline.

Table 6. Estimated dynamic groundwater resources availability, utilization and stage of development in the Indo-Gangetic Basin (Indian part) CGWB (2006).

State	Annual replenishable GW resources (bcm)	Natrual discharge (bcm)	Net GW avail. (bcm)	Annual groundwater draft (bcm)			Stage of GW development (%)
				Iriig.	Domestic & industrial	Total	
Ganga Basin							
Bihar	29.19	1.77	27.42	9.39	1.37	10.76	39
Chattisgarh	5.70	0.45	5.25	0.89	0.22	1.11	21
Delhi	0.30	0.02	0.28	0.20	0.28	0.48	171
Haryana	6.79	0.51	6.28	6.94	0.24	7.18	114
Himachal Pradesh	0.08	0.01	0.07	0.007	0.003	0.01	14
Jharkhand	3.53	0.20	3.33	0.54	0.23	0.77	23
Madhya Pradesh	27.50	1.37	26.13	13.04	0.80	13.84	53
Rajasthan	7.73	0.74	6.99	7.9	0.915	8.815	126
Uttar Pradesh	76.35	6.17	70.18	45.36	3.42	48.78	70
Uttaranchal	2.27	0.17	2.10	1.34	0.05	1.39	66
West Bengal	22.85	2.18	20.67	8.83	0.62	9.45	46
Total	182.29	13.59	168.70	94.44	8.15	102.35	61
Indus Basin							
Haryana	4.15	0.28	3.87	3.69	0.12	3.81	98
Himachal Pradesh	0.35	0.03	0.32	0.08	0.017	0.10	31
Jammu & Kashmir	2.70	0.27	2.43	0.10	0.24	0.34	14
Punjab	23.78	2.34	21.44	30.34	0.83	31.17	145
Rajasthan	2.33	0.23	2.10	2.18	0.39	2.57	122
Total	33.31	3.15	30.16	36.39	1.60	38	126

Table 7.   Estimated groundwater availability and use in the Indus and Ganga Basins (CGWB (2006), BADC (2007), BMDA (2004), Hossain et al., (2002), WECS (2004)).

Basin name	Groundwater available (bcm)	Annual groundwater draft (bcm)			Stage of GW development (%)
		Irrigation	Domestic, industrial & others	Total	
Ganga Basin					
India	168.7	94.4	8.2	102.4	61
Nepal	11.5	0.8	0.3	1.1	10
Bangladesh	64.6	25.2	4.1	29.3	45
	244.8	120.4	12.6	132.8	54
Indus Basin					
India	30.2	36.4	1.6	38.0	126
Pakistan*	55.1	46.2	5.1	51.3	93
	85.3	82.6	6.7	89.3	105

* It is assumed that 90% of groundwater use is consumed by irrigation sector.

Table 8.   Water resources in the Yellow River and Huang-Huai-Hai plain in 2002[1] (Jin et al., 2006).

Catchments/ regions	Rainfall (billion m³/yr)	Surface water (billion m³/yr)	Ground water (billion m³/yr)	Water deducted[2] (billion m³/yr)	Total water resources (billion m³/yr)	Per capita water res. (m³)
Hai-river Basin	127.381	6.408	14.609	9.491	15.899	121
Huai-river Basin	237.591	44.536	34.366	25.647	70.183	343
Yellow River Basin	322.487	35.766	33.401	11.574	47.340	428
Total	6261.029	2724.329	869.718	101.701	2826.130	2200

Notes: [1] Data from the Annual report of 2002 by the Ministry of Water Resources PR China.
[2] The water deducted is the water volume that has to be subtracted from the sum of surface and groundwater due to the interrelation between surface water and groundwater.

Geomorphologically the basin covers the Qinghai plateau, Loess Plateau and Huang-Huai-Hai plains. These control the occurrence of groundwater in the basin (Lin & Wang 2006; Chen & Cai 2000). Areas covered by the three geotectonic units in the basin are in the raio of 6:3:1. Considering the characteristics of geotectonic units, the north China platform has the best conditions for groundwater occurrence. The Qinling-Qilian-Kunlun geosyncline and Yunnan-Tibet geosyncline underwent fierce tectonic movements and generated many folds and faults, and aquifers tend to be small in area. Figure 7 and Table 9 show regional availability of groundwater in the Yellow River Basin and the Huang-Huai-Hai plain.

In the Yellow River Basin groundwater may be intergranular, contained in secondary porosity, and there is some karst. Intergranular storage occurs mainly in the Huang-Huai-Hai plain, alluvial plain of the main branches of the Yellow River and the intermountain basins. The main aquifers consist of sand, gravel and cobble grade material of Quaternary age and occur. Fracture storage is widespread in the hilly regions and Cenozoic basins of upstream and middle reach of the Yellow River Basin. Karst rocks are mainly distributed in the caves and fractures of Cambrian and Ordovician carbonates in Luliang mountain, Taihang mountain, Zhongtiao mountain, Tai Mountain and eastern part of Qinling. Fracture storage in magmatic and metamorphic rocks mainly occurs in weathering material in the mountain and hilly regions of the basin.

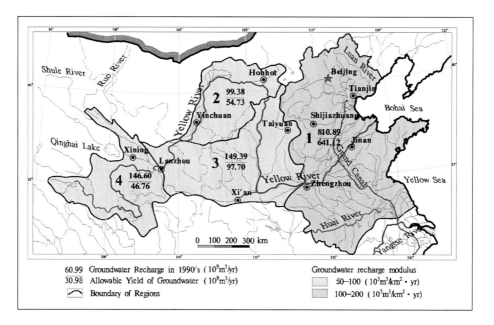

Figure 7.   Groundwater resources in the Yellow River basin and the Huang-Huai-Hai plain (See colour plate section).

Table 9.   Available groundwater resources in Yellow River basin and Huang-Huai-Hai plain[1].

		Groundwater recharge		Allowable yield of groundwater	
Regions		(bcm/yr)	(10^3 m^3/km^2·yr)	(bcm/yr)	(10^3 m^3/km^2·yr)
	1. Huang-Huai-Hai plain	63.533	114.6	51.210	101.8
Yellow River catchments	1. Lower reaches of Yellow River[2]	4.045	162.2	4.053	162.1
	2. Ordos plateau and Yinchuan plain	7.285	56.1	3.958	31.4
	3. Loess plateau	13.054	54.2	9.375	64.0
	4. Upper reaches of Yellow River	14.144	62.5	4.378	20.9
	Sum	38.528	61.1	21.764	43.0

Notes: [1] Data from Ministry of Land and Resources, China, 2003; [2] The number of Huang-Huai-Hai plain includes 4.045 bcm/yr of the lower reaches of Yellow River; [3] Location see Figure 7.
Source: Jin et al., 2006.

3.1.1 *Hydrogeology of the Ordos Basin*

The Loess Plateau has an area of 614,700 km^2, and covers seven provinces of China including most of the Yellow River Basin. It is a comparatively water scarce area. The Ordos Basin is typical of the Loess Plateau.

The Ordos Basin is located in the east of northwestern China with an area of 275,000 km^2 and covers five provinces, i.e. Shaanxi, Gansu, Ningxia, Inner Mongolia and Shanxi (Wang et al., 2005). The total water resource in the basin is about 28.9 billion m^3/year. Allowable groundwater abstraction is 12.91 billion m^3/year. Current water use is about 22.96 billion m^3/year, in which contribution from groundwater is 5.23 billion m^3/year. Water demand in 2010 is estimated at 35.36 billion m^3/year (Wang et al., 2005).

There are four kinds of aquifers in the Ordos Basin: the Quaternary intergranular aquifer, Cretaceous fracture aquifer system, Carboniferous-Jurassic fracture aquifer system and Cambrian-Ordovician Carbonate Karst aquifer system (Fig. 8). All of these aquifer systems are superimposed vertically and connected laterally. There is a close hydraulic connection among these aquifer systems, which makes the Ordos Basin a semi-open groundwater basin (Hou et al., 2006).

Quaternary intergranular aquifer system overlies the Jurassic-Carboniferous rocks consisting of Pleistocene sand which is 60–80 m thick. It is generally in hydraulic connection with the underlying Jurassic aquifer. Water quality is generally good with TDS less than 1000 mg/l and pumping rates up to 40–125 m^3/hour.

Cretaceous fractured aquifer system in the middle-west part of the Ordos Basin is a sub-sedimentary basin overlying the Jurassic and pre-Mesozoic basin. It is artesian and is recharged by surface water and lateral recharge and flows towards rivers. There are no regional aquicludes in the system which is about 1,000 m thick.

Carboniferous-Jurassic fracture aquifer system lies between the Cambrian-Ordovician Karst aquifer system and the Cretaceous aquifer system covering an area of 64,700 km^2. It mainly comprises alternate layers of sandstones and mudstones. Due to cementation, the well yield rate is low. Direct rainfall recharge takes place in the shallow or weathered zones (50–100 m depth). The Carboniferous-Jurassic aquifers in thew river valleys are interconnected with the Quaternary system with good quality and high yielding rates in river valleys.

The Karst aquifer system of Cambrian-Ordovician Carbonate rocks mainly outcrops along the margins of the Ordos basin and is confined by the overlying Carboniferous aluminous shale. More than 70% of the recharge is from direct rainfall recharge and groundwater flows westward parallel to the outcrop. The discharge is in the form of large springs in the sections where the aquifer is cut by the Yellow River and its tributaries. The higher yielding locations are mainly in the lower reaches in the west where most of wells are artisan with yield rates of 1,000 and 10,000 m^3/d, exceptionally up to 50,000 m^3/d. The water quality is good with total dissolved solids less than 1000 mg/l.

3.1.2 *Hydrogeology in Huang-Huai-Hai plain*

Huang-Huai-Hai plain is one of the largest food production regions in China, with an area of 310,000 km^2 (the north of the Yellow River of the plains is also called North China plains, with an area of 134,780 km^2). Groundwater in the Quaternary aquifer system is the principal source for water supply in the plains due to the scarcity of surface water (Fig. 9).

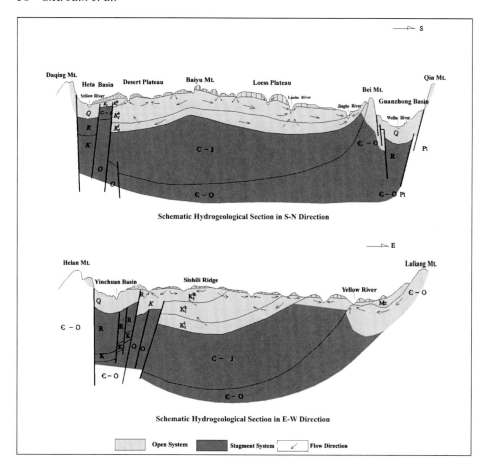

Figure 8.   Hydrogeological section of the Ordos Basin (Wang et al., 2005; Hou et al., 2006).

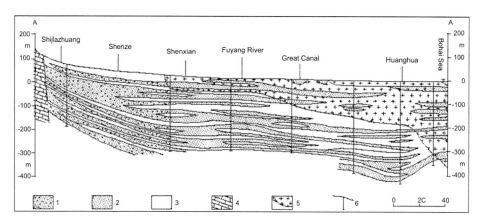

Figure 9.   Profile of the North China plain (Zhang et al., 1994).
1-gravel and sand; 2-sand; 3-clayey soil; 4-carbonate rock; 5-saline water (TDS>2 g/L); 6-borehole.

Groundwater in the west piedmont plain is fresh water (TDS is les than 2000 mg/l). The aquifers are mainly composed of sand and gravel with clayey and loam layers. The fluvial-lacustrine plain is mainly clayey with some sand units. There is a widespread layer of saline water between 50 and 120 m in thickness and of 2000–10,000 mg/l TDS. The groundwater above the saline water is called the shallow fresh water and below the saline water is deep fresh water.

In the east fluvial-marine plain there are clayey beds with intermediate sand layers. There is a marine saline layer 100–300 m in thickness with the TDS reaching a maximum of 35,000 mg/l. There is little shallow fresh water in the fluvial-marine plain, but some fresh water occurs at depth.

Long-term over-extraction of groundwater has resulted in serious environmental degradation such as a continuous decline in groundwater levels resulting from over-exploitation. The rate of decline is 1–2 m/yr for many deep freshwater aquifers and the bottom of the saline water is moving down towards the deep fresh water; land subsidence and degradation of water quality (harmful ions are released during consolidation of the aquitards (clay).

3.2 *Groundwater resources in the Yellow River Basin*

The total water use in the Yellow River Basin was 51.2 billion m^3 in 2006. Total groundwater abstraction in the YRB in 2006 was 13.7 billion m^3, of which 6.9 billion m^3 (50.2%), was for irrigation purposes (Table 10).

4 COMPARATIVE ANALYSIS OF INDUS, GANGETIC AND YELLOW RIVER BASINS

Table 11 presents the overall comparative picture of Indus, Gangetic and Yellow River Basins based on some common parameters. Both the Indus and Gages drain vast areas of more than 1 million sq. km. and as such support well developed surface and groundwater based irrigation systems. Availability of water resources both on a per capita basis and per hectare cultivated is on the decline in all the three basins with its impact most severely felt in the Yellow River Basin followed by the Indus and Gangetic basins. Complex interactions lead to widespread distribution of salinity and fluoride problems in the Indus Basin and the release of arsenic in the lower parts of the Ganges Basin. The aquifers in the Yellow River Basin are threatened by industrial and domestic pollutants. Vast regions of rice and wheat are threatened due to the continuous decline of the water table in the Indus, the western Ganges and Yellow River basins. The eastern Ganges basin still provides an opportunity to harness the available surface and groundwater resources for a more stable, improved and diversified agricultural production system. Decreasing water supplies, increasing demand and a rapidly growing economy have added new challenges to the management agenda. Urgent supply and demand interventions are needed to mediate this gap in sustainability. These include integrated use of surface and groundwater resources, development of groundwater in less exploited areas, selected and need based development of deeper aquifers, safe use of saline and brackish aquifers, enhanced and coordinated efforts for recharge and conservation projects and practices, aligning cropping patterns with water resource availability and improving water productivity at all levels- field, farm, project and basin.

Table 10. Composition of groundwater use (billion m^3) in YRB in 2006 (YRCC, 2007).

Regions			Groundwater use					
Users	Total water use	GW use	Agriculture irrigation	Forest, livestock	Industry	Urban public	Urban domestic	Eco-environment
Up Longyangxia	0.21	0.014	/	0.002	0.003	0.004	0.005	/
Longyangxia-Lanzhou	3.69	0.627	0.141	0.003	0.306	0.046	0.110	0.021
Lanzhou–Toudaoguai	18.53	2.602	1.268	0.271	0.677	0.095	0.257	0.034
Toudaoguai-Longmen	1.75	0.709	0.356	0.064	0.178	0.015	0.090	0.006
Longmen-Sanmenxia	10.13	5.472	2.618	0.390	1.418	0.208	0.799	0.039
Sanmenxia-Huayuankou	3.85	1.769	0.658	0.096	0.672	0.044	0.245	0.054
Down-Huayuankou	12.72	2.319	1.725	0.172	0.159	0.019	0.237	0.007
Inland drainage area	0.33	0.206	0.123	0.060	0.011	/	0.010	/
Total	51.21	13.718	6.889	1.058	3.424	0.431	1.753	0.161
Percentage	100	26.79	50.22	7.71	24.96	3.14	12.78	1.17

Table 11. Comparison of Indus, Gangetic and Yellow River Basins.

Parameters	Ganga	Indus	Yellow
Length (km); Drainage area (km²)	Length–2,525; D.A.–1,089,370	Length–3,199; D.A.–1,029,089	Length–5,464; D.A.–794,712
Country	India, Nepal, Bangladesh, Tibet (China)	Pakistan, India, Afghanistan, China	China
Population (million)	600	150	172
Altitude range	7,010 m to the sea	5,182 m to the sea	5,400 m to the sea
Climate	Hot and Subhumid	Semi-arid	Arid, semi-arid
Rainfall (mm)	up to 1000 (Western part), above 1800 (Eastern and lower part)	400–800	360–380 (Upper), 456–570 (Middle), 614–733 (Lower)
Mean inflow (bcm)	525	175	58
Irrigated area (million ha)	19.5	14.6	7.5
Sediment load (billion ton/year)	0.73 (at Farakka)	0.2	1.6
Erosion processes	*Himalaya*: Landslides, glaciation, channel, sheet and rill, glacier lake-burst floods (GLOFs), and landslide lake-burst floods (LLFs); *Siwalik*: Landslides, sheet and rill, channel; *Plain*: Sheet and rill, channel, wind	Landslides, glaciation, rill, gully, stream bank, pinnacle, wind	*Loess plateau*: Sheet, gully, Debris flow, Landslide, Funnel, wind blowing, Ice-melt (mainly concentrated in middle reach of YRB)
Lithology	*Himalaya*: Crystalline, sedimentary, and meta sedimentary rocks; *Terai Bhabar belt*: Boulder, cobble, pebble, gravels, sands, silt and clays; *Siwalik*: Sedimentary rocks, often partly lithified; *Plain*: Unconsolidated alluvium and aeolian deposits (Clays & silts, gravels & sands, lenses of peat & organic matter, carbonate and siliceous concretions (Kankar); *Floodplain & Delta*: Aquifer of grey medium to course sands capped by 10–60 m surface clay, Silty clay, peat	Young (Quaternary) Alluvial and deltaic deposits, consisting mainly of fine-medium sand, silt and clay; Older (palaeozoic) sediments and crystalline basement rocks are restricted to north (Gilgit and J&K)	North China platform (60%) followed by Qinling-Qilian-Kunlun geosyncline and Yunnan-Tibet geosyncline; Large area of unconsolidated Quaternary (downstream); sedimentary formation of geosynclinal clastic rock and volcanic rock, along with metamorphic rock (upstream)

(Continued)

Table 11. (*Continued*)

Parameters	Ganga	Indus	Yellow
Aquifer characteristics	*Central plain:* Transmissivity (m²/d): 520–3530; Storage coefficient: $1.13 \times 10^{-3} - 1.5 \times 10^{-3}$; Hydraulic conductivity (m/d): 37–55 *Floodplain & Delta:* Transmissivity (m²/d): 1000–5000; Storage coefficient: $1.13 \times 10^{-3} - 1.5 \times 10^{-3}$; Hydraulic conductivity (m/d): 11–114	Plains between Ravi and Chenab rivers: Hydraulic conductivity: 47–120 m/day; Specific yield: 0.01–0.13; Groundwater salinity: 500–1000 μS/m	Transmissivity (m²/d): 100–5000; Storage coefficient: 0.01–0.25; Hydraulic conductivity (m/d): 0.1–50
Groundwater yield (m³/hr)	*Bhabar* : 100–300; *Tarai*: 100–200; *Central plain:* 90–200; *Floodplain & Delta:* 50–200	*Alluvial plain:* 100–300; *Coastal area:* up to 50	40–210
Net annual GW available (mcm/km²)	0.225	0.083	0.027
Net annual groundwater draft (mcm/km²)	0.122	0.086	0.017
Stage of groundwater development (%)	54	105	63
Dependency of irrigation on GW (%)	91	92.5	50
GW quality	*Salinity*: Haryana, Rajasthan (western Ganga) *Arsenic*: West Bengal and Bengal delta plains in Bangladesh	*Salinity*: Punjab and part of lower Indus basin *Arsenic*: Part of Punjab and Sindh	*Salinity*: Ningxia and Inner Mongolia areas (middle reach) *Arsenic*: Shan Xi Province and Inner Mongolia

Groundwater in the Indus-Gangetic and Yellow River basins offer great opportunities for sustaining the livelihoods of about one billion people. Better use of the resource can be achieved by improved domestic socio-economic and policy environments in the respective countries (India, Nepal and Bangladesh and China) and also through a shared modern, transparent and real-time data management system on resource availability, abstraction, quality, and other hydrogeological variables.

REFERENCES

Aggarwal, P.K., A. Basu, J.R. Poreda, K.M. Khukarni, K. Froelich, S.A. Tarafdar, M. Ali, N. Ahmed, A. Hossain, M. Rahman. and S.R. Amed (2000). Isotope hydrology of groundwater in Bangladesh: Implications for characterization and mitigation of arsenic in groundwater. IAEA-TC Project: BGD/8/016.

BADC (2007). Minor Irrigation Survey Report 2006–2007. Bangladesh Agricultural Development Corporation, Ministry of Agriculture, Govt. of Bangladesh.

BMDA (2004). Data Bank: Groundwater and Surface Water Resources Bangladesh. Barind Integrated Area Development Project (BIADP). Barind Multipurpose Development Authority, Ministry of Agriculture, Govt. of Bangladesh.

BWDB (1989). Report on the compilation of aquifer test analysis results as on June, 1988, BWDB Water Supply Paper No. 502.

BWDB (1994). Report on the compilation of aquifer test analysis results as on June, 1993, BWDB Water Supply Paper No. 534.

BWDB-UNDP (1982). Groundwater Survey: The Hydrogeologic Conditions of Bangladesh. Technical Report. Bangladesh Water Development Board.

CGWB (1984). Project Findings and Recommendations. Ground Water Studies in Upper Yamuna Basin, Technical Series-P Bulletin No. 4, Central Ground Water Board, Government of India, Faridabad.

CGWB (1999). Ground Water Resources of India, Central Ground Water Board, Govt. of India, New Delhi.

CGWB (2006). Dynamic Ground Water Resources of India, Central Groundwater Board. Government of India.

CGWB (2008). State Ground Water Profile, Central Groundwater Board. Government of India. http://cgwb.gov.in/gwprofiles.html

Chen M. and Cai Z., (2000). Groundwater resources and the related environ-hydrogeologic problems in China. Beijing: Seismological Press.

Fahlbusch H., Schultz B. and C.D. Thatte (2004). In: The Indus Basin-History of Irrigation, Drainage and Flood Management, ICID Publication, ISBN: 81-85068-77-1, New Delhi.

Geological Survey of Pakistan (2000). Hydrogeological Map of Pakistan. http://www.gsp.gov.pk/pakistan/ground_water. html (accessed on 20 August, 2008).

GWTF (2002). Report of the Ground Water Task Force, Ministry of Local Govt. Rural Dev. and Co-operatives, Govt. of Bangladesh.

Hossain M., Faruque, H.S.M. and Alam S., (2002). Sustainable Groundwater Management: Irrigation Economics, Quality, Energy and Poverty Alleviation in Bangladesh, Country Paper Presented in the Seminar on Forward Thinking Policies for Groundwater Management, Energy, Water Resources and Economic Approaches, New Delhi, India, 2–6 September 2002.

Hou G., Liang Y., Yin L (2006). Groundwater system and resources of the Ordos basin, China. Proceedings of the 34th Congress of International Association of Hydrogeologists, October 9–13, 2006. Beijing, China.

IUCN (2005). Pakistan Water Gateway, www.waterinfo.net.pk (accessed on 20 August, 2008). International Union for the Conservation of Nature and Natural Resources (IUCN).

IWMI (2008). http://dw.iwmi.org/ (accessed on 20 August, 2008).

Jain, S.K., Agarwal, P.K. and Singh, V.P. (2007). Hydrology and Water Resources of India, Springer Publisher, The Netherland.

Jin M., Liang X., Cao Y. and Zhang R. (2006). Availability, Status of Development, and Constraints for Sustainable Exploitation of Groundwater in China. In: Sharma, B.R., Villholth, K.G. and Sharma, K.D.

2006 (Eds.). Groundwater Research and Management: Integrating Science into Management Decisions. Malhotra Publishing House, New Delhi. 47–61.

Karanth, K.R. (1987). Ground Water Assessment, Development and Management, Tata McGraw-Hill Publishing Company Limited, New Delhi, pp. 576–657.

Khan, A.R. (1999). An Analysis of Surface Water Resources and Water Delivery Systems in the Indus Basin, International Water Management Institute (IWMI), Lahore, Pakistan.

Khan, A.R., Ullah M.K. and Muhammad, S. (2002). Water Availability and Some Macro Level Issues Related to Water Resources Planning and Management in the Indus Basin Irrigation System in Pakistan.

Lin, X. and Wang, J. (2006). Groundwater resources in Yellow River Basin and their renewable capability. Zhengzhou:Yellow River Conservancy Press. (in Chinese).

Ministry of Land and Resources of the People's Republic of China (2003). Annual report, 2003, Beijing.

Ministry of Water Resources of the People's Republic of China (2002). Annual report, 2002. Beijing.

NBS (1999). China Statistical Yearbook 1998. National Bureau of Statistics of China (NBS), China Statistical Press, Beijing. (in Chinese).

NESPAK. (1991). Contributions of Private Tubewells in the Development of Water Potential. Ministry of Planning and Development, Islamabad, Pakistan.

Revenga, C., Murray S., Abramovitz J. and Hammond, A. (1998). Watersheds of the World: Ecological Value and Vulnerability. Washington, DC: World Resources Institute.

Rogers P., Lydon, P. and Seckler, D. (1989). Eastern Waters Study: Strategies to Manage Flood and Drought in the Ganges-Brahmaputra Basin, ISPAN, Virgina, USA.

Shah, T., Deb Roy, A., Qureshi, A.S. and Wang, J. (2003). Sustaining Asia's groundwater boom: An overview of issues and evidence. Natural Resources Forum, 27 (2): 130–140.

Shamsi, R.A. (1989). Hydrogeological map of Pakistan. 1:2,000,000 scale. WAPDA, Lahore, Pakistan.

Sharma C.K. (1984). Ground Water Resources of Nepal, Sangeeta Publication.

Sharma C.K. (1995). Shallow Aquifers of Nepal, Sangeeta Publications.

United Nations Educational, Scientific and Cultural Organisation (UNESCO) (2001), Courier, Internet Newsletter, October 2001 Ed.

Valdiya, K.S. and Bartarya, S.K. (1989). Diminishing discharges of mountain springs in a part of Kumaun Himalaya, Current Science, 58 (8): 417–426.

Wadia, D.N. (1990). Geology of India. Tata McGraw-Hill Publishing Company Ltd., New Delhi.

Wang R., Ren H. and Ouyang Z. (2000). China Water Vision, China Meteorological Press, Beijing, China.

Wang D., Liu Z. and Yin L. (2005). Hydro-geological characteristics and groundwater system of the Ordos Basin, Quaternary Science, 25 (1): 6–14. (in Chinese).

WAPDA/EUAD 1989. Booklet on hydrogeological map of Pakistan, 1:2,000,000 scale. Water & Power Development Authority, Lahore and Environment & Urban Affairs Division, Govt. of Pakistan, Islamabad.

Water and Energy Commission Secretariat (WECS) 2004. "National Water Plan." Draft report, Water and Energy Commission Secretariat, Kathmandu.

WWF (2005). An Overview of Glaciers, Glacier Retreat, and Subsequent Impacts in Nepal, India and China. WWF Nepal Program. http://www.freewebs.com/climatehimalaya/WWF2005/WWF-Country%20case%20 study-India. pdf (accessed on 20 August, 2008).

YRCC (2007). Annual report on water resources in the Yellow River Basin. http://www.yellowriver.gov.cn/other/hhgb/2006.htm. (in Chinese) (accessed on 20 August, 2008).

Zahid A. and Hassan MQ. (2007). Arsenic distribution and characterization of multi layer aquifer system in Bengal delta for sustainable use of groundwater. Pre Conference paper Volume on Water and Flood Management, IWFM, Bangladesh University of Engineering And Technology (BUET), Dhaka, Bangladesh, Vol. 1: 19–27.

Zhang R., Jin M., Sun L. & Gao Y. (1994). Systems analysis of agriculture-water resources-environment in Hebei Plain, Proceedings of the Water Down Under 1994 Conference. Vol. 1, Adelaide, Australia, 453–458.

Region specific case studies

CHAPTER 4

Groundwater resource issues and the socio-economic implications of groundwater use: Evidence from Punjab, Pakistan

S.M. Kori
Mehran University of Engineering and Technology, Sindh, Pakistan

A. Rehman
International Waterlogging and Salinity Research Institute (IWASRI), Lahore, Pakistan

I.A. Sipra
Irrigation Research Institute, Old Anarkali, Lahore, Pakistan

A. Nazeer & A.H. Khan
International Water Management Institute (IWMI), Lahore, Pakistan

ABSTRACT: Irrigated agriculture in Pakistan was once dependent upon surface water although the large contiguous Indus Basin Irrigation System (IBIS) is now heavily dependent on groundwater resources. The number of private tubewells in the country, particularly in the province of Punjab grew exponentially over the last two decades. The overall objective of this study was to investigate the present groundwater issues and water use patterns and their impact on farmers' income in the Punjab province of Pakistan. Results of the study revealed that informal groundwater markets were better developed in the mixed-wheat zone as compared to rice-wheat zone. Water productivity of the water-buyer was slightly lower than pump-owner suggesting that pump-owners over-irrigate their fields as compared to water buyers. Farmers reported three major problems associated with groundwater irrigation; high diesel cost, difficulty in getting new electricity connections and falling water tables.

1 INTRODUCTION

Agriculture is the largest sector of Pakistan's economy. It serves as one of its principal engines of growth by contributing almost one-fourth of the country's Gross Domestic Product (GDP), employing 44–50% of the labour force, supporting 75% of the population and being responsible directly or indirectly for 60–70% of export earnings (GoP 2002; World Bank, 2004). Agriculture is the major and primary user of water and it uses about 95% of available water resources. 80% of the agricultural production comes from 17 million ha (Mha) of irrigated land (World Bank, 2004). The vital role of water resources for increasing agricultural and economic growth in Pakistan cannot be overemphasized.

Pakistan has the largest contiguous irrigation system in the world, serving more than 17 Mha in the alluvial plains of the Indus Basin. An additional 5 Mha is under rainfed agriculture. A major source of surface water in Pakistan is the Indus River and its tributaries (Jhelum, Chenab, Ravi, on the east and Kabul River on the west). Because of high variability in weather during the year, flows into the Indus River and its tributaries are also very variable. The average annual flows into the Indus System are about 192 billion (10^9) cubic meter (BCM) of which around an average of 128 BCM is diverted for irrigation through a well integrated irrigation system (PWP, 2004).

The total groundwater endowment of Pakistan is estimated at 68 billion cubic meters (BCM) of which about 59 BCM is currently exploited (WAPDA, 1997). The density of private tubewells per 1000 ha in Punjab has increased from 3 in 1965 to 46 in 2002 (Qureshi and Mujeeb, 2003). A decline in groundwater levels is occurring in many fresh groundwater regions due to intensive exploitation of the groundwater resources mainly through the rapidly increasing number of private pumps. Before the introduction of weir controlled widespread surface irrigation in the early twentieth century, the average groundwater table in the Indus Basin varied from 12 m in Sindh to about 60 m in the Punjab. The water table started rising due to poor irrigation management, lack of drainage facilities and additional recharge from canals, secondary channels, tertiary water courses and farms. At some locations, the water table rose to the ground surface or very close to the surface thereby causing water-logging and soil salinity, which in turn led to reduced agricultural productivity.

The government of Pakistan embarked upon a programme of Salinity Control and Reclamation Projects (SCARPS) in the late 1950s when high capacity deep tubewells were installed to control groundwater table rise. Over a period of about 30 years, some 13,500 public tubewells were installed. These proved to be quite effective in lowering the water table and supplementing supplies in many parts of the Indus Basin system. The declining performance and the increasing operation and maintenance cost of the SCARP deep tubewells led to their disinvestment by the government during the late 1980s and 1990s. The lack of timely availability and reliability of water to the farmers from the public tubewells led to private investment in shallow tubewells for irrigation. The availability of locally manufactured inexpensive diesel engines resulted in the rapid growth of these tubewells in most of the fresh groundwater areas in Punjab. The biggest increase took place over the 1990s when the tubewell density almost doubled from the preceding decade (Shah et al., 2003). Based on recent statistics, there are around 0.89 million tubewells in the Punjab, out of which 90% are operated with diesel (GoP, 2005).

The exploitation of fresh groundwater resources provided an opportunity for farmers to supplement their irrigation requirements and cope with the vagaries of the surface supplies thus increasing accessibility and reliability of water while increasing crop production. However, due to uncontrolled and unregulated use of groundwater, the problems of over-draft of the aquifer and saline water intrusion have emerged in many areas of the Indus Basin (Kijne, 1999). In Punjab as a whole, 25% more groundwater is being pumped out than is being recharged leading to declining water-table in many areas (Meinzen Dick et al., 1997). The increasing water table depth and high diesel cost is making groundwater use quite uneconomical. Secondary salinisation associated with the use of poor quality groundwater for irrigation has further compounded the problem. Therefore, salt affected soils are becoming an important ecological entity in the Indus Basin of Pakistan. Generally, the major reason of emerging groundwater problems is that the management of groundwater resources could not keep pace with its development. The major issues pertaining to groundwater governance in Pakistan are high population density, exceedingly large number of groundwater users, low levels of resource management capacity, high share of

Table 1. Sample size by Zone/District and villages.

Zone/District	Village	Sample size	Pump owners	Water buyers
Zone I (District Gujranwala)	Chadhiala kalan	20	20	0
	Kot Bilal	20	12	8
	Sub-total	40	32	8
Zone II (District Jhang)	Chak 125/JB	20	12	8
	Chak 132/JB	20	11	9
	Sub-total	40	23	17
Total sample		80	55	25

agriculture in GDP, widespread poverty and dependence of rural livelihoods on tubewell irrigation, poor institutional arrangements and lack of information on groundwater use.

The objectives of this paper are as follows:

1. to investigate groundwater issues in irrigated agriculture of Punjab, Pakistan;
2. to investigate the trends in water use patterns for crop production in the study area; and
3. to evaluate the impact of increasing groundwater use on agricultural profitability of the pump owners and water buyers.

2 METHODOLOGY

This study was conducted in two zones (district Gujranwala as Zone I and district Jhang as Zone II) of the Rechna Doab located in the Punjab province of Pakistan. Rechna Doab is a part of the IBIS and represents the land located between the River Chenab and River Ravi (Figure 1). Four villages in the Doab were selected for detailed case study, of which two villages (Chadhiala Kalan and Kot Bilal) were selected in the high rainfall zone with rice—wheat cropping pattern while the other two (Chak No. 125-JB and Chak No. 132-JB) were selected in the medium rainfall zone representing the sugarcane—wheat cropping pattern (Figure 2). Key parameters on groundwater use, existing groundwater technology, engineering and socio-economic aspects were investigated through pre-designed questionnaires and interviewing schedules.

The village level detailed information was gathered from key informants through a pre-designed village profile questionnaire and 20 individual farmers were selected for in-depth interviews in each village. In all, 80 farmers were interviewed in the four selected villages (Table 1). In addition to this, one well-driller working in the vicinity of each selected village was also interviewed to obtain information on groundwater history, water table trends and soil and water quality. The survey team also met with the officials of the local water departments and conducted informal interviews to get information on the groundwater situation in the area.

3 GENERAL DESCRIPTION OF THE STUDY AREA

Rechna Doab lies between longitude 71°48' to 75°20' east and latitude 30°31' to 32°51' north and has a gross area of 2.98 Mha, out of which 2.32 Mha is the cultivatable land. Surface elevation of this area varies from 138 m (west-south) to 302 m (north-east). The physiography of the Rechna Doab consists of (a) active flood plains, (b) abandoned flood

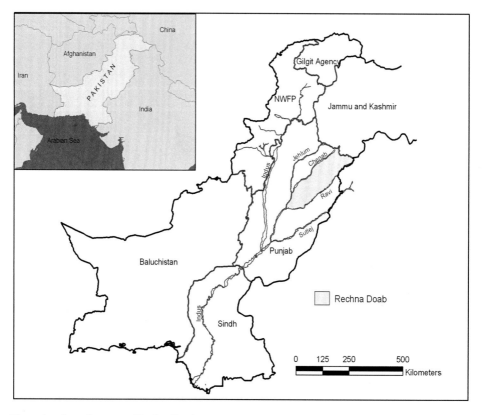

Figure 1.   Location map of Rechna Doab.

plains, (c) Bar Uplands and (d) Kirana Hills (longitudinal across the doab). The soil texture of Rechna Doab is clay, clay loam, silty, silty-clay, silty-loam, fine sandy loam, loamy-sand, sandy clay loam, and sandy clay. There is a southwesterly slope with a gradient of 0.38 and 0.29 m/Km in the upper part and the lower part, respectively. Surface salinity is found in more than 20% of the cultivated area in the Rechna Doab (about 1.17 Mha).

The climate of Rechna Doab is sub-tropical and categorized as semi-arid, indicating large seasonal fluctuations in temperature and rainfall. Summers are long and hot, lasting from April through September with temperature ranging from 21°C to 49°C. The winter season lasts from December through February with maximum temperature ranging between 25°C and 27°C. Mean annual precipitation is about 650 mm in the upper Doab, falling to 375 mm in the central and lower area. Nearly 75% of the annual rainfall occurs during the monsoon season—from mid-June to mid-September (Ahmed et al., 2007).

4   GROUNDWATER RESOURCE CONDITIONS

The gross area of Rechna Doab is 2.98 Mha, out of which 2.39 Mha is cultivated and classified as the irrigated crop land (Jehangir et al., 2002). Three types of irrigation sources are generally used by the farmers in the Rechna Doab, i. e., surface water (canal irrigation) only, groundwater (tubewell irrigation) only and use of surface and groundwater. Most of

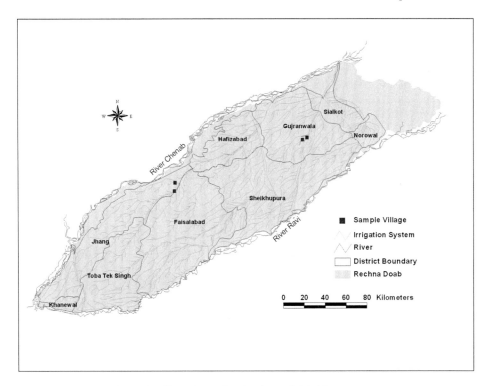

Figure 2.   Location of sample villages in Rechna Doab, Punjab Pakistan.

the present irrigated land in the Rechna Doab is under the command area of two major canals: (i) the UCC, which covers the bulk of the upper one-third of the Doab; and (ii) the LCC, which covers most of the lower two-thirds of the Rechna Doab. Irrigated agriculture started in the Rechna Doab in 1892 with the construction of the LCC, while UCC started operating in 1915. Out of total 2.39 Mha of cultivated area in Rechna, about 1.42 Mha is under the command area of the LCC and 0.59 Mha is under the command area of the UCC with a designed discharge of 275 and 467 cumecs, respectively (Ahmed, 1988; Rehman et al., 1997; Ashraf & Khan, 1984). The surface irrigation system was designed for a lower cropping intensity of about 60–70% per annum which has now increased to more than 120 percent by using groundwater from more than 200,000 tubewells (GoP, 2005) in Doab for feeding increasing population. In general the canal water supplies are inadequate and insufficiently reliable to meet the overall crop-water requirements. A recent study found that due to the scarcity of surface water and conveyance losses, the present surface water supply is not adequate for meeting the irrigation requirement. The same study also found that only 2.5% of the sample farmers reported receiving adequate canal water, while only six % reported receiving reliable and timely canal water supplies in the Rechna Doab (Nazeer et al., 2007). According to Federal Bureau of Statistics, the surface water availability reduced by 26% during the cropping year 2000–01 as compared to 1996, while the growth of private tubewells during this period increased by 59% in the Punjab province alone.

Groundwater has become the most reliable source of irrigation in the Rechna Doab. The growth of tubewell coverage during the last two decades has led to falling water tables in many locations. The exploitation of usable groundwater provided an opportunity to the farmers of these areas to supplement their irrigation requirements but excessive lowering of the water table is making the pumping more expensive and wells are going out of production (Qureshi et al., 2003). Deterioration of groundwater quality in the form of secondary salinization is another serious issue in the Doab. The groundwater quality in the Doab is divided into three distinct zones (i) Fresh Water Zone: the upper Rechna with high rainfall and rice cultivation has good quality water (TDS <1000 mg/l) covering an area of 1.36 Mha. (ii) Mix Zone (TDS 1,000–3,000 mg/l) and (iii) Saline Zone covering the central part of lower Rechna has poor quality groundwater (TDS >3,000 mg/l) (Jehangir et al., 2002). Figure 3 presents an overall picture of groundwater quality status in the Doab. A review of literature shows that both fresh as well as saline water zones are present within the LCC command area, whereas the UCC command area is mostly underlain by fresh water layers (Table 2).

4.1   *Characteristics of aquifer*

The inter-fluvial region of the Rechna Doab is flat with little natural drainage and is underlain by a deep, unconfined, high yielding aquifer that is relatively homogeneous and highly anisotropic. Bennett et al. (1967) and Jehangir et al. (2002) provided detailed hydrologic

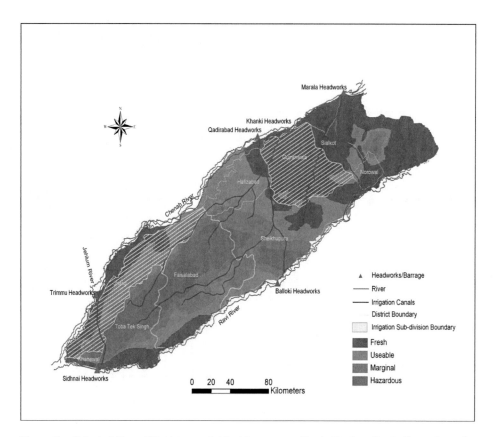

Figure 3.   Selected Zones/Districts overlaid with water quality in Rechna Doab (See colour plate section).

Table 2.   Distribution of fresh and saline groundwater area in Rechna Doab, Punjab.

Canal circles	Fresh water area (ha)	Saline water area (ha)	Total (ha)
Upper Chenab canal circle	0.65	–	0.65
Lower Chenab canal west circle	0.38	0.09	0.47
Lower Chenab canal east circle	0.60	0.16	0.76
Total	1.63	0.25	1.88

Source: Jehangir et al., 2002.

Table 3.   Aquifer characteristics by canal commands in the Rechna Doab.

Irrigation canal circles	Number of test well	Hydraulic conductivity (m/s)		Specific Yield (%)	
		Range	Mean	Range	Mean
Upper Chenab canal circle	16	3.05E-04–1.52E-03	9.14E-04	0.01–0.22	0.082
Lower Chenab canal east circle	21	6.10E-04–3.05E-03	1.22E-03	0.06–0.33	0.175
Lower Chenab canal west circle	10	3.05E-04–2.13E-03	1.22E-03	0.06–0.29	0.129

Source: Jehangir et al. (2002) adopted from Khan (1978).

description and the general characteristics of the aquifer in Rechna Doab. Groundwater in the alluvial aquifer is generally unconfined, however, because of the random distribution of clayey strata, the aquifer is anisotropic and lateral permeability is generally much greater than vertical permeability. The water-bearing characteristics of the alluvial aquifer were evaluated by about 140 pumping tests in the Punjab province. Table 3 below presents statistics on the range and mean for hydraulic conductivity and specific yield values in three irrigation circles of Rechna Doab.

4.2   *Geological formation of the study region*

The geological formations in both surveyed zones (Zone I: District Gujranwala and Zone II: District Jhang) are presented in Tables 4 and 5. The geological formations in Zone I most commonly consisted of four lithological strata at different approximate depths, which included clay (0–3.7 m), sand (3.7–19.8 m), clay (19.9–22.9 m) and gravel plus sand (>22.9 m). Whereas, in Zone II, most commonly three layers were noted consisting of loamy-clay (0–7.6 m), sand (7.6–24.4 m) and gravel plus sand (>24.4 m). The survey area of Zone II also consisted of intermediate thin impervious layers at some areas.

4.3   *Historical trend in groundwater level*

Our survey team visited the related government offices in the study area but records of any historical trends in water table depth were not available at the village level. Neither the well drillers nor the farmers had been monitoring levels. However, the well drillers provided estimates of how much depletion has been occurring for the last 25 years. Based on well drillers' perceptions, the water table depth in Zone I was about 0.9 to 1.2 m below ground level in 1980, whereas in Zone II, the water table depth was varying between 2.1 to 3.0 m for the two villages during the same period. Substantial depletion was noted over 25 years mainly due to the continuous extraction of groundwater which has exceeded annual average

Table 4.   Geological formation of Zone I/District Gujranwala.

Layers	Depth range (m)	Local name	Scientific name	Colour
STRATA-1	0–3.7	*Mati*	Clay	Reddish/blackish brown
STRATA-2	3.7–19.8	*Rati*	Sand	Brown
STRATA-3	19.8–22.9	*Mati*	Clay	Blackish brown
STRATA-4	Above 22.9	*Rati + kanker*	Gravel + sand	Blackish brown

Source: Data collected through well drillers' perception survey during November 2006.

Table 5.   Geological formation of Zone II/District Jhang.

Layer	Depth range (m)	Local name	Scientific name	Colour
STRATA-1	0–7.6	*Mati*	Loamy clay	Brown; some times yellowish-brown or reddish-brown
STRATA-2	7.6–24.4	*Rati*	Course sand	Wheatish-brown
STRATA-3	24.4–54.9	*Rati + kanker*	Gravel + sand	Blackish brown
Intermediate layers (some cases)	Intermediate different thicknesses	*Chikni Mati*	Hard Clay	Blackish- or Redish-brown

Source: Data collected through well drillers' perception survey during November 2006.

Table 6.   Historical trend of water-table depletion in study area (m) based on well drillers' perceptions.

	Zone-I		Zone-II	
Year	Chadhiala Kalan	Kot Bilal	Chak No.125	Chak No.132
1980	0.9	1.2	2.1	3.1
1985	1.5	1.5	3.1	4.6
1990	3.7	2.1	4.6	7.6
1995	4.9	2.4	6.4	9.1
2000	5.5	3.1	8.5	12.2
2005	6.1	4.0	10.7	15.2

Source: Through local well-drillers' interviews (December 2006).

recharge. Based on the local well drillers' perception, the water table depletion data for the last 25 years is listed in Table 6.

4.4   *Evolution of groundwater extraction technologies*

The use of groundwater for irrigated agriculture in the study area has a long history. In the early days, groundwater abstraction was by means of open wells with rope and bucket and Persian wheels. Large-scale pumping and use of groundwater for irrigation started during 1960s with the launching of SCARPS The farmers in the study area had previously enjoyed abundant water supply through public tubewells (SCARPS tubewells) installed at the head of the outlet delivering water in conjunction with canal water for

Table 7.   Average drilling cost of tubewell in the study area (Pakistan Rupees PKR[1]).

Year	Zone-I		Zone-II	
	Chadhiala Kalan	Kot Bilal	Chak No.125	Chak No.132
1980	1000	700	500	1600
1985	1500	1000	600	2700
1990	1500	1000	800	3000
1995	1500	1500	1000	3500
2000	2000	2000	2000	4000
2005	2500	4000	5000	6000

[1] USD ~ PKR 62.7 as in March 2008.
Source: Through local well-drillers' interviews (December 2006).

crop production during the 1960s through the late 1980s. Qureshi et al. (2003) reported that a large number of high-capacity tubewells (0.08 to 0.14 m^3/sec) were installed to lower the water table and supplement irrigation supplies. Later on, these public tubewells were discontinued mainly due to very high maintenance cost. Closure of these public tubewells led the farmers to install their own tubewells. Initially, public tubewells were substituted by electric pumps in private tubewells where farmers were given a subsidy in the form of free electricity connections. However, due to frequent electricity fluctuations and shutdowns farmers switched to diesel tubewells. Qureshi et al. (2003) indicated that disinvestment of public tubewells led to a proliferation of private tubewells with a capacity of about one cusec (0.03 m^3/sec and even less) by farmers in the 1970s and 1980s. Subsidized power supply and the introduction of locally manufactured diesel engines and cheap Chinese engines provided the motivation to increase the number of private tubewells. The estimated number of groundwater users in Punjab province is now over 2.5 million farmers (Qureshi et al. 2003), exploiting groundwater directly or buying water from neighbours.

An overwhelming majority of the farmers located in the study area were irrigating their fields through diesel pumps. In shallow groundwater areas the borehole depth ranged from 24.4 to 36.6 m while in deep water table areas it ranged between 45.7 to 76.2 m. The majority of the farmers in the study area had diesel centrifugal pumps using locally manufactured or Chinese pumps. However, there were few cases where farmers had only boreholes and were operating these through tractor engines. Almost all the farmers in Zone I were extracting groundwater through diesel centrifugal pumps where groundwater was shallow and rice cultivation was high, while in case of Zone II, there were cases of electricity operated centrifugal pumps (about 2% of total sample tubewells). Operational costs of electric tubewells are lower than diesel pumps.

The local professional well drillers were asked to provide information on the changing drilling cost over last two decades. Data presented in Table 7 shows that the cost of drilling wells varied from village to village. On an overall basis, drilling cost over the period increased by 60 to 90 percent in the selected villages of the study area. The variation in cost depended on the boring depth and the drilling technology, which has been changed from manual percussion to diesel percussion.

Well-drillers were generally not aware of any registration requirements and regulations in the survey region. They were not registered with any groundwater authority and were

Table 8.   Average discharge of tubewells in the study area (m^3/hr).

Village name	Tubewell discharge (m^3/hr)
Chadhiala Kalan	69
Kot Bilal	75
Chak No. 125/JB	76
Chak No. 132/JB	90

Source: Field measurements.

Table 9.   Ownership and access to groundwater irrigation.

Parameters	Zone I	Zone II	Total
Number of farmers interviewed	40	40	80
Number of pump owners	32	23	55
Number of non-pump owners	8	17	25
Shared well owners (%)	3	13	7
Water sellers (% of pump owners)	6	17	11
Groundwater market participation rate (% of all respondents who participate in groundwater markets either as water seller or buyer or both)	8	23	15

free to install tubewells without considering the technical aspects of water quality and groundwater depletion.

Discharge of selected tubewells in the study region was measured. The tubewells were selected on the basis of representation of different categories of water extraction mechanisms (WEMs) in terms of capacity, type of prime mover and source of energy. Results of discharge measurement indicated that the average discharges of the selected tubewells ranged from 69 to 90 m^3/hr (Table 8).

5   SOCIO-ECONOMIC IMPLICATIONS OF GROUNDWATER USE

5.1   *Pump ownership status and access to groundwater irrigation*

An important issue relating to management and regulation of groundwater is the question of ownership of the resource. The sample farmers were asked to give their perceptions on the ownership of groundwater resources. On an overall basis, 20% of the sample farmers were of the opinion that groundwater resource is the "property of the landowner"; while 76% of the respondents thought it as a "common property/state property", and 4% reported that they did not know about the ownership of the resource. The results in Table 9 illustrate the ownership of private tubewells in the study area.

Water from electric tubewells is sold on relatively lower per hour rate as compared to diesel tubewells (see Table 17). The incidence of water selling is less in the rice-wheat zone (Zone I) because almost all farmers in this zone have their own tubewells since the rice-wheat cropping pattern needs an assured water supply, which water markets may not provide.

Table 10.   Average number of tubewells per farmer in the study area.

	Zone I			Zone II		
Parameter	Diesel WEM	Electric WEM	Total WEM	Diesel WEM	Electric WEM	Total WEM
Number of WEM	122	0	122	36	3	39
Number of farmers	32	0	32	20	3	23
Average Number of WEM per farmer	3.8	0.0	3.8	1.8	1	1.9

Table 11.   Landholdings and tubewell ownership in study area.

		Farmer's land holding categories											
		Total sample		Sub-marginal <0.5 ha		Marginal 0.51 to 1.0 ha		Small 1.01 to 2.0 ha		Medium 2.01 to 4.0 ha		Large >4.01 ha	
Survey villages	Farmer's status	No.	(%)	No.	(%)	No.	(%)	No.	(%)	No.	(%)	No.	(%)
Zone I	Owners	32	80	0	0.0	0	0.0	5	15.6	8	25.0	19	59.4
	Buyers	8	20	1	12.5	2	25.0	3	37.5	2	25.0	0	0.0
Zone II	Owners	23	58	1	4.3	1	4.3	1	4.3	9	39.1	11	47.8
	Buyers	17	42	3	17.6	5	29.4	7	41.2	2	11.8	0	0.0
Overall	Owners	55	69	1	1.8	1	1.8	6	10.9	17	30.9	30	54.5
	Buyers	25	31	4	16.0	7	28.0	10	40.0	4	16.0	0	0.0

An overwhelming majority of farmers in the study area owned diesel pumps. These results are also comparable with the official statistics reported by the government agencies (GoP, 2005). 91% of tubewells in Punjab province are diesel operated while this figure is 92% for the Rechna Doab. Table 10 indicates that average number of tubewells in Zone I is significantly higher (almost double) than Zone II. Generally the larger farmers owned 3 to 4 tubewells each as compared to small farmers who owned only one tubewell each.

Data on land holding patterns of the sample farmers was classified into five categories ranging from sub-marginal land holding size (<0.5 ha) to large farmers (>4.01 ha) and was analyzed by pump ownership (Table 11).

5.2   *Cropping pattern and yields of major crops*

There are two distinct cropping zones in the Punjab Province, namely, the Punjab Rice-Wheat (PRW) zone and the Punjab Sugarcane-Wheat (PSW) zone (WAPDA, 1997). However, Rechna Doab can fairly be divided into three cropping zones of Upper, Middle, and Lower Rechna mainly based on existing cropping patterns. Upper Rechna is characterized by rice-wheat cropping pattern, middle Rechna as sugarcane-wheat and the lower Rechna as the mixed-wheat cropping pattern having the mix of wheat, sugarcane and cotton (Nazeer et al., 2007). Similar to other parts of the Punjab and all other regions, there are two major cropping seasons in the Doab; i. e., Kharif (summer) and Rabi (winter) season. Kharif refers to a cropping season in Pakistan extending from April to September in which rice, cotton, sugarcane, maize and kharif fodder are the major crops. Rabi refers to a cropping season from October to March in which wheat and gram are the major crops (Nazeer et al., 2007).

Zone I represents rice-wheat cropping pattern (upper Rechna) while Zone II is characterized by sugarcane-wheat pattern (middle Rechna). Table 13 presents the cropping pattern of pump owners and non-owners in both the zones. On an average rice and wheat were the major crops sown during *kharif* and *rabi* seasons, respectively in Zone I while wheat and sugarcane emerged as the major crops of Zone II during the corresponding seasons. Table 13 indicates that in Zone I, pump owners grew 10% more rice than non-owners mainly due to widespread tubewell ownership and reliable access to groundwater. Contrary to rice crop that has high water requirement, there were no large differences in area under wheat cultivation in both the zones between pump owners and non-owners as the water requirement of the wheat crop is much lower and this could easily be met through buying water from others. In Zone I, the remaining soil moisture after harvest was adequate for growing a wheat crop. Pump owners were estimated to devote eight percent more land to sugarcane than non-owners. The water buyers cultivated less sugarcane mainly due to high cost and unreliability of groundwater irrigation. However, water buyers devoted a larger share of their cropped area to *kharif* and *rabi* fodder as compared to pump owners in both the zones. This is because fodder crop requires relatively less water and water buyers can meet these requirements through buying water from their neighbours. Fodder is used for feeding their livestock and extra fodder is also sold in the nearby markets. Another important finding was that large area in Zone I was covered by staple foods (wheat and rice crops) and the incidence of growing other cash as well as low water consuming crops was less (3% of the gross cropped area only) whereas, the cultivation of other crops like, oilseeds, pulses, gram, maize grain, and vegetables was significantly higher (36% of gross cropped area) in Zone II. This was partly a response to deep water table in zone II, soil type and government policy of sugarcane pricing *vis-à-vis* rice and wheat pricing.

On an overall basis, cropping intensity in the study area was 160%; this was 155% for tubewell owners and 178% for water buyers (Table 12). The higher cropping intensity of

Table 12.   Cropping pattern by selected zones in the study area.

Zone	Average farm size (hectares)	Cropping pattern (percent cropped area)						Cropping intensity
		Wheat	Rice	Sugarcane	Fodder Rabi	Fodder Kharif	Others*	
Zone I	17.3	42.0	43.0	0.04	5.0	7.0	3.0	144
Pump owners	20.6	42.0	45.0	0.05	5.0	5.0	3.0	139
Water buyers	4.3	39.0	35.0	0	8.0	15.0	3.0	158
Zone II	5.6	31.0	1.0	14.0	10.0	8.0	36.0	182
Pump owners	8.1	29.0	1.2	17.0	7.0	9.0	37.0	177
Water buyers	2.3	33.0	0	9.0	15.0	7.0	36.0	190
Overall	11.5	36.0	22.0	7.0	8.0	8.0	19.0	160
Pump owners	15.4	37.0	27.0	7.0	6.0	7.0	16.0	155
Water buyers	2.9	35.0	11.0	6.0	13.0	9.0	26.0	178

*Other crops include, maize grain, gram, pulses, orchard, vegetables, oilseeds, water melon, musk melon, etc.

Table 13. Yield of major crops for pump owners and water buyers (in kg/ha).

Category/Crop	Yield (Kg/ha)				
	Wheat	Rice	Sugarcane	Fodder Rabi	Fodder Kharif
Zone I	2628	3026	49400	19580	11066
Pump owners	2579	2991	49400	19357	10937
Water Buyers	2836	3176	NC	20913	11733
Zone II	3137	3088	56896	30351	18179
Pump owners	3260	3088	61157	26918	15944
Water Buyers	2962	NC	49793	33098	24330
Overall	2870	3036	54678	24549	13733
Pump owners	2848	3008	57428	21682	12659
Water buyers	2917	3176	49793	30285	18031

NC = Not cultivated.

water buyers was mainly due to smaller average farm size (1.2 ha) as compared to pump owners (6.2 ha). It is a general understanding that large farmers usually keep a larger area fallow in a cropping season as compared to small size farmers. Also, farmers owning small plots have to eke out a living from their limited landholding and hence often crop intensively.

Table 13 presents statistics on per ha yield of major crops by pump owners and water buyers. Results show that on an overall basis, per ha yield of three major crops; wheat, rice, and sugarcane were estimated at 2870 kg/ha, 3036 kg/ha, and 54678 kg/ha, respectively. For rice, wheat and sugarcane crops, pump owners achieved the highest yield, while for fodder crops (grown in both *kharif* and *rabi* season), water buyers achieved the higher yield. Fodder crops need minimal irrigation, while rice, sugarcane are high water consuming crops. The difference in yields may be attributed to difference in reliability of access to groundwater between the tubewell owners and water buyers.

5.3 *Groundwater use and applied groundwater productivity*

The dominant source of irrigation in the study area was groundwater, as groundwater met 75 to 81% of the total irrigation demand of different crops. However, farmers reported that 85% of their farm area was irrigated by both surface and groundwater. Since the sample sites were selected within a fresh groundwater area, the majority of the farmers were practicing cyclic use of tubewell and canal water whereas mixing of canal and tubewell water is a common phenomenon in poor quality saline groundwater areas. Though every farmer had canal water allocations, they did not receive adequate and reliable canal water during their allocated turns. Results of a recent study in Rechna Doab (Nazeer et al., 2007) showed that the allocated time for canal water supplies during each weekly turn was just over an hour per hectare. Under the present canal water supply situation, the sample farmers in Rechna Doab anticipated consumption at an average of 7.6 hours for irrigating one ha of land; showing a wide gap of more than 6 hours. The results of the same survey showed that the farmers were able to irrigate only 18.6% of their farm area on a weekly basis (during each water turn) with the present canal water supplies. Thus groundwater contributes as the major source of irrigation in study area as well as in the province of Punjab.

Tables 14 a and 14 b indicate number of total as well as canal and tubewell irrigations for major crops for the pump owners and water buyers. Results show that on an overall

Table 14a.   Amount of water applied, percentage share of groundwater and groundwater productivity of pump owners in the study area.

Crop/Parameter	Total number of irrigations	Number of canal irrigations	Number of tubewell irrigations	Amount of tubewell water applied (m³/ha)	Percentage share of ground-water	Groundwater productivity (Kg/m³)
Wheat	5.1	1.3	3.8	5022	75	1.43
Rice	25.0	3.9	21.1	20869	84	0.20
Sugarcane	28.3	8.0	20.3	20778	72	4.57
Fodder Rabi	9.3	1.9	7.4	8452	80	4.56
Fodder Kharif	8.7	1.8	6.9	11219	79	5.02

Table 14b.   Amount of water applied, percentage share of groundwater and groundwater productivity of water buyers in the study area.

Crop/Parameter	Total number of irrigations	Number of canal irrigations	Number of tubewell irrigations	Amount of tubewell water applied (m³/ha)	Percentage share of ground-water	Groundwater productivity (Kg/m³)
Wheat	5.1	1.9	3.2	3016	63	2.20
Rice	25.8	13.4	12.4	13034	48	2.24
Sugarcane	26.8	11.3	15.5	12469	58	3.56
Fodder Rabi	9.5	3.3	6.2	6899	65	6.25
Fodder Kharif	7.2	2.2	5.0	5236	69	6.04

basis, groundwater is the most dominant source of irrigation for major crops in the study area. The amount of tubewell water applied in cubic meter per ha was also estimated for major crops. Comparison of water use patterns from major irrigation sources between pump owners and water buyers shows that water buyers, applied fewer irrigations from tubewell water compared to pump owners. The percentage share of groundwater in total irrigation applied by the water buyers was also less than the pump owners. This does not mean that water buyers had access to larger share of canal water but simply that they use less amounts of water per irrigation.

Water productivity is the mass of crop production per unit volume of water applied. Data on the number of canal and tubewell irrigations, time needed to irrigate one ha of land, average tubewell discharges and crop yields was collected to compute groundwater productivity of major crops grown on the selected farms. Four representative tubewells were selected from each village for discharge measurement using volumetric or the trajectory method depending upon the situation in the field.

Results (Tables 14a and 14b) indicate that groundwater productivity of pump owners was lower for all crops except sugarcane compared to that of the water buyers. This was mainly due to the higher proportion of groundwater share in total water applied by the pump owners. This implies that water buyers produce more from one cubic meter of groundwater. There may be two reasons of higher groundwater productivity of water buyers as against pump owners; (i) use of more surface water irrigation (ii) groundwater irrigations applied

Table 15. GVP and CoP (PKR/ha) of pump owners and water buyers.

Crop	Pump owners				Water buyers			
	GVP	CoP	GM	CB ratio	GVP	CoP	GM	CB ratio
Wheat	29218	17485	11733	1:1.7	27869	15465	12404	1:1.8
Rice	39495	31241	8255	1:1.3	39826	25466	14361	1:1.6
Sugarcane	95745	68404	27340	1:1.4	77126	40812	36314	1:1.9
Fodder Rabi	18194	13017	5177	1:1.4	19486	16277	3209	1:1.2
Fodder Kharif	9038	10762	−1724	1:0.8	8771	17564	−8793	1:0.5

by the water buyers are optimally used by the crop whereas pump owners over-irrigate their land leaving higher amounts of water lost in the system.

5.4   *Crop economics*

Data on cropped area, total crop production and output prices for the 2006 cropping year were gathered from the sample farmers. The gross value of product (GVP) was computed for five major *rabi* and *kharif* crops. Similarly, data on major cost components of crop production, including land tillage, fertilizer, seeds, pesticides, irrigation and labour was also gathered for these crops. Table 15 shows the GVP, cost of crop production (CoP), gross margins (GM) and cost-benefit ratio (CB ratio) for five major crops for pump owners and water buyers. On an overall basis, per ha GVP of wheat, rice, and sugarcane was estimated at PKR 28822, PKR 39545 and PKR 89041, respectively, where wheat and rice were seasonal (six months) crops and sugarcane was an annual crop but usually counted in *kharif* season. *Rabi* and *kharif* fodders were mainly grown for feeding livestock. The GVP of these fodders was calculated based on current market prices.

Gross margins from each crop were simply calculated by deducting total cost of production (variable costs) from GVP. The highest GM (PKR 29833/ha) was noted in case of sugarcane, followed by wheat (PKR 11930/ha) and rice (PKR 9184/ha). However, it is to be noted that GVP of sugarcane was on annual basis. Table 16 shows that the water buyers were getting higher GVP and GM from wheat, rice and sugarcane while incurring less cost. Cost-benefit analysis shows that water buyers were obtaining slightly higher benefits in case of wheat, rice and sugarcane than pump owners.

Among various cost components, the share of irrigation cost (surface plus groundwater) in total cost of production was the highest for all reported crops, ranging from 30.7% for wheat to 59.7% for rice. It is important to note that irrigation cost is mainly the groundwater pumping cost as canal irrigation charges in the Punjab province of Pakistan are fixed; i.e., only PKR 250/ha. The second highest cost component was fertilizer for wheat, rice and fodders, and hired labour for sugarcane.

Tables 16a and 16b presents comparative data on the share of various cost components for pump owners and water buyers. It shows that the percentage share of irrigation cost (mainly groundwater cost) in total CoP in the case of wheat and rice was higher for pump owners as compared to water buyers. It is evident from Table 16b that pump owners use more groundwater irrigation cycles than water buyers. Regarding per hour

Table 16a.   Share of cost components in total cost of production of pump owners.

Percentage share of cost components	Percentage share of various cost items for major crops				
	Wheat	Rice	Sugarcane	Fodder Rabi	Fodder Kharif
Cost of production (PKR/ha)	17485	31241	68404	13017	10762
Land tillage	16.7	12.1	7.6	15.5	16.6
Seed	6.7	0.9	19.2	10.9	11.3
Fertilizer	24.2	14.3	9.5	24.6	23.6
Chemicals	5.2	4.1	2.4	2.1	4.8
Irrigation	32.9	58.6	18.7	43.1	40.2
Hired labour	14.3	10.0	42.6	3.8	3.5

Table 16b.   Share of cost components in total cost of production of water buyers.

Percentage share of cost components	Percentage share of various cost items for major crops				
	Wheat	Rice	Sugarcane	Fodder Rabi	Fodder Kharif
Cost of production (PKR/ha)	15465	25466	40812	16277	17564
Land tillage	16.8	20.9	6.9	16.2	13.4
Seed	7.8	0.9	20.7	9.4	6.6
Fertilizer	24.8	18.2	9.1	14.5	16.6
Chemicals	6.4	5.2	2.0	7.8	8.0
Irrigation	21.8	40.9	47.3	52.1	55.4
Hired labour	22.4	13.9	14.0	0.0	0.0

rate of tubewell water, the pump owners reported consuming an average of 2.1 l of diesel in an hour while the average per litre cost of diesel was PKR 37.7. Therefore, the owner of the pump in the study area was bearing a cost of PKR 79.2 per hour while the water buyer reported to purchase tubewell water on an average rate of PKR 139 per hour. However, the purchasing rate was higher in Zone II (PKR 148/hour) as compared to PKR 128/hour in Zone I for diesel WEM. It was only PKR 65/hour for electric WEM that was only practiced in Zone II. Thus, in the case of diesel WEMs, the water buyers were consuming PKR 59.8/hour (or 43%) more than the pump owners.

Overall CoP of pump owners was higher than the water buyers for all major crops except fodders. This was mainly because the pump owners hired more labour, applied more groundwater irrigations and fertilizer and used good quality and expensive seed as compared to the water buyers. Water buyers who were marginal small farmers were using less groundwater irrigations due to their limited financial resources (Table 15).

5.5   Economics of groundwater extraction

The average cost of operation and maintenance for running the diesel as well as electric tubewells was estimated (Table 17). The number of hours of extraction for electric pumps

Table 17. Cost of extraction of groundwater in study area.

Parameter	Zone I		Zone II	
	Diesel WEM	Electric WEM	Diesel WEM	Electric WEM
Average operational and maintenance cost in PKR/WEM/year	42,641	NA	37,902	69,211
Average no. of hours of extraction/WEM/year	556	NA	403	3,087
Hourly cost of water extraction (PKR/hour/WEM)	77	NA	94	22
Average discharge (m^3/hr.)	92	NA	97	98
Cost of extraction of 1 m^3 water in PKR	0.84	NA	0.97	0.23

NA = Not available.

in Zone II was the highest, because these were being operated on share basis and were also used for selling additional water to other farmers. The number of hours of extraction for diesel pumps was higher in Zone I compared to Zone II. The reason may be the maximum utilization of groundwater for the rice crop. The cost of extracting water was 3–4 times higher for diesel WEM compared to electric WEM.

5.6 *Emerging groundwater problems*

In the survey region the groundwater quality was good and the extent of groundwater pollution was moderate (as reported by the sample farmers and well drillers). However, the most serious problem was the lowering of the water table (see Table 6). The average depth to the water table has increased from 3.6 m in 1988 to 7.0 m in 1996 and 5% of the area in Punjab has already gone beyond the reach of poor farmers due to the high cost of tubewell installation in those areas (Qureshi and Mujeeb, 2002; 2003). This figure is expected to reach 15 % in Punjab province during the next decade. Qureshi et al. (2003) reported that dug-wells in Pakistan, particularly in the Punjab province, are constructed according to the prevailing water table conditions and even a small decline in water table depth is enough to abandon a "dug well". Continuous drought conditions during the last decade have also caused about 40% of the dug wells to either be deepened or reconstructed or simply abandoned (Qureshi et al., 2003).

Pakistan is a water scarce country and the drought phenomenon (the dry year) has occurred in 4 out of the last 10 years. Precipitation during 1997–2000 has been exceptionally low i.e. 50% normal. Agricultural growth also suffered a severe setback during 2000/01 due to the unprecedented drought situation and shortage of irrigation water, causing a decline of 2.5% in agricultural growth as against an impressive growth of 6.1% in the previous year (PWP, 2001).

The problems faced by well irrigators in Punjab can be prioritized into seven major problems based on farmers' perceptions in rank order as; (i) high energy costs, (ii) falling water tables, (iii) salinity, (iv) high rate of failure of wells/tubewells, (v) pumping of wells for shorter periods/low life of pumping wells, (vi) poor water quality and expensive electricity connections, and (vii) unreliable power supply.

6 POLICY AND INSTITUTIONAL PERSPECTIVE ON GROUNDWATER AGRICULTURE

Groundwater governance is a major concern in Pakistan, particularly in the Punjab where irrigated agriculture heavily depends upon groundwater. The groundwater crisis in the country was ignored until recently because governments were under pressure to produce more food for the growing population and groundwater generated prosperity. Groundwater exploitation was, in any case, mainly privately financed. Government is only now beginning to develop a management policy to control groundwater issues in Pakistan.

Steenbergen & Gohar (2005) have summarized the history of groundwater legislation and regulations. A number of acts and laws have been issued in the past for supporting groundwater management but they were not fully implemented. The Punjab Soil Reclamation Act of 1952 was the first in this regard under which a Soil Reclamation Board was established to control water-logging and salinity through the development and operation of drainage tubewells. This Board was supposed to issue a license to land owners to install private tubewell for the designated land reclamation areas. Later on, the Board was suspended and its responsibilities were transferred to the Provincial Irrigation and Power departments (PIPD). Licensing rules were framed under the PIPD but never implemented. In 1958, another Act, namely the Pakistan and Power Development Authority Act was introduced. This later formed the legal basis for the establishment of the Water and Power Development Authority (WAPDA). In 1978 the government of Baluchistan introduced legislation to control groundwater mining and an administration ordinance was spelled out. In 1996, the Environmental Protection Agency Act was formalized to address water quality issues but again had difficulty in getting implemented. Finally, the Provincial Irrigation and Drainage Authority Acts, 1997 were framed to address the groundwater management issues, mainly groundwater monitoring. Currently, the Directorate of Land Reclamation (dlr) is involved in groundwater monitoring. For this purpose they have divided Punjab into eight administrative zones.

Considering the emerging complexity of groundwater issues, there is a need to develop a holistic groundwater management policy framework that addresses the complex groundwater issues of multiple uses and at the same time lends itself to phased implementation in the particular socio-economic and physical resources of Pakistan. The Government of the Punjab (Pakistan) is developing a strategy that would include the implementation of groundwater management tools with immediate effect in problem areas, and a regulatory framework for groundwater management, launching a province-wide programme on groundwater recharge and water saving, a programme with special focus on the protection of groundwater quality, a comprehensive drainage programme to manage shallow aquifers, and extend monitoring and research programmes for the sustainable use of groundwater resources.

7 CONCLUSIONS AND RECOMMENDATIONS

It is increasingly realized that the water crisis in Pakistan, like in most other countries of South Asia, is a crisis of governance, and not just of water scarcity. However, canal water supplies have reduced over the years due to population growth and poor maintenance of the canal networks putting more pressure on groundwater resources for meeting the agricultural water requirements in Pakistan.

This study suggests that depletion of groundwater is the most serious problem in the study areas. This is true all over the Punjab. The discharge from the aquifer was, generally, more than the recharge potential and this has created groundwater depletion. The sustainability of groundwater resources in the Punjab province of Pakistan is now under threat. Installation of turbine engines to supply drinking water to major cities has also put enormous pressure on groundwater resources thereby limiting future agricultural water use. This reflects a weak and unplanned integrated water management policy at the government level that needs immediate attention. The absence of regulations to control groundwater over-exploitation has already led to dangerous mining that has adversely affected not only socio-economic benefits to the poor but is seriously undermining groundwater quality in many parts of the province as well as at basin level. It is necessary to put in place a regulatory framework in the provinces for optimal development and management of this important resource. There is an urgent need to develop and test a groundwater management model and to initiate pilot studies for implementing a groundwater regulatory framework; development of a database for groundwater use and quality; assessment of groundwater recharge and re-use; re-allocations of canal irrigation water to minimize pumping of groundwater, particularly in saline areas; to evaluate the optimum depth of groundwater in different regions/cropping zones, and to review and evaluate groundwater rules and regulations existing in different regions of the country to create an effective groundwater regulatory framework.

The study suggested that pump owners, generally, over-irrigate their crops and are not fully aware of crop water requirements. There is a need to develop awareness among the farming community towards optimal use of this precious resource. There is also a need to introduce high efficiency irrigation system and water conservation technologies at farm level, which will increase productivity gains and profitability of the farmers. These tech-nologies must be accessible and affordable, particularly for the poor and marginal farmers. This should be implemented on area priority basis, i.e., should start first in sensitive areas areas with very low water tables, and poor water quality regions. The role of agricultural extension services agencies in the province has deteriorated and the farmers have lost their trust in this service.

Increasing energy prices and lack of an energy pricing policy is an important issue for groundwater use and management. The existing energy pricing policy in the water sector suffers from the lack of dialogue between the service provider and service user. Policies were in any case designed to reduce the expense of revenue collection and not to regulate groundwater use. All groundwater stakeholders still need to know more about the problem and its resolution according to the prevailing conditions. Presently, at the farm level, high diesel prices are a great concern. Farmers are paying overmuch for irrigation water, reducing their income and profitability. Government subsidizing energy costs for irrigation is not a sustainable option for the energy utility or the groundwater sector.

There remains a lack of clear definition of groundwater management objectives, required levels of decision making, roles of stakeholders, data collection and monitoring and filling the capacity gaps. The main elements of the groundwater management policy and legal framework include rights definition, a neutral negotiating and dispute resolution forum, institutional arrangement for groundwater management, a system of information generation and dissemination and coordination at local and regional level and all of these are lacking at present.

REFERENCES

Ahmed, N. (1988). *Irrigated agriculture of Pakistan*, Mimeo, International Water Management Institute (IWMI), Lahore, Pakistan.

Ahmed, M.D., H. Turral, I. Masih, M. Giordano, and Z. Masood, (2007). Water saving technologies: Myths and realities revealed in Pakistan's rice-wheat systems. *IWMI Research Report 108* Colombo, Sri Lanka: International Water Management Institute. 44 p.

Ashraf, M. and M.A. Khan, (1984). *Irrigation Directory Punjab*. Lahore, Pakistan. Planning and Investigation Organization, Planning Division. Pakistan Water and Power Development Authority.

Bennett, G.D., A. Rehman, I.A. Sheikh, and S. Ali, (1967). *Analysis of aquifer tests in the Punjab region of West Pakistan*. US Geological Survey Water Supply Paper 1608-G, 56 p.

GOP (Government of Pakistan). (2002). *Pakistan Economic Survey 2001–02*. Finance Division, Economic Advisors Wing, Islamabad, Pakistan.

GOP (Government of Punjab). (2005). *Punjab Development Statistics 2005*. Bureau of Statistics, Government of the Punjab, Lahore.

Jehangir, W.A., A.S. Qureshi, and N. Ali, (2002). *Conjunctive water management in the Rechna Doab: An overview of resources and issues*. Working Paper: 48. Lahore, Pakistan: International Water Management Institute.

Khan, M.A. (1978). *Hydrological Data, Rechna Doab, Vol. 1, Publication Number 25*, Lahore, Pakistan: Project Planning Organization (NZ) and Pakistan Water and Power Development Authority.

Kijne, J.W. (1999). *Improving the productivity of Pakistan's irrigation: The importance of management choices*. Colombo, Sri Lanka: International Water Management Institute.

Meinzen-Dick, R.S., Brown, L.R., Feldstein, H.S. and Quisumbing, A.R., (1997). "Gender, property rights, and natural resources," FCND discussion papers 29, International Food Policy Research Institute (IFPRI).

Nazeer, A., M. Ahmad, M. Giordano, H. Turral, A. Hussain, and Z. Masood. (2007). "Land and Water Productivity Patterns across Irrigation Sub-divisions of Rechna Doab", Draft paper submitted to IWMI, Colombo.

Pakistan Water Partnership (PWP). (2001). Supplement to the Framework for Action (FFA) for Achieving the Pakistan Water Vision 2025.

Pakistan Water Partnership (PWP). (2004). Draft National Water Policy, Islamabad. Pakistan.

Qureshi, A.S. and A. Mujeeb, (2002). Groundwater Economy of Pakistan. Paper presented at the Workshop on "Water, Livelihoods and Environment in India: Frontline Issues in Water and Land Management and Policy", Annual Partner's Research Workshop-IWMI-Tata Water Policy Program, Institute of Rural Management Anand, Gujrat, INDIA, January 26–29.

Qureshi, A.S. and A. Mujeeb, (2003). The impact of utilization factor on the estimation of groundwater pumpage. Journal of irrigation and Drainage, PARC, Islamabad, Pakistan.

Qureshi, A.S., Shah, T., Akhtar, M. (2003). The groundwater economy of Pakistan. Working Paper 64. Lahore, Pakistan: International Water Management Institute.

Rehman, G., W.A. Jehangir, A. Rehman, M. Aslam, and G.V. Skogerboe (1997). Principal findings and implications for sustainable irrigated agriculture: Salinity management alternatives for the Rechna Doab, Punjab, Pakistan. Research Report No. 21.1. Lahore, Pakistan: Pakistan National Program, International Irrigation Management Institute.

Shah, T., A. Debroy, A.S. Qureshi, and J. Wang (2003). Sustaining Asia's groundwater boom: An overview of issues and evidences. Natural Resource Forum, 27 (2): 130–141.

Steenbergen, F.V., and M.S. Gohar, (2005). *Groundwater development and management in Pakistan*. Country Water Resources Assistance Strategy. Background Paper No. 11, March 2005. The World Bank. Islamabad. Pakistan.

WAPDA. (1997). Revised action plan for irrigated agriculture. Main Report, Vol. 1. Lahore, Pakistan: Master Planning and Review Division, WAPDA.

World Bank (2004). Pakistan Public Expenditure Management, *Accelerated Development of Water Resources and Irrigated Agriculture*, Volume II, Report No 25665-PK, Environment and Social Sector Development Unit, Rural Development Sector Unit, South Asia Region.

CHAPTER 5

Groundwater resources and the impact of groundwater sharing institutions: Insights from Indian Punjab

V. Selvi
Central Soil and Water Conservation Research & Training Institute,
ICAR, Tamil Nadu, India

D. Machiwal
Soil and Water Engineering Department, CTAE, MPUAT, Rajasthan, India

F. Shaheen
Sher-i-Kashmir University of Agricultural Sciences and Technology, Kashmir, India

B.R. Sharma
International Water Management Institute (IWMI), New Delhi, India

ABSTRACT: Hoshiarpur district in the north-eastern part of state of Punjab, India is a region with ample groundwater potential. Variability in groundwater resources along with the existing caste system has given rise to two distinct patterns of groundwater access—shared wells and private, informal water markets. The sharing of groundwater in one study village has helped farmers to achieve equitable access to groundwater as well as increase their crop and water productivity. Even a very competitive groundwater market in the two neighboring study villages did not allow the water buyers to realize the same levels of benefits as obtained by the well owners. The government policy of providing free electricity for the farm sector has provided incentives to the farmers to install additional tubewells. However, this has also led to a decline in the groundwater table. A more sustainable approach would be to propagate small water harvesting measures for utilization of torrents in rainy season and micro irrigation techniques for plantation crops.

1   INTRODUCTION

India occupies the first place in the world's irrigated agriculture with 21.7 per cent of the total global irrigated area. Groundwater plays an important role in Indian agriculture. Gravity systems have dominated irrigated agriculture until the 1970s, but by the early 1990s, groundwater irrigation had surpassed the use of surface irrigation (Debroy & Shah, 2003 and Shah et al., 2003). According to Moench (1996), groundwater contributes to 9% of the country's GDP and it accounts for as much as 70 to 80% of the value of irrigated farm produce. Energy, especially electricity, has contributed significantly to the development and exploitation of groundwater resources thereby improving productivity and providing livelihood and food security to millions of rural poor (Shah et al., 2003). While groundwater development has played an important role for the economy, over-exploitation of groundwater is emerging as a major concern (Kaushal & Sondhi, 2006) particularly in the western Indo-Gangetic basin (IGB) states of Punjab and Haryana.

Punjab, a north-western state of India occupies only 1.5% of the total geographical area of the country but produces about two-thirds of the total food grains. Food surpluses from Punjab meet the food deficits of several states of India. Groundwater availability at shallow depth and well-developed surface water irrigation systems paved the way for the Green Revolution in the mid 1960s. As in other parts of the country, groundwater irrigation slowly became the most important source of irrigation and at present 72% of the total irrigated area in the state is dependent on groundwater. Hoshiarpur district lying in the northeastern part of Punjab has a predominantly agrarian-based economy with very little industrialization. However, agricultural development in this district is less well developed than the rest of the state. In this district, unlike many of other districts of Punjab, canal networks are not extensive and hence groundwater is often the sole source of irrigation. This has encouraged local groundwater institutions to be formed, which are governed not only by the social fabric but also by the state of groundwater resources prevailing in that area. With this background, a study was initiated in Hoshiarpur district to get a broad perspective on groundwater irrigation with the following specific objectives.

1. To characterize the hydrogeology of the study area.
2. To study of the evolution of tubewell technology and local groundwater institutions.
3. To understand the role of groundwater institutions in the agrarian economy.
4. To investigate the role of government policies and print media in groundwater related issues.

2   METHODOLOGY

Three contiguous villages, Rampur, Bilron and Bhajjal in Garhshankar block of Hoshiarpur district, Punjab, India (Fig. 1) were selected as study sites. These villages were selected as they depended entirely on groundwater irrigation, yet they had developed different water sharing institutions, e.g. private water markets in Rampur and Bilron and kin based shared well ownership in Bhjajjal village.

A general overview of the land use pattern, cropping status, demographic details, caste breakup, literacy rates and infrastructural facilities was obtained for the respective villages using a specially developed village questionnaire. Mapping of the various resources in the study area was also carried out. A special questionnaire which had been developed to gather information on the individual farmers land holding size, cropping pattern and intensity, water selling buying or sharing pattern, intensity of water use and crop economics along with other socio economic details was give to 75 respondents selected from the three villages.

For hydrogeological investigations, lithologic samples from six sites in and around the study area were collected at every 3 m depth where drilling was in progress during the study period, well logs were drawn for each of them and aquifer characteristics like hydraulic conductivity and transmissivity were studied. Three well drillers who working in the study area were contacted and interviewed using a questionnaire to get an insight into the well drilling technologies employed in the region, the growth pattern in the number of wells and the change in the groundwater levels observed during the period of operation.

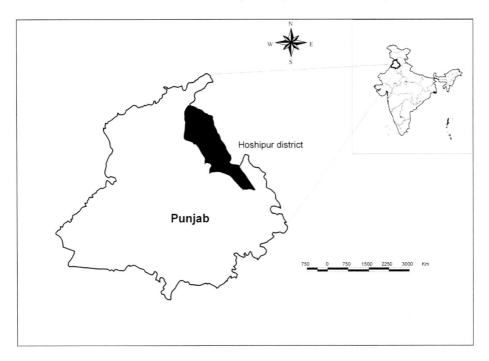

Figure 1.   Location map of Hoshiarpur district, Punjab, India.

To understand the views of formal groundwater institutions regarding groundwater governance, officials from major groundwater authorities such as the Central Ground Water Board, Water Resources and Environment Directorate of the Irrigation Department and Ground Water Cell of the State Agriculture Department, Govt. of Punjab were contacted.

3   LOCATION AND GENERAL DESCRIPTION OF THE STUDY AREA

The state of Punjab receives a long term average annual rainfall of 508 mm (1994–2004) with nearly 65 per cent of it occurring during the south-west monsoon i.e. from July to September (Fig. 2). Hoshiarpur district, being adjacent to the Shivaliks, receives higher average annual rainfall of 768 mm (1994–2004).

The soils of Punjab can be classified into eight major types: flood plain soils, loamy soils, sandy soils, desert soils, *Kandi* (foothills) soils, sierozems, grey- brown podzolic & forest soils, and sodic and saline soils.Hoshiarpur district has a predominance of *Kandi* soil and texture of these soils varies from sandy, sandy loam, silt loam and clay-silt to gravelly nature. The texture becomes coarser and rougher eastward of the Shiwalik hills where gravel, pebbles and conglomerates predominate and which have been deposited by numerous *choes* (torrents) coming from these hills. Forest soils are also found in the Shiwalik hill zone of the district (Manku, 2007).

According to 2001 Census, Punjab has a total population of 24.4 million and a density of 482 persons/km^2. Punjab is dominated by its rural population, which forms nearly 66.1% of the total. Hoshiarpur district with a population density of 439 persons per km^2 is a region

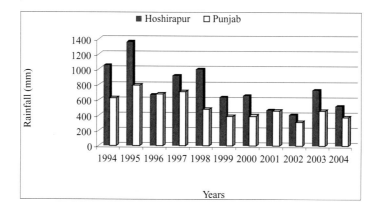

Figure 2.   Average annual rainfall for Punjab State and Hoshiarpur district (1994–2004).

Table 1.   Geographical area of the study region.

Name	Status	Area (sq. km)
Punjab	State	50,332
Hoshiarpur	District	3,394
Garhshankar	Block	391
Rampur, Bilron and Bhajjal	Study villages	37

Table 2.   Area under irrigation in the study region.

Irrigated area	State Punjab	District Hoshiarpur
Gross irrigated area, (GIA '000 ha)	7,692.2	297.5
Net irrigated area (NIA '000 ha)	4,035	162.0
Ratio of GIA to Gross Cropped Area (GCA)	96.97	83.10
Ratio of NIA to Net Cropped Area (NCA)	96.07	81.00

GIA: is total area under crops, irrigated once and/or more than once in a year. It is counted as many
    times as the number of times the area are cropped and irrigated in a year;
NIA: is the area irrigated through any source of water once a year for a particular crop;
GCA: represents the total area sown once and/or more than once in a particular year, i.e. the area is
    counted as many times as there are sowings in a year;
NCA represents the total area sown with crops and orchards. Area sown more than once in the same
    year in counted only once.

of moderate density where agriculture is the predominant occupation. The geographical
area of the study region is presented in Table 1.

The land use pattern of the state of Punjab is predominantly agricultural (83.5%) and
the same trend is observed in Hoshiarpur district. However, the land used for cultivation
purpose in Hoshiarpur district is only around 60% with forests occupying a little more than

Table 3.   Source wise irrigated area in study region.

Source of irrigation	Punjab		Hoshiarpur	
	Area ('000 ha)	% of total	Area ('000 ha)	% of total
Govt. canals	1101	27.29	14.0	8.64
Private canals	7	0.17	0	0
Tube wells & wells	2919	72.34	146	90.12
Other sources	8	0.19	2	1.23
Total	4035	100	162	100

30% of the total area of the district (Indiastat, 2007). The cropping intensity for the state is higher (190%) than that of Hoshiarpur district (179%).

Punjab, also known as the "Granary of India" leads in the production of food grains like wheat and rice in India. Agriculture is carried out mostly under irrigated conditions. Table 2 shows that close to 96% of the net and gross sown area in the state has irrigated agriculture while for Hoshiarpur district, it is slightly lesser with 83 and 81%, respectively. The district depends largely on groundwater for irrigation purpose (90%) with a small area coming under canal irrigation (Table 3). Among the irrigated crops, wheat occupies 44.3% of the gross irrigated area followed by rice (34.2%) in the state. In Hoshiarpur district, the maximum area under irrigation is again for the wheat crop (42.9%) followed by rice (19.5%). However, the acreage for the maize crop grown under irrigated condition in the district at 15.9% and is far higher than that of the state (1.26%).

4   CHARACTERIZATION OF GROUNDWATER RESOURCES

4.1   *Hydrogeology*

The study area lies in the southwest region of the Shiwaliks which has steep slopes and forms the origin of several streams. On the foothills of the Shiwaliks lies the *Kandi* belt with undulating topography and milder slopes. The study villages are situated in the border of the *Kandi* belt where the slopes are milder with a relief of 30 m. The streams originating in the Shiwaliks and passing through the *Kandi* belt disappear in the study area giving way to underground *choes* (*torrents*).

The hydrogeology in and around the study villages was explored by collecting and analyzing localized lithologic samples from six well drilling sites (Fig. 3). The aquifers can broadly be divided into three layers: good, medium and poor with respect to their water conducting properties. Medium aquifer (i.e., very fine sand) with thickness varying from 20 to 35 m can be encountered at depths ranging from 2 to 13 m beneath the ground level. Below the medium aquifer layer, lies a good aquifer (i.e., fine sand and fine sand with pebbles) with an average thickness of 30 m at a depth of 37 to 42 m for sites near Bhajjal village. This good aquifer with the same thickness exists at a deeper depth (i.e., 42 to 80 m) at sites in and around Rampur and Bilron. A 3 m thick concretion layer with poor aquifer properties occurs almost randomly within the good aquifer layers.

Hydraulic conductivity of all these layers was computed by using the Hazen model. The variation of hydraulic conductivity with depth for the sampling site at Satnor near Bilron

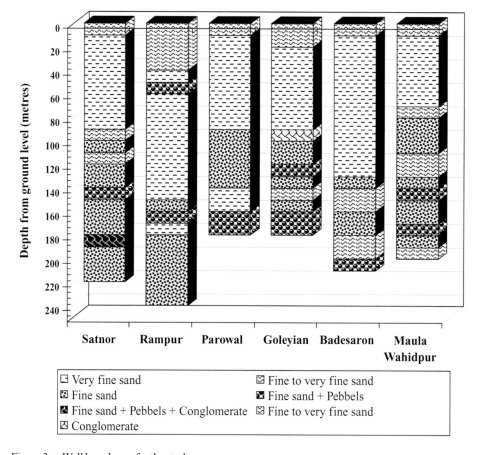

Figure 3.   Well logs drawn for the study area.

Table 4.   Characteristics of private tube-wells in the study area.

Village name	Irrigation TW density (Nos./km²)	Well depth (m)	Depth to water table (m)	Pump hp	Discharge (m³/hr)
Bhajjal	23.60	30–120	25–39	5–10	12.6–46.4
Bilron	4.19	60–150	35–70	7.5–20	20.1–59.4
Rampur	4.66	65–215	35–100	10–25	77.8–301.3

village is shown in Fig. 4. The hydraulic conductivity is 27–42 m/day at a depth range of 50 to 70 m.

Figure 5 depicts the topographic contours of the study villages and the contours of equal depth to water table (isobath) for the study area. The depth to water table in general reflects the topographic contours. The depth to water table in the villages of Rampur and Bilron vary between 33 and 100 m while in Bhajjal, it hovers between 33 and 37 m. The change in the water table between pre and post monsoon season is around 3 to 5 m in all the

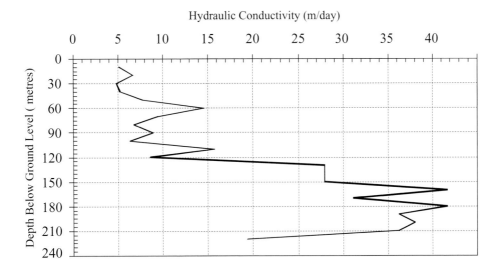

Figure 4.   Variation of hydraulic conductivity with subsurface lithology (Satnor site).

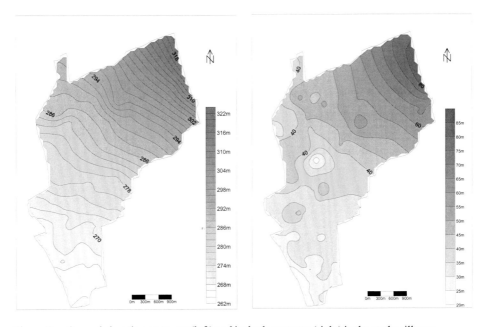

Figure 5.   Ground elevation contours (left) and isobath contours (right) in the study villages.

three villages. Bhajjal village, which is relatively low lying, has shallow water table and
the highest density of tubewells. Rampur and Bilron have a relatively low tubewell density.

Singh (1993), in a Central Ground Water Board (CGWB) report, states that aquifers
with good yields occur at depths 90 to 120 m bgl in Garhshankar block of Hoshiarpur
district. This is one of the areas where extraction is high because of the need to irrigate

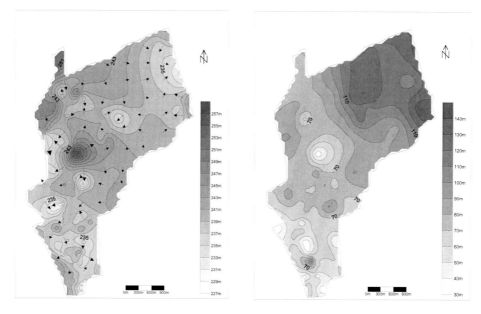

Figure 6.   Direction of groundwater flow (left) and well depth contours (right) in the study villages.

large plantations. The water table elevation contours shown in Fig. 6 show the direction of flow in the study area. The flow directions reveal two main patterns; a general flow towards the north-east boundary is observed which could be attributed to the cone of depression arising from heavy pumping in that area. Another pattern is the recharge zone, i.e., the village pond in Bilron from which flow paths radiate. Even though a similar recharge zone can be expected from the pond in Bhajjal village, no such pattern is detected. In fact, no general direction of flow is seen, possibly due to a high density of wells and heavy pumping in this village. In general, wells are deeper in Rampur and Bilron villages than those in Bhajjal (Fig. 6). Shallow dug-cum-bore wells are present in Bhajjal village (33 to 35 m depth). These wells have a lower yield of about 12 m^3/hr than that of the tubewells found in the same village. The experience of the people in Bhajjal is that these shallow wells become operational only when extraction stops from the tubewells in their vicinity.

4.2   *Groundwater quality*

Quality of irrigation water plays a major role in deciding the quantity and quality of crop produce. Major soil properties under irrigated conditions are influenced by the quality of water used for irrigation. In order to ascertain of the quality of groundwater used for irrigation across the study area, parameters such as pH, Electrical conductivity (EC), Soluble Sodium percentage (SSP), Sodium Absorption Ratio (SAR), Residual Sodium Carbonate (RSC), Chloride (Cl) were determined (Muthuvel and Udayasoorian, 1999) and based on this the gypsum requirement needed for each well was estimated. Table 5 presents the groundwater quality parameters for wells in the study area.

A large percentage of wells in the study villages had good quality irrigation for the most parameters except SAR. Irrigation water with high values of SAR entails poor quality as

Table 5.   Salient groundwater quality parameters for the study areas.

Parameter	Range	Quality rating	No. of wells	% of total
pH	7.25–7.50	Excellent	3	2.3
	7.50–8.00	Good	121	93.8
	8.0–8.50	Fair	5	3.9
EC (dS/m)	0.5–0.75	Good	23	17.8
	0.75–1.00		100	77.5
	1.00–1.50		6	4.7
Cl (meq/litre)	<2.5	Excellent	66	51.2
	2.5–5.0	Good	63	48.8
SAR	4–8	Poor	50	38.8
	8–15	Very poor	79	61.2
SSP	30–45	Good	7	5.4
	45–60		93	72.1
	60–75	Fair	29	22.5
RSC (meq/litre)	<1.0	Excellent	106	82.2
	1.0–1.25	Good	4	3.1
	1.25–2.0	Fair	7	5.4
	2.0–2.5	Poor	6	4.7
	2.5–3.0	Very poor	4	3.1
	>3.0	Unsuitable	2	1.6

this water may produce harmful levels of exchangeable sodium in most soils. Based on the water quality parameters studied, it can be concluded that gypsum is required for certain wells of the study villages in order to improve their suitability for irrigation by reducing their sodium hazard.

5   HISTORY OF GROUNDWATER IRRIGATION AND INSTITUTIONS

Groundwater extraction and its use for irrigated agriculture have largely been shaped by the policies of state and central governments in India. In Punjab, during the early 1950s, tube-wells did not yet exist and only 52.3% of the total cultivated area was under irrigation (GOP, 2007). The '*Morababandi*' or consolidation of landholdings was popular during this period. This was considered to be a pre-requisite for utilization of canal water that became available through the Bhakra–Nangal canal system. Hoshiarpur district, however, could not receive much water through the canal irrigation projects and as a result had to depend more on rainfed farming. The advent of the green revolution in the 1960s brought in the high yielding crop varieties which needed an assured supply of irrigation and the agricultural community in this region responded by drilling tube-wells for groundwater extraction.

The evolution of well technology in Bhajjal village is shown in Fig. 7. In this village, where *jat* sikh, a predominantly agricultural community was in the majority, the importance of irrigated agriculture was recognised. When land consolidation took place in 1952, the holdings with an open well as an irrigation source fetched a better price. The adjoining landowners became share well partners who got access to water according to their land holding size. This system of open wells using a Persian wheel for abstraction continued

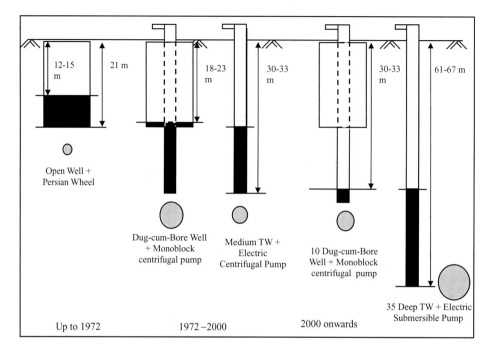

Figure 7.   Timeline diagram showing evolution of tubewell technology in Bhajjal village.

until the 1970s after which the depleting water table made further abstraction difficult. The open wells were then converted to dug cum bore-wells with either diesel or electricity operated mono-block pumps. After the year 2000, most of the mono-block pumps gave way to submersible pumps driven by electricity as the water table got deeper and power became completely free for the farm sector. It is imperative to note that in spite of the technological changes, the share-owners of wells in Bhajjal remained steadfast in their groundwater sharing mechanism. As the number of shareowners increased due to division of land holdings, the farmers in this village adapted their cropping pattern so that excess water was not drawn from the wells. For example, only one partner (on a rotation basis) among the many is allowed to grow paddy, which is a water intensive crop. The investment in tube-well and operation and maintenance cost is apportioned between the partners as per their holding size irrigated by that particular well. Tiwary & Sabatier (Chapter 12, this Volume) have described the various aspects of water sharing found in this village.

The neighbouring villages of Rampur and Bilron were not dominated by the agricultural community and reacted slowly to irrigated agriculture. Till the 1960s, only 25% of the total geographical area of these villages was cultivated with practically most of it under rainfed farming. The land holdings even today are fragmented and the depth to water table higher than that of Bhajjal village. Until 15 years ago, landowners who had a tube-well enjoyed a higher status in the society. The concept of water markets was then introduced and water was sold at a rate of Indian rupees (INR) 10–12 per hour (1 USD ~ INR 40 as in May 2008). Conflicts frequently arose in buyer group regarding their turn, which have to be settled through the *gram panchayat*. The populist policy of the state government in doing away with the electricity charges for agricultural purposes and assured supply of three phase electricity, especially during the peak rabi season, had considerably eased the pressure in the water market as many farmers now developed their own tube-wells. The cost of drilling a new well has considerably reduced with the introduction of percussion drilling and increased competition among the local drillers. Another reason for the increased number of wells was the reclamation of large area under wasteland by big farmers who converted them into plantations as well as the introduction of paddy cultivation, especially after 1995, in these two villages. In spite of all this, there is still a flourishing water market as many of the small farmers who had fragmented land holdings were of the opinion that purchasing water is cheaper for them. The number of water buyers accessing different wells in these two villages ranges from two to forty which is greatly dictated by the location of a particular well and its pump horse power (10 hp to 35 hp). This has given rise to a situation whereby a tube-well owner whose well is in a very strategic location enjoys a monopoly and dictates the terms and conditions to the water buyers. It also happens that a farmer whose land is fragmented purchases water from a large number of well owners at different rates.

The rotation cycle for access to water among the buyers is generally fixed on a first come first served basis. Similarity can be discerned in adjustment of the irrigation rotation calendar between the water buyers in these two villages and shared well owners in Bhajjal village. The mode of payment varies. Purchasers in some cases have to pay advances for irrigation water especially where the water seller has monopolistic advantage. An attempt to create a cooperative society of farmers who access water from such a well would be obstructed by the well owner.

The pond in Bilron village, which has become a sewage water disposal lagoon, currently functions as an irrigation source for around fifteen marginal farmers of the village. In the year 2006, State Soil Conservation Department laid underground pipelines from the pond to the fields of these farmers covering a total of 9.2 ha. A diesel engine operated mono-block pump has been installed and these farmers access the pond water for the price of the diesel used. This water helps to provide irrigation for around 250 hours per year. A seven-member committee constituted by the *gram panchayat* undertakes the operation and maintenance of this system. New common water harvesting ventures such as this are attractive.

6   IMPLICATIONS OF GROUNDWATER INSTITUTIONS ON THE AGRARIAN ECONOMY

Groundwater institutions played a significant role in shaping the agrarian systems of the study area. The implications of the functioning of these institutions on irrigated agriculture are important.

6.1   *Land and groundwater ownership access*

Table 6 shows the pattern of groundwater ownership among the sampled farmers of the study villages. The farmers were divided under six categories. Among the 75 sampled respondents, there are 27 water extraction mechanisms (WEM) or tube-well owners in Rampur and Bilron villages. The rest of the respondents were share well owners from Bhajjal village. The groundwater market participation rate was estimated at 55% for the sampled farmers, which indicates the important role of groundwater markets in sustaining the livelihood of water buyers.

Table 6.   Classification of respondents as per access to groundwater.

Pump ownership characteristic	No. of farmers	(%)
WEM owner who does not sell water but buys water from others	3	4
WEM owner (who neither sells nor buys water)	12	16
WEM owner who sells as well as buys water from others	4	5
Shareholder of a group of owned WEM	22	29
Non-WEM owner who buys water from others	26	35
WEM owner who sells water to others but does not buy	8	11
Total number of sampled farmers	75	100
GW market participation rate (%)		55%

Table 7.   General information about farmers in different categories of groundwater ownership.

Particulars	Pump owners	Share well owners	Water buyers	Over all
Number of farmers	27	22	26	75
Per cent living in *pucca* houses	100	100	100	100
Agriculture as main source of income	20 (74.1)	13 (59.1)	9 (34.6)	42 (56.0)
Additional sources of income	22 (81.5)	15 (68.2)	24 (92.3)	61 (81.3)
Average holding size (ha)	5.0	2.0	1.6	3.0
Average number of plots owned	4.5	2	3	3
Category of farmers as per land holding				
Sub-marginal (less than 0.5 ha)	1	1	3	5
Marginal (0.51–1.0 ha)	1	2	7	10
Small (1.01–2.0 ha)	3	11	11	25
Medium (2.01–4.0 ha)	9	7	4	20
Large (more than 4.01 ha)	13	1	1	15

Note: Figures in parenthesis is percentage to total number of farmers in that group.

Based on groundwater ownership and access, farmers of the study area can be categorized under three groups, pump owners, shareholders and non-pump owners or water buyers (Table 7). The pump owners mostly belonged to the category of large and medium farmers while shareholders fall mostly in category of small farmers. The majority of non-pump owners were from small and marginal categories.

From the equity perspective, the access to groundwater resource is clearly skewed as per land ownership in Rampur and Bilron villages where large and medium farmers mostly own WEMs. However, the benefits of sharing the WEMs among the farmers is clearly evident in the case of Bhajjal village, where all farmers, irrespective of holding size, have a share in WEM as proportional to their land holding. Similar pattern of equitable access to the groundwater resource among the farmers of company tubewells was found by Shaheen and Shiyani (2005) in the North Gujarat, India.

6.2   Gross cropped area and income

The reliable access to groundwater, choice of crops and input application has direct implications on the gross income realized from crop cultivation. Well owners have the highest average gross cropped area (GCA) of 7.3 ha followed by share well owners (3.9 ha) while water buyers have the least land (2.4 ha) (Table 8). The average gross income follows a similar pattern, and is directly proportional to the land holding size. The results substantiate the fact that reliable access to groundwater resource has a direct bearing on the income realized from crops as uncertainty in the adequacy and timeliness of supply is overcome.

6.3 *Economics of groundwater irrigated crops*

The economics of groundwater-irrigated crops in the study area irrespective of well ownership are shown in Table 9 Among the food grain crops, gross income per ha was highest in case of paddy crop followed by wheat, though paddy is grown by only a small number of farmers and in a limited area owing to its high water consumption. The major share of total cost of crop production for all the crops was for tractor hiring charges, labour wages and fertilizer, in that order. Pesticide use was found to be quite limited in the study villages.

6.4 *Investment on water extraction mechanism and amortized cost of groundwater extraction*

The investment in water extraction mechanisms (WEM) can be categorized under three major heads, (a) tube-well construction which includes the cost of drilling and casing, (b) the pumpset and its installation, and (c) conveyance or pipe line distribution system. The investment per well at historical prices worked out to be INR 146,431 in the

Table 8.   Gross cropped area and income from different well ownership categories.

Well ownership classes	Average GCA (ha)	Yearly gross income (INR)	Gross income/ha (INR)
Well owners	7.3	215137	39638
Shared well owners	3.9	102738	27205
Water buyers	2.4	51570	21178
. Over all	4.6	125464	29593

Table 9.   Economics of some important groundwater irrigated crops (INR/ha).

Particulars/Crop name	Barseem fodder	Maize	Paddy	Wheat
Gross income	35083	21105	38628	34608
Costs				
Fertilizer	1568	2473	3835	2718
Seed	2190	1505	338	1558
Labour		2033	4650	2930
Farm machinery	2093	2283	2995	2688
Pesticide	65	495	1028	563
Irrigation	4238	743	10945	1490
Total cost	10153	9530	23790	11945
Net return	24930	11575	14838	22663

study area (Table 10). Well investment at current prices was computed by compounding the historical cost (year of construction for surveyed wells) at reasonable interest rate of 2%. A similar interest rate was applied by other researchers in compounding the well investment costs (Chandrakanth et al., 1998 and Shaheen and Shiyani, 2005). The well investment at current prices was estimated to be INR 169,254, showing an increase of 15.6% over that of the historical cost. A major share of this cost (58.1%) was tube-well construction, while the cost of pump-sets accounted for about 30% of the total costs.

Valuation of the cost of the groundwater resource is crucial for rational decision making with respect to well investment. The cost of irrigation was taken as a proxy for the value of groundwater used. For computing the cost of irrigation wells, the historical investments in wells was compounded at an interest rate of two % from the year of construction to the year 2006 (survey year). The capital cost of the well was amortized over an average life span of 15 years. The variable costs comprised of expenses incurred on repairs-maintenance and replacement of working parts.

From the study on sampled tubewells, it was estimated that the annual amortized cost per well per year was INR 15,548 (Table 11). The repair and maintenance costs were quite low, as very few farmers had such expenses during the survey year. The energy cost is currently zero as farmers are supplied free electricity since 2004, while previously the electricity tariff was based on very low flat rate. The total irrigation cost worked out at INR 16,258 per year. The average number of hours for which well owners had operated their tube-wells in the agricultural year 2005–06 was estimated as 1120 hours. This also includes the hours of water sold by the well owners to water buyers. The amortized cost per hour and per cubic metre of groundwater extraction was found to be INR 14.47 and 0.37, respectively.

Table 10. Investment on sampled tube-well at historical and current prices (Amount in INR).

Particulars	Historical	(%)	Current	(%)
Tubewell construction cost	84,303	57.57	98,350	58.11
Cost of pumpsets and installation	43,228	29.52	49,485	29.24
Conveyance Cost	18,900	12.91	21,419	12.65
Total Fixed cost	1,46,431	100.00	1,69,254	100.00

Table 11. Amortized irrigation cost per well.

Particulars	Cost (INR)
Annual amortized cost of well investment	15,548.00
Variable cost per well	
(a) Operation and maintenance cost	710.00
(b) Electricity charges	0
Total irrigation cost per well	16,258.00
Average number of operation hours per year per well	1123.20
Amortized cost per hour of groundwater extraction	14.47
Amortized cost per cubic metre of groundwater extracted	0.37

6.5 *Economics of groundwater markets*

Ground water markets were found to play an important role in sustaining the agricultural livelihood economy of marginal and small farmers in Rampur and Bilron. The sample size of 33 water buyers also included seven well owners who were also purchasing water for their crops. A total area of 56.7 ha was irrigated by the water buyers through purchased groundwater irrigation in agricultural year 2005–06. The area irrigated through purchased groundwater was high in rabi season (38.4 ha) than that of kharif season (16.5 ha) and on an average 1.7 ha was irrigated by water buyer in these villages. However, during kharif season, water buyers cultivate mostly rainfed crops. Groundwater is sold on per hour basis in the study region with the price ranging from INR 15 to INR 50. The water rates were found to increase with pump horsepower and well yield. Water buyers on an average bought 142 hours of water in a year and paid INR 3762 as water charge to the well owner. More than 86% of the total hours have been purchased for crops grown in the rabi season.

Water sellers' income was computed with a sample size of fifteen pump owners. These water sellers sold a total of 7223 hours with an average of 481 hours per water seller and through this, an area of 47 ha was irrigated. A total amount of INR 1,66,265 was realized through water selling. The results clearly show that water sellers recover the amortized cost of the tube-well by water selling as the average income per water seller was INR 11,085 a value close to the annual amortized cost of well investment. Results also show that water sellers are making a net profit of INR 10 per hour sale of groundwater if the average water price per hour sold/bought is taken as INR 25.

6.6 *Crop and water productivity*

Water productivity can be measured in two ways. One method of estimating water productivity is in terms of yield per unit of applied water, which is also known as physical water productivity or applied water productivity. The second way is by estimating the returns realized per unit of applied water, which is known as economic water productivity (Kijne, 2003). More emphasis is nowadays given on increasing the total value of output per unit or drop of water in order to utilize the scarce water resources more efficiently. Water productivity was estimated for selected crops among the well owners, shared well owners and water buyers.

The crop and water productivity for different crops realized across categories of farmers having different access to groundwater are presented in Figs. 8 and 9. For irrigated crops like paddy and wheat, water productivity was maximum in the case of shared well owners followed by individual well owners. The same trend was observed for fodder crops. Shaheen and Shiyani (2005) found that shared well owners achieved higher water productivity as they try to optimize use of their shared resource. Crop productivity for paddy was more for well owners than shared well owners which can be attributed to the fact that reliable and assured water supply for the crop is available in the case of the former while share owners have to wait generally for their turn which may some times cause time lag in irrigation during critical crop stages. This is in spite of the fact that only one among the share well owners of a well is usually allowed to grow paddy. The crop productivity for wheat, however, follows the same trend as that of water productivity. The maize crop is chiefly grown in the kharif season and is largely rainfed. Water buyers resort to buy only to fulfill one or two supplemental irrigations for their crop. Hence, applied groundwater productivity for

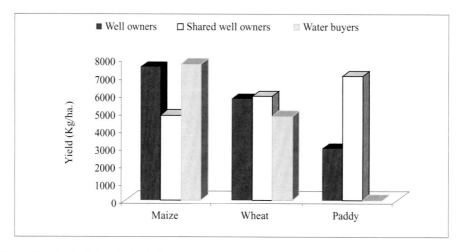

Figure 8.   Productivity of principal crops in study area.

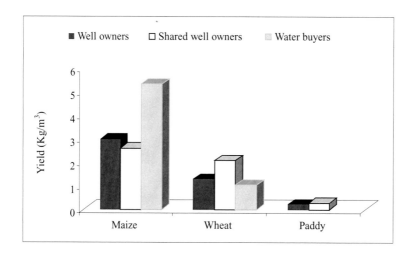

Figure 9.   Water productivity of principal crops.

maize is highest. This can be substantiated by the observation that the maize crop grown by well owners or share well owners is invariably over-irrigated.

7   ROLE OF GOVERNMENT POLICIES AND PRINT MEDIA

Populism by state governments offering free or subsidized power to farmers is damaging the aquifers across the country. According to a report (Anonymous, 2006) between 1995 and 2004, 75% of the blocks in Punjab state have been overexploited. The state government of Punjab incurs a huge cost in supplying free electricity to the farmers but as the political parties come to power by virtue of support from the peasant lobby, they are reluctant to impose policies that will be resented by their electorate. The statistics of PSEB reveal that on an average, 12% of the consumers served in a year (2001–06) belong to the farm sector and 18% of the power consumed is for tubewell irrigation (PSEB, 2006).

There are several central and state groundwater authorities functioning in the state of Punjab. These agencies have a wide network of monitoring wells from which the level of groundwater exploitation in a block or district can be calculated. The PSEB does not restrict the number of connections awarded in a year based on these reports.

The state water policies formulated by these agencies have started to advocate the use of artificial recharge methods to replenish the aquifers. Micro-irrigation techniques, which promote sustainable use of water, were not propagated much by the state government though some subsidies are now available. However, the villagers of the study area lack awareness on these technologies. A new proposal by the Soil Conservation Department allows a farmer with a certificate from the Horticulture/Soil Conservation Department showing that he is adopting sprinkler/drip irrigation for his plantation or vegetable crops, will provide him with priority access for a tubewell connection by the PSEB.

Much of the media reported emphasis is laid on the current issues of agriculture, including special economic zones, shortage of farm inputs and strength of farmers' lobbies. Whenever groundwater related issues are highlighted, they are depicted in headlines (e.g. suicide of farmers due to well failure) on issues which evoke emotional response. The media can play a more positive role by spreading the message of sustainable groundwater use.

8   CONCLUSION

The location of study area in Hoshiarpur district is strategic with respect to groundwater potential as numerous streams originating in the Shiwaliks assist recharge of the aquifers. Hoshiarpur district is classified as a safe area for groundwater extraction has no immediate depletion problem even though the water table has declined in the recent past. The causes for the decline are primarily the increased use of tubewells. The ease with which farmers in the study villages can get an electricity connection for their tubewell has made sinking of tubewells an attractive proposition, and the free power provided to the farm sector is an added incentive. Former waste lands have been reclaimed, converted to big plantations and increased the use of groundwater.

The variability in the physical resources, particularly the depth to the water table, coupled with the social caste system seen in the study villages has given rise to two distinct patterns of access to groundwater. The role that an agrarian based community can play in evolving a sharing mechanism for the available groundwater, which has withstood the test of time in Bhajjal village, is an example to be emulated. On the other hand, small and fragmented landholdings have led to a flourishing water market in the neighbouring villages of Rampur and Bilron as substantiated by a groundwater market participation rate of 55% found among the respondents of the study area. Access to groundwater is clearly skewed towards the land holding size in the cases of Rampur and Bilron villages, whereas, in a sharing village, every farmer has access to the resource in proportion to his land holding. Water productivity of shared well owners is greater than that of the sole well owners and water buyers, indicating the efficiency of this informal institutional mechanism. The adequacy of water in the study area had not entirely satisfied the cropping needs of farmers who fall in the category of water buyers. These small and marginal farmers usually face delay in irrigating their crops, especially when the number of purchasers in a group is large. This constrains their crop development and water productivity which can only be overcome by having more common water harvesting structures in the villages by which access to irrigation water can be improved for the marginal and small farmers who have highly fragmented land holdings.

It is not only the amount of water applied that controls crop and water productivity but also the quality of water and soil fertility within a region. The quality of water drawn from the tubewells in the study region is found to be quite suitable for irrigation purpose except that gypsum has to be added in certain wells to improve suitability. However, the soil fertility status of the soils in the three villages is lacking in the nutrients nitrogen and potash. Addressing these concerns through integrated nutrient management can go a long way to improving the water productivity, since the availability and quality is not a constraint for farmers in this region. However, the fertilizer recommendations usually made for the whole state need to be more location specific.

The government policy of appeasing the farmers lobby with zero electricity tariff can be a double edge to sword as it can lead to unsustainable groundwater exploitation. The current need is the promotion of artificial recharge methods, which can replenish the aquifers, and adoption of micro irrigation techniques, especially for the plantation crops.

REFERENCES

Anonymous (2006). Free power for tubewells rings alarm bells—Massive overdrawing of groundwater due to subsidies to farmers: Report. *Yojana*, 50, August 2006: p. 50.

Chandrakanth, M.G., S. Adhya, and K.K. Ananda (1998). 'Scarcity of Groundwater for Irrigation: Economics of Coping Mechanisms in Hard Rock Areas'. Monograph, Department of Agricultural Economics, University of Agricultural Sciences, GKVK, Bangalore, India.

Debroy, A. and Shah, T. (2003). Socio-ecology of groundwater irrigation in India, in *Groundwater Intensive Use: Challenges and Oppurtunities*, pp. 307–335 Llamas, R&E. Custodio (ed). The Netherlands: Swet and Zetlinger Publishing Company.

GOP (2007). http://www.punjab.gov.in (accessed on 17 February 2007).

Indiastat (2007). www.indiastat.com (accessed on 1 March 2007).

Kaushal, M.P. and Sondhi (2006). Simulation Modeling and Optimization Studies for the Groundwater Basins of Northwest India: Case Studies and Policy Implications. In: Sharma, B.R., Villholth, K.G., and Sharma, K.D. (eds.). *Groundwater Research and Management: Integrating science into management decisions*. International Water Management Institute. Colombo, Sri Lanka 282 p.

Manku, D.S. (2007). Geography of Punjab. www.Punjabenvironment.com (accessed on 17 February 2007).

Moench, M. (1996). Groundwater policy: issues and alternatives in India, International Irrigation Management Institute, *IIMI Country paper*, India, No.2, Colombo, pp. 61.

Muthuvel, P. and Udayasoorian, C.U. (1999). The laboratory manual "soil, plant, water & agrochemical analysis", Tamil Nadu Agricultural University, Coimbatore.

Kijne, J.W., R. Barker, and D. Molden (2003). *Water productivity in agriculture: limits and opportunities for improvement*, CABI Publishing, UK.

PSEB (2006). Electricity Statistics of Punjab ending 31.3.2002. Planning Organization, Punjab State Electricity Board, Patiala.

Shah, T., Scott, C., Kishore, A. and Sharma, A. (2003). Energy-irrigation nexus in South Asia: improving groundwater conservation and power sector viability. Research Report 70, Colombo, Sri Lanka: International Water Management Institute.

Shaheen, F.A. and Shiyani R.L. (2005). Water use efficiency and externality under the groundwater over-exploited and energy subsidized regime. *Indian Journal of Agricultural Economics*, 60 (3).

Singh, T. (1993). Hydrogeology of Hoshiarpur District based on Reappraisal Hydrogeological Survey-FSP-1992-93. Central Ground Water Board, Chandigarh.

CHAPTER 6

Groundwater resource conditions, socio-economic impacts and policy-institutional options: A case study of Vaishali District of Bihar, India

A. Islam
Indian Council of Agricultural Research, Research Complex for Eastern Region, Patna, India

R.S. Gautam
The Livelihood School (BASIX), Indore, India

ABSTRACT: The tubewell irrigated area in Bihar accounts for 30% of the gross cropped area and around 81% of the net irrigated area in the state. In Vaishali district, where groundwater is available at shallow depths, the number of shallow tubewells has greatly increased in the last two decades. The number of pump sets has not increased in proportion to the number of tubewells because of the high cost of pumps and low investment capacity of small and marginal farm units. Diesel operated pumps are common and escalating diesel costs and high demand for irrigation has led to a rapid increase in water prices. Because of the scattered landholdings, water is purchased even by the tubewell owners for irrigating lands located away from their own tubewell. Though community tubewells are preferred by villagers, but various socio-political influences and lack of a conflict resolution mechanism, means that some community tubewells are not working as envisaged. An attempt is made to explore groundwater utilization scenarios in the three villages in the Vaishali District.

1 INTRODUCTON

Irrigation has contributed significantly towards increasing India's food production and to creating grain surpluses, but there are some irrigated areas where agricultural productivity continues to remain low. The state of Bihar, which has an area of 94,200 km^2, is rich in resources but has low productivity. The soils are predominantly thick alluvial (Gangetic) deposits, or swamp and Terai soils, which are rich in nutrients. The state has surplus water resources with the potential for double and multiple cropping. In spite of the rich soil and abundance of easily accessible water, agriculture in Bihar has remained stagnant due to its colonial legacy, ecological conditions, demographic pressure, the land tenure system, and a lack of infrastructure and economic incentives (Banerjee and Iyer, 2002, Ballabh and Sharma, 1990, Kishore, 2004). Land holdings in Bihar consist predominantly of small farms (average size of each holding is around 0.6 ha), many with a high degree of fragmentation. Fragmentation of land holding impedes productivity, and subsistence farming continues to predominate. Hayami (1981) considered technological options as the only practical approach for agrarian transformation in this region, where land redistribution/consolidation is politically difficult. The Reserve Bank of India

(RBI, 1984) stressed that a machine based approach for exploiting the region's abundant groundwater resources using shallow tubewells was essential for increasing agricultural productivity.

The gross cropped area of Bihar is estimated at 79,460 km^2 and the total irrigated area is 48,400 km^2 (60.9%). The tubewell irrigated area is 30% of the gross cropped area and less than 50% of the total irrigated area. Net groundwater availability and annual groundwater draft of the state has been estimated as 28,456 Mm^3 and 9,472 Mm^3, respectively (CGWB, 2006), with only a third of the available groundwater resources so far developed. Groundwater is generally available at shallow depth, but varies considerably depending upon the geology, terrain and the season. During pre-monsoon (May) the groundwater level ranges from 1.2 to 18.1 m bgl, but in the post-monsoon (November) it varies from 0.5 to 9.2 m bgl (CGWB, 2006). Kishore (2004) reported one bore well for every 2.5 ha of cultivated land in nine villages of six districts in Bihar.

Three villages Chakramdas, Amritpur and Jaitipur in the Vaishali and Lalganj blocks of Vaishali district, were selected for investigation because groundwater is locally their main source of irrigation. The three villages differ in terms of socio-economic conditions, groundwater use, cropping pattern and the economic value of the irrigation water. An informal private groundwater market exists in all three villages. In addition, in Chakramdas community tubewells co-exists with private tubewells.

Data gathering included 65 interviews with individual farmers (both water seller and buyer), regarding cropping patterns and how they access groundwater. Data were also collected at village level and from five well drillers. Village level data were mainly gathered through discussion with the villagers and the *gram panchayat* (elected village council) representatives. Group discussions were also conducted with members of the community tubewell groups, and members of self-help groups.

Data on rainfall, groundwater status, irrigation, cropping patterns, various government programmes/schemes, and demographic profiles and other information pertaining to the Lalganj and Vaishali blocks were collected from a variety of secondary sources including reports, published papers, web sites and NGOs working in the area.

2   LOCATION AND GENERAL DESCRIPTION OF THE STUDY AREA

2.1   *Vaishali district*

Vaishali district is located at latitude of 25°28′–26°00′N and longitude of 85°03′–85°38′E. The district is bordered by the rivers Ganga in the south and Gandak in the west (Fig. 1). The district lies within the semi-tropical Gangetic plain and covers an area of 2,036 km^2.

The maximum and minimum temperatures recorded in the district are 44°C and 6°C respectively and the district typically receives rainfall of 1160 mm per year. Around 87% of the total rainfall occurs during the monsoon season (June to September) (Figs. 2a and 2b). The soil of the area is fertile and suited for both food and cash crops. The topsoil is characterized by a high percentage of silt and is classified as silty loam.

Agriculture is the main occupation but fragmented land holdings are major constraints for the development of agriculture. Some 92.5% of the holdings are less than 1 ha in size (Table 1). There are various types of share cropping mechanisms, typical for lower and middle income farmers.

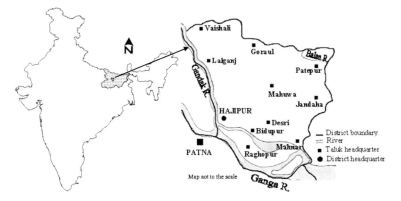

Figure 1.   Location map of the Vaishali district.

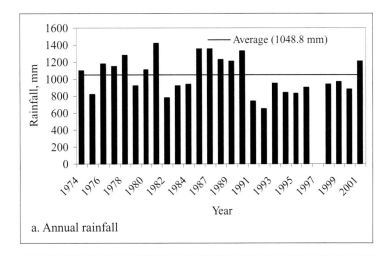

a. Annual rainfall

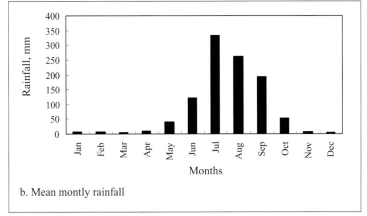

b. Mean montly rainfall

Figure 2a & b.   Rainfall distribution pattern in Vaishali district.

Table 1.   Details of landholdings in Vaishali district.

Size of holdings	Number	Area in ha
Less than 1 ha	439,643 (92.5%)	84,401 (55.1%)
Between 1 and 2 ha	24,214 (5.1%)	32,252 (21.1%)
Above 2 ha	11,263 (2.4%)	36,532 (23.9%)
Total	475,120 (100%)	153,185 (100%)

Source: Bihar through figures (2001).

Table 2.   Landuse details of Vaishali district.

Land use type	Area (ha)	Percentage of total area
Total geographical area	203,600	NA
Total reported area	199,119	100
Barren and uncultivable land	23,984	12.1
Land put to non-agricultural use	38,828	19.5
Land under miscellaneous trees and groves	9,550	4.8
Current fallow	3,712	1.9
Other fallow	457	0.2
Cultivable waste land	325	0.2
Permanent pasture	359	0.2
Net area sown	121,903	61.2
Area sown more than once	73,098	36.7

Source: Bihar through figures (2001), NA: Not applicable.

Net area sown accounts for 61.2% of the total geographical area, of which nearly 73,000 ha (36.7%) is sown more then once per year (Table 2). The most common crops grown in the area are paddy and maize in *kharif* (*monsoon*), and wheat, winter maize, gram, mustard and tobacco in rabi (*winter*) During the summer the land is either kept fallow or cropped with maize, green gram (moong), or vegetables. Some perennial crops such as sugarcane are also grown. The production and yield of the major crops are shown in Table 3.

The 3rd Minor Irrigation Census 2000–01, reported that groundwater is the major source of irrigation in the Vaishali district, accounting for 91% of the net irrigated area (Table 4) (MoWR, 2007). The number of shallow tubewells (depth less than 60–70 m) in the district was estimated at 25,267, and most (46%) of these tubewells are owned by marginal farmers (0–1 ha) (Fig. 3). Diesel pumps are the most common water lifting devices in this district.

2.2   *The study blocks*

The village of Jaitipur is located in the Lalganj block, and Amritpur and Chakramdas are located in the Vaishali block. In the Lalganj block most of the farmers have land holdings of less than 1 ha, but in the Vaishali block most of the farmers are medium farmers with land holding 2–10 ha (Table 5). Agriculture is the main source of income in both the blocks. Maize is the preferred crop during *kharif* (*monsoon*) in the Lalganj block, but in the Vaishali block paddy is the main crop during *kharif*.

Table 3.   Area, production and yield of major crops in Vaishali district.

| | 2002–2003 | | |
Crop	Area (ha)	Production (MT)	Yield (kg/ha)
Paddy			
Garma (*summer*)	188	356	1,894
Bhadai (*Auf*)	168	134	798
Agahani (*Aman*)	47,937	60,804	1,268
Wheat	43,369	91,786	2,116
Maize			
Rabi (*winter*)	5,677	16,282	2,868
Garma (*summer*)	12,105	48,981	4,046
Pigeon pea	1,522	1,289	847
Potato	6,861	45,570	6,642
Tobacco	6,252	7,027	1,124
Mustard	2,816	1,760	625

Source: Bihar through figures (2001).

Table 4.   Irrigated area in Vaishali district by source.

Details	Area (ha)
Net irrigated area	81,495
Irrigation through groundwater	73,864
Irrigation through surface water	7,621
Irrigation through major/medium irrigation schemes	10

Source: MoWR (2007).

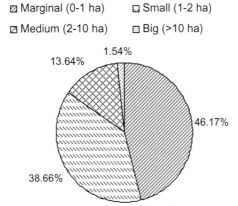

☒ Marginal (0-1 ha) ☐ Small (1-2 ha)

☒ Medium (2-10 ha) ☒ Big (>10 ha)

1.54%
13.64%
46.17%
38.66%

Figure 3.   Distribution of shallow tubewells among different land holdings classes in Vaishali district (MoWR, 2007).

Table 5.   The Lalganj and Vaishali blocks.

	Lalganj	Vaishali
Geographical area (ha)	22,194	19,973
Total population (1991 census)	209,723	138,628
Sex-ratio (Females per 1000 males)	890	1,038
Literacy rate (%)	27.3	24.9
Households involved in agriculture (no.)	22,907	20,100
Agricultural labourer households (no.)	19,925	19,283
Households in household & cottage industries (no.)	1,923	563
Others occupations (no.)	2,634	2,010
Net cultivable land (ha)	16,154	17,021
Land holdings (no.)		
Marginal farmers (upto 1 ha)	40,814	3,305
Small farmers (1–2 ha)	7,623	6,000
Medium farmers (2–10 ha)	722	9,100
Big farmers (>10 ha)	–	–
Cropping pattern (ha)		
Maize	11,000	2,000
Paddy	1,245	12,000
Wheat	5,000	6,300
Oilseeds	1,198	300
Vegetables/Tobacco	1,754	600

Source: District credit plan (2005–06).

The study villages have many features in common. These include diverse caste system with one or two dominant castes and the majority of land resources owned by the dominant caste group, subsistence farming dependent on groundwater, seasonal migration among weaker sections, and non-availability of electricity. The major differences (Table 6) are that the density of tubewells is greatest in Jaitipur and least in Amritpur, the number of pumps per unit of cultivable land is maximum in Chakramdas, and least in Jaitipur. In Amritpur most tubewells and pumps are owned by middle caste groups in contrast to the other two villages where upper caste groups posses the majority of the pumps and tubewells. Maize is the major crop of *kharif* in Amritpur (Table 6).

3   RESOURCES

The geological formation (aquifer) comprises various unconsolidated alluvial sediments. Transmissivity lies in the range 400 to 700 m^2/day with specific yield and coefficient of storage of 12 to 20% and 0.15, respectively, and is described as a 'good aquifer' (CGWB, 1993). The tubewells operating in the area are mostly privately owned. Generally, the private tubewells have an irrigation capacity of 2 to 5 ha each (CWC, 1995) with a discharge of 10–25 m^3/hr. The CGWB's hydrogeological map of Bihar describes the region as "Large Yield Prospects (more than 150 m^3/hr)". However, only 58% of the total potential has been developed (Table 7) and there is scope for expansion.

Interviews with the local well drillers provided details on well construction, depth to water table, water quality and drilling methods. Well drilling is largely carried out by

Table 6.   Profile of three study villages.

	Chakramdas	Amritpur	Jaitipur
	Vaishali	Vaishali	Lalganj
Distance from district headquarters (km)	35	27	25
Total Area (ha)	130	194	211
No. of households	251	330	499
Population	2,320	2,500	3,086
Cultivable land (ha)	109	162	147
Irrigated land (ha)	97	97	118
Chaur land/Low land (ha)	13	16	7
Tubewells density (number of tubewells per 100 ha of cultivable land)	36.6*	22.2	40.0
Tubewell to pumps ratio	1.6	1.2	2.2
Caste distribution in the village (%)			
Low	26.7	37.9	54.7
Middle	38.7	51.5	14.2
High	34.7	10.6	31.1
Caste wise land-ownership pattern			
Low	5.0	15.0	7.0
Middle	12.7	32.6	9.7
High	82.3	52.4	83.3
Cropping pattern			
Kharif	Paddy	Maize, paddy, and cauliflower	Paddy, Maize
Rabi	Wheat, oilseed, maize and potato	Wheat, maize, cauliflower and Brinjal	Wheat, Oilseeds
Wage rate (INR[1]/day)			
Normal	INR 15/day + meal	INR 20/	INR 20/
Peak season	INR 60/	INR 50/	INR 80/
Cost of land (INR/ha)			
Best quality agricultural land	1,235,000	741,000	988,000
Poor quality unirrigated agricultural land	494,000	432,250	370,500

* Includes four community tubewells.
[1] USD ~ Indian Rupees (INR) 40.00 as in May 2008
Source: Primary survey, 2007.

the *Mallah* (fishermen caste) in the area. The groundwater is available at shallow depth (Table 8) and one farmer in Chakramdas village uses manually operated treadle pumps to extract groundwater from shallow tubewells (6 m deep) (Figure 4). The land is divided into three topographic levels with an elevation difference of about 6 m between the upper and lower lands. Depth of boring varies with elevation, the depth to the water table is maximum in the northeast where the elevation is highest and reduces towards the river Gandak. There are a large number of small depressions locally known as *Chaurs*, where rainwater remains accumulated each year until January or February, and these promote infiltration to the aquifer. There is not much change in water level from one year to the next although a steady long term decline of up to 4.5 m in 15 years has been reported by the farmers and well drillers.

Table 7.   Groundwater resources availability and use in Vaishali district.

Total groundwater resources (ha m)	63,293
Quantity of groundwater used for irrigation (ha m)	9,494
Available groundwater for irrigation (ha m)	53,799
Utilizable groundwater for irrigation (ha m)	48,419
Net groundwater draft (ha m)	27,970
Irrigation potential created as on 01.04.1998 (ha)	43,031
Ultimate irrigation potential (ha)	74,491
Irrigation potential to be created (ha)	31,460
Stage of groundwater development in April 1998	5,848

Source: CGWB (2006).

Table 8.   Depth to water table during different months.

	Depth to water table (m)	
Months/Seasons	Chaur /Low lands	Up lands
July–Aug.	0 (at surface)	2
Dec.	2	4
April–May	5	5–6

Source: Primary survey, 2007.

Figure 4.   Treadle pump for water extraction in Chakramdas village.

The quality of water is generally potable. However, in Jaitipur village excessive iron is present at shallow depths (10 to 15 m) at some places. Deeper boreholes cased to depths below 30 m avoid the problem. Drilling is mainly undertaken by percussion method using hand boring sets and the tubewells with galvanized iron pipe is common although PVC pipe is also used now.

4 SOCIO-ECONOMIC IMPLICATIONS OF GROUNDWATER USE

4.1 *Groundwater irrigation technology*

During focused group discussions with village elders/heads, it was reported that before the advent of tubewell technology *dhekul* (counterpoise lift) and *ghirni* (mohte) systems were used for lifting water from dug wells for irrigation by some farmers in north Bihar. As these devices were manually operated, only a very limited parcel of land could be irrigated and a large portion of agriculture land remained rainfed. Tubewells were introduced in the mid-1960s, especially after the famine of 1966–67. The head of Amritpur village constructed the village's first tubewell in 1965 with government support. At the same time, the government promoted *kuchcha* (earthen) open wells in order to bring more agriculture land under irrigation. However, this scheme failed, because the soil condition did not favour earthen open wells and they usually collapsed after a few seasons or after heavy rainfall. Five community tubewells were constructed for irrigation between 1970 and 1975 at Chakramdas. The tube-wells constructed in 1960s and 70s were generally deep tubewells (up to 80 m) and were drilled by government drilling machines. Over time, local drilling technology evolved and shallow tubewells flourished. Initially, diesel pumps were used for water extraction and later on they were replaced by electric centrifugal pumps, but by the 1980s these were replaced by diesel operated pumps because of the irregular electricity supply, the so-called 'de-electrification' of eastern India (Shah et al., 2006). In recent years lightweight kerosene/diesel operated pumps such as Honda and Chinese pumps are also becoming popular due to their low cost, low fuel consumption and small size and ease in transportation.

Diesel operated 5 hp centrifugal pumps are common in all the three villages. Most of the water sellers prefer light weight low hp kerosene operated pumps, but non-availability of kerosene in the open market is forcing them to shift towards light weight diesel operated centrifugal pumps.

4.2 *Water transaction*

The number of the pump-owner selling water to other is maximum in Chakramdas (33%) followed by Amritpur and Jaitipur, where as number of pump owner buying water is maximum in Jaitipur (Table 9). The proportion of gross irrigated area is maximum in Chakramdas (98%) followed by Jaitipur and Amritpur (Table 10).

4.3 *Cost of water extraction*

The operational cost of pumping in the three villages ranges from 42.3 INR/hr to 54.2 INR/hr with minimum in Chakramdas and maximum in Amritpur. This may be due to variation in tubewell density. Hourly rates charged by water sellers varies from 45 to 70 INR/hr. As

Table 9.   Water transaction status in study villages.

Description	Chakramdas	Amritpur	Jaitipur
Total number of pump owners	15	11	10
Total number of non-pump owners	10	10	9
Total number of shared well owners	8	1	0
% of pump owners who sell water to others	33	27	20
% of pump owners who buy water from others	13.5	18	40
% of pump owners who sell as well as buy water	40	55	0
% of pump owners who neither buy nor sell water	0	9	40

Source: Primary survey, 2007.

Table 10.   Irrigated area and mode of water transactions.

Description	Chakramdas	Amritpur	Jaitipur
Gross cropped area (GCA) (ha)	44.8	20.9	37.6
Gross irrigated area (GIA) (ha)	44.0	15.1	32.0
% GIA to GCA	98.2	72.2	85.2
% GIA irrigated by own water	60.0	63.5	67.5
% GIA irrigated by purchased water	20.5	36.0	32.5
% GIA irrigated by canal and other surface source of irrigation	5.6	0	0
% GIA irrigated by shared tubewell	14.0	0.5	0

Source: Primary survey, 2007.

lands are fragmented, water is conveyed through plastic pipe up to a distance of 300 m, and in Chakramdas water is also conveyed through lined channels from community tubewells. It was also observed that the water buyer has to pay to use the plastic pipes. Pump owners not having their own tubewell have to pay INR 5/hr to the tubewell owner for abstracting water. A pump owner with no tubewell could get water from a tubewell owner free of cost in exchange for the pump-owner allowing the tubewell owner to use his pump without any charge other than for fuel. Generally water sellers irrigate their own land first and then supply water to others on a first come first served basis. As water availability is not a constraint in these villages, the water allocation generally depends on the availability of pumps.

4.4   *Cropping pattern and crop productivity*

Crop productivity (yield per unit of land) is high in Chakramdas compared to the two other villages (Table 11). Crop productivity is higher for the pump-owner than the non-pump-owner in all the three villages. Similarly, groundwater productivity (yield per unit of groundwater used) of the pump owner is comparatively higher than that of non-pump owner for both paddy and wheat due to the timely availability of irrigation water in the case of pump owner.

Table 11. Productivity of major crops.

	Paddy Productivity (kg/ha)	Wheat Productivity (kg/ha)
Chakramdas		
PO	2610	2895
NPO	2015	2750
ALL	2360	2840
Amritpur		
PO	2075	2430
NPO	1300	2030
ALL	1730	2240
Jaitipur		
PO	2080	2615
NPO	1590	1920
ALL	1880	2220
Overall		
PO	2375	2680
NPO	1740	2240
ALL	2115	2480

Source: Primary survey, 2007, PO: Pump owners, NPO: Non-pump owner.

5 POLICY AND INSTITUTIONAL PERSPECTIVES

5.1 *Community tubewell*

During a period 1970–1975 thirty six community wells were installed in 16 villages in Vaishali and Saraiya. The Vaishali Area Small Farmer's Association was created to support and facilitate the smooth functioning of community tubewells with the financial support from Danish International Development Agency (DANIDA). Five community tubewells were constructed in Chakramdas village, from where this project was started (Pant, 1984).

The community tubewells paved the way for land consolidation and rapidly transformed the rainfed areas into irrigated areas. Though the concept of community tubewell was envisaged for small and marginal farmers, inclusion of some of the bigger farms was unavoidable due to their location. Bank loans were taken for construction of the tube-wells, purchasing of pump sets, construction of pump houses and concrete field channels for water distribution. Most of the farmers, particularly small farmers, repaid their loans but those who could not pay were exempted from their loans by the government in 1989.

Many of the bigger farmers managed to install tube-wells on their land, and came to control and dominate the decision-making and group functioning and influenced water allocation and collection of water charges. The small farmers, who belonged to lower caste, had little say in decision-making became reluctant to attend group meetings. As a result of irregularity in organizing group meeting, the organisation could not easily be changed, and lack of transparency and accountability in financial activities as well as elements of corruption created conflicts and impeded the functioning of the community tube-wells. By the 1980s three tube-well groups had failed to pay the electricity bills and the electricity board stopped their supply of electricity.

New diesel pumps were brought under a government scheme that gave a subsidy of 67% over market price. The members of the two operational community tubewells managed to collect the required one-third cost of the diesel pump set, while the three defunct groups failed to do so. The big farmers, who had paid one-third the cost of the pump, became owners of their respective community tubewells and continued operation and collection of water charges for several years without maintaining any transparency and accountability. This again raised conflicts between the big farmer and other group members.

The two group tubewells were functional in Chakramdas during this study. The key to success was: active participation of group members in group processes, transparency and accountability, strong leadership, a comparatively big command area comprising large number of small and marginal farmers, cultivation of cash crops (vegetables) by farmers throughout the year, regular collection of water charges, easy access to market, intensive irrigation leading to higher revenue collection, and selling of water to non-members on higher rates. Some of the problems had been:

- Many group members have sold their land but are still members of the group. Though the new landowners are getting water on the ownership right of previous owner they cannot participate in the decision making.
- Over time, agricultural lands get divided within families, but the siblings (sons) are not provided with membership of the community tube-well group, and if any of the siblings fail to pay the water dues, water is not provided to any of the family members the next season.
- The big farmers lease out their lands to people who sometimes leave without paying their water dues, as the water dues are paid the next season.
- Many big farmers have constructed their own tube-wells with government support under various subsidy schemes and are selling water to others in competition to the community groups.

Despite all these difficulties, most of the group members were in favour of the system owing to higher water discharge and lower water charges compared to the private tube-well.

5.2   *Groundwater utilization and government policies*

The government promoted tubewell technology in Bihar through various subsidy schemes. The Planning Commission of the Government of India implemented the "Million Shallow Tubewell Programme (MSTP)" in 38 districts (including Vaishali) during 2001–02 and provided credit, subsidy and required only a marginal contribution from eligible farmers. So far 217111 shallow tubewells with pumps have been installed (NABARD, 2004–05).

6   CONCLUSIONS AND POLICY OPTIONS

Small and fragmented landholdings, low productivity of rain fed agriculture and burgeoning population forced the farmers to increase agriculture production and productivity through irrigation development. The number of tubewells has increased dramatically in Vaishali region in the last two decades. The number of pump sets (diesel/kerosene) however, has not increased in proportion to the number of borings because of the high cost of pumps and low investment capacity of small and marginal farmers. The pump sets and the numerous borings, therefore, play a prominent role in groundwater extraction and its marketing.

An informal groundwater market exits in the villages. The water is sold at the price of INR 45–70/per hour and the price of water is increasing by an average rate of INR 10/- per year. The price of water is comparatively low in Chakramdas because of the presence of community tubewells. In the case of Amritpur it was observed that with increase in distance from tubewell to farm plot, the water buyer has to pay INR 10 per hour extra for conveying water using plastic pipes. Availability of groundwater is not a major constraint in these villages and groundwater related disputes are few. Water is available as and when needed depending upon the availability of the pump.

Community tubewells are preferred by villagers but they are not always performing satisfactorily. Increasing water prices forced the farmers, especially the small and marginal farmers, to change their cropping pattern in favour of vegetables or other cash crops which are grown all year round as the return from these crops is higher than the traditional rice-wheat crops.

Recommendations to help boost the agriculture production through groundwater irrigation include:

- Provision of rural electricity for lowering the cost of pumping.
- Provision for subsidy and credits for pumpset purchase.
- Availability of low cost pump sets for small and marginal farmers.
- Encouraging community/shared tubewells with effective and transparent governance.

REFERENCES

Ballabh, V. and Sharma, B.M. (1990). HYV Adoption, Productivity Gaps and Production Adjustment Mechanisms in Flood Prone Areas of Utter Pradesh: A Synthesis, *Research Report 5*, IRMA, Anand.

Banerjee, A. and Iyer, L. (2002). *History, Institutions and Economic Performances: The Legacy of Colonial Land tenure System in India'*. http://info.worldbank.org/etools/docs/voddocs/183/367/land_tenure_India. pdf (accessed on June 15, 2007).

Bihar through Figures.(2001). http://bih.nic.in (accessed on December 22, 2006).

CGWB (1993). Hydrogeology and groundwater resources of Vaishali district, Bihar, *Technical Report No 103*, Series D. Central Groundwater Board, Patna.

CGWB (2006). http://cgwbbihar.nic.in/ (accessed on December 22, 2006).

CWC (1995). Guideline for planning conjunctive use of surface and groundwater in irrigation projects, Central Water Commission, New Delhi.

District Credit Plan (2005–2006). District—Vaishali (Bihar), Central Bank of India, Hajipur, Vaishali.

Hayami, Y. (1981). '*Agrarian problems in India from an Eastern and South Eastern Asia perspective*', Institute of Economic Growth, New Delhi.

Kishore A. (2004), Understanding Agrarian Impasse in Bihar, *Economic and Political Weekly*, 30 (31): 3484–3491.

MoWR (2007). http://mowr.gov.in/micensus/mi3census/ (accessed on June 15 2007).

Shah, T., O.P. Singh, and A. Mukherji (2006). 'Groundwater irrigation and South Asian agriculture: Empirical analysis from a large scale survey of India, Pakistan, Nepal and Bangladesh', *Hydrogeology Journal* 14 (3):286–309.

NABARD. (2004–05). Million Shallow Tube Wells Programme in Bihar- a quick impact evaluation study. *Evaluation Study Series Patna No 7*, National Bank for Agricultural and Rural Development, Bihar Regional Office, Patna.

Pant, N. (1984). 'Community tubewell: An organizational alternative to small farmers' irrigation', *Economic and Political Weekly*, 19 (Review of Agriculture) June 1984.

RBI (1984). *Agricultural productivity in Eastern India*, Reserve Bank of India, Mumbai, India.

CHAPTER 7

Groundwater resource conditions and groundwater sharing institutions: Evidence from eastern Indo-Gangetic basin, India

K.H. Anantha
Institute for Social and Economic Change (ISEC), Bangalore, India

D.R. Sena
*Central Soil and Water Conservation Research and Training
Institute (CSWCRTI), Gujarat, India*

A. Mukherji
International Water Management Institute (IWMI), Colombo, Sri Lanka

ABSTRACT: Based on socio-economic survey, well driller's survey and secondary data, this paper investigates three perspectives of groundwater use, namely, the resource, the socio-economic and policy perspectives. The study region in Murshidabad district of West Bengal state is representative of the eastern Indo-Gangetic basin and is characterized by abundant groundwater at relatively shallow depths that gets adequate recharge due to high rainfall. While there is an abundant supply of groundwater, access to it is determined by capability to construct a tubewell or to buy water from other pump owners, government tubewells and tubewells operated by water users committees. All these options are available in the region and each has somewhat different implications for water buyers, who for the most part, are small and marginal farmers. Lack of electricity and rising cost of diesel are the two major impediments for further development of groundwater.

1 INTRODUCTION

Irrigation, in general and groundwater irrigation in particular is the backbone of Indian agriculture. The two basic advantages of groundwater irrigation over surface water irrigation are that it is less sensitive to rainfall variability and it offers the potential of obtaining water more or less on demand. These advantages lead to more secure agricultural planning and lower levels of risk and thus encourage investment in the inputs necessary to utilize new agricultural technologies (Kahnert and Levine 1989). Various scholars have also shown that groundwater irrigation can be deployed for poverty alleviation especially in water abundant regions such as the eastern Indo-Gangetic basin (Chambers 1986; Chambers et al. 1989; Shah 1993; Vaidyanathan 1996, 1999; Deb Roy and Shah 2003). Groundwater provides irrigation for the whole year and opens up new farm employment opportunities for improving the income level of farming households (Hussain and Biltonen 2001). Also, unlike canal irrigation, groundwater development mostly took place in the private arena, and in many locations access to groundwater is chiefly determined by informal water markets operating at a local level (Shah 1993; Mukherji 2004). As per the 54th round of the National Sample

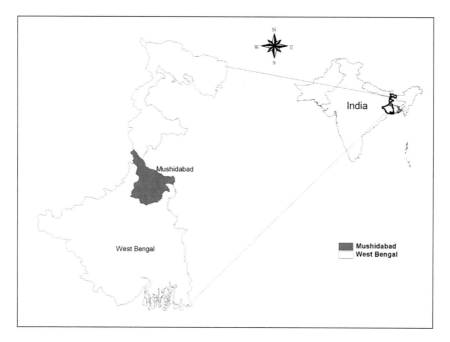

Figure 1.   Location of Murshidabad district, West Bengal, India.

Survey (NSSO 1999), of the 85 million farming households in India, as many as 21 million households own WEMs while another 24 million households buy water from them (also see Mukherji 2008).

While there are several studies that look into physical characteristics of groundwater there are relatively few studies that look into socio-economic and policy perspectives of groundwater use. This paper integrates the three perspectives. Field work was carried out in Murshidabad district of West Bengal. West Bengal located in the eastern part of the Indo-Gangetic basin.

2   DESCRIPTION OF THE STUDY SITE

This study is based on the primary survey conducted in four villages of Berhampore block of Murshidabad district in the state of West Bengal (Fig. 1). West Bengal is an eastern of India and receives an average rainfall of 1500 to 2500 mm annually. The study villages belong to the sub-humid sub-tropical zone of the Indian peninsula and are situated in the eastern part of the river Bhagirathi in the Indo-Gangetic Basin (IGB). The annual rainfall of the area is about 1389 mm mostly received from the south-east monsoon (June to September).

3   DATABASE AND METHODOLOGY

The village selection was made on the basis of groundwater development status, energy use for water extraction mechanism (WEM) operation, prevalence of different types of water sharing institutions, socio economic condition of the respondents and cropping pattern. The aim was to choose villages that offered sufficient contrasts in terms of groundwater sharing

Table 1.   Type of motive power, ownership and mode of water sharing.

Name of the village	Type of WEMs (e.g. deep tubewell, shallow tubewell medium deep tubewell)	Motive power of pumps (electric/diesel)	WEMs owned and operated by (individuals/govt./ CADC/govt. TW turned over to farmer's coop)	Mode of water sharing (private GW market, Non-market water sharing arrangements
Kaya	STWs, MDTWs	Electric and diesel	Individuals and farmers groups	Private water markets
Paschim Gamini	DTWs, MDTWs and STWs	Electric and diesel	Govt., CADC and individuals	Private GW markets as well as CADC
Mankara	DTWs, MDTWs and STWs	Electric and diesel	Govt., CADC and individuals	Private as well as government and CADC
Baninathpur Dakshin Para	STWs and DTWs	Only diesel	Individuals	Private water markets
Baninathpur Uttar Para	STWs and MDTWs	Electric and diesel	CADC and individuals	Private as well as CADC

Source: Primary survey.

institutions. Village of Kaya was chosen because this village has agricultural electricity connection. Both diesel and electric operated WEMs are in use in this village. There is no government irrigation infrastructure in this village and private informal water markets operate and flourish here. Paschim Gamini village has electricity operated medium deep government tubewells run by either West Bengal Comprehensive Area Development Corporation (WBCADC in short CADC) or farmers' cooperatives. It has a mixed groundwater economy with private water market totally dependant on diesel WEMs and non-market water sharing arrangements from WBCADC or cooperative electric WEMs. The village of Mankara has complete dependency on government owned deep tubewells with stray cases of diesel pump owners selling to needy farmers in case the government infrastructure fails to meet their irrigation requirement. Baninathpur, a part of which inhabited by Hindus (*Dakshin pada*) does not have electricity connection and is completely governed by private market with diesel WEMs. However Baninathpur (*Uttarapada*) with majority of Muslims have ready access to government operated (DTWs) and CADC run medium duty tube wells (MDTWs) and water sharing arrangements are on a cooperative basis.

A primary survey was conducted in all the four villages during December 2006 and early January 2007, with a questionnaire schedule covering a representative cluster of 80 farm households using a stratified random sampling method. Representative samples of 17 WEMs operated by various motive powers were chosen covering the four villages for tubewell discharge[1] measurement. The discharge was estimated with the volumetric method

---

[1] The discharge rates of the measured wells have been scaled to fit into the historical time frame by extrapolating the discharge data to the data collection period. Individual crop scaling coefficients have been estimated based on the water table condition existing during the crop growth period. This methodology gives a realistic estimation of volume of water applied from groundwater source to a particular crop.

and was used for computing irrigation water needs. Apart from this, secondary data on the hydrogeology, agro-climatic parameters, and census data on groundwater infrastructures have also been collected from organizations like the State Water Investigation Department (SWID), Agri-irrigation department in Berhampore block and WBCADC. Time series data on water table fluctuation in pre- and post-monsoon season from an observation well maintained by SWID was also collected. In order to harness local knowledge on hydrogeology, information was collected from the well driller's. Interviews with key officials from SWID, WBCADC, Central Groundwater Board (CGWB), West Bengal State Electricity Board (WBSEB) and Agri Irrigation Dept, were carried out. All these organization have a direct or indirect role in groundwater management in the state. Some of the informal institutions such as water user beneficiary members of the CADC tubewells were also interviewed to study the different dimensions of water trading in the study villages.

4   GROUNDWATER RESOURCE POTENTIAL

4.1   *Monthly water balance of study site*

The long term monthly water balance (Thornthwaite and Mather 1955) has been computed for the study site (Table 2). Rainfall, temperature and potential evapo-transpiration data have been derived from IDIS (http://dw.iwmi.org/dataplatform/Home.aspx). The maximum soil moisture storage is adequate during July only, reflecting the high potential evaporation. Rainfall is confined to four months of the year (June to September). This shows that rainfed agriculture alone cannot sustain the production system.

4.2   *Subsurface geology and aquifer characteristics of the study area*

Number of exploratory and production tube wells installed by the SWID and test drilling by the Agri-Irrigation Department shows that clay, sand, silt and gravel with pocketed cases of cemented sand are predominant in the upper 300 m. The thickness of recent alluvium is more than 300 m. A granular zone can be seen from the surface to a depth of 90 m. A fine sand and clay zone is often found in the surface up to a depth of 3 m. Below the depth of 90 m clay is predominant. However, intercalated zones of micaceous sand of varied thickness are also found within the clay zones. The granular materials occur at an average depth of 8 m from surface with an average thickness of 70 m within one to two aquifers (see the Bhakuri gram panchayat bore well log Fig. 2 after SWID, 1998).

The State Water Investigation Department (SWID) has done several aquifer performance tests in this district. Deep tubewells yield about 150 m^3/hour and shallow tubewells 30 m^3/hour for drawdowns not exceeding 4 m. The transmissibility of the aquifer is about 7.4×10^6 m^2/d and storage coefficient is 2.26×10^{-3}.

4.3   *Status of groundwater irrigation development*

Surface lift irrigation and irrigation from groundwater are widespread in the district using deep tubewells, shallow tubewells and medium duty tubewells. In the eastern part of Bhagirathi and the associated flood plains in Berhampore block that contain the study villages, shallow tubewells fitted with centrifugal pumps are predominant. In contrast, in the western part (*Rahr* region), submersible pumps in shallow tubewells are more predominant.

Table 2.   Monthly water balance of the study site (Based on meteorological data Average Monthly Rainfall, Temperature and Evapo-transpiration data derived from IDIS http://dw.iwmi.org/dataplatform/ClickandPlot.aspx. (data period 1902–2002)).

Month	J	F	M	A	M	J	J	A	S	O	N	D
P	1.0	2.1	0.0	0.8	2.0	108.3	309.4	266.4	129.4	13.5	0.8	1.7
PET	76.0	93.0	146.1	172.3	164.3	125.7	108.0	106.4	101.7	106.5	84.8	71.8
AET	1.0	2.1	0.0	0.8	2.0	108.3	108.0	106.4	101.7	26.5	0.8	1.7
S_t	0.0	0.0	0.0	0.0	0.0	0.0	13.0	0.0	13.0	0.0	0.0	0.0
?S	0.0	0.0	0.0	0.0	0.0	0.0	13.0	-13.0	13.0	-13.0	0.0	0.0
W	0.0	0.0	0.0	0.0	0.0	0.0	188.4	173.0	14.8	0.0	0.0	0.0
D	-74.9	-90.9	-146.1	-171.5	-162.3	-17.4	0.0	0.0	0.0	-80.0	-84.0	-70.1
SMI	0.01	0.02	0.00	0.00	0.01	0.86	2.99	2.50	1.40	0.13	0.01	0.02
MAI	0.014	0.022	0.000	0.004	0.012	0.862	1.000	1.000	1.000	0.249	0.010	0.023

P   = Rainfall (mm),   PET = Potential Evapotranspiration (mm),
AET = Actual Evapotranspiration (mm) S_t = Plant available water (mm)
S   = soil moisture change (mm)   W   = Runoff or Surplus water (mm)
D   = Deficit (mm)   SMI = Soil Moisture index
MAI = Moisture Adequacy Index (AET/PET).

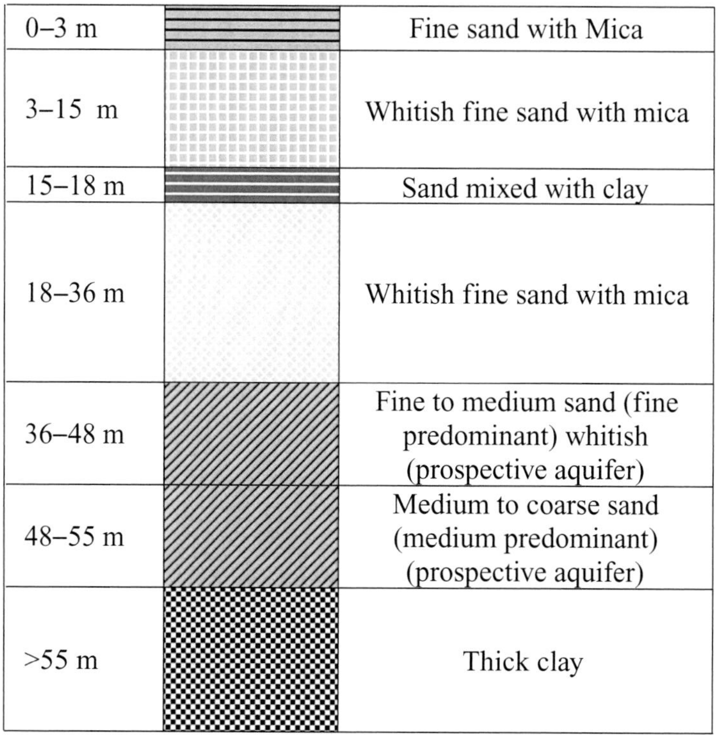

0–3 m		Fine sand with Mica
3–15 m		Whitish fine sand with mica
15–18 m		Sand mixed with clay
18–36 m		Whitish fine sand with mica
36–48 m		Fine to medium sand (fine predominant) whitish (prospective aquifer)
48–55 m		Medium to coarse sand (medium predominant) (prospective aquifer)
>55 m		Thick clay

Figure 2.   Representative well log chart of the bore well at Bhakuri gram panchayat.
Source: SWID log chart.

This reflects the depth limitation of centrifugal pumps. The gross annual draft from deep tubewell, shallow tubewell fitted with centrifugal pump and shallow tubewell fitted with submersible pump is about 18.5 ha-m, 1.8 ha-m and 6.0 ha-m respectively.

The numbers of tubewells in Murshidabad district have increased by 40% between 1991 and 1997. The Berhampore block shows a similar trend. Shallow tube wells (STWs) are increasing at a higher rate as compared to deep or medium duty tube wells (DTWs/MDTWs). From 1988 to 2001, the numbers of STWs have increased from 3,921 to 5,169 in 2001 (Figure 3). As per the Minor Irrigation Census 2000–01 (GOI 2001), the gross irrigated area served by shallow tubewells is about 12,315 ha of which 4,245 ha is under *boro* paddy alone. The numbers of deep tube wells have increased only by 54 in 1988 and 69 in 2001).

The rising trend in STWs is because the STWs are less capital intensive than the MDTWS/DTWs. With the arrival of the cheap Chinese motors (Fig. 4) which are fuel efficient (about 500 ml diesel per hour as compared to 1 l/h) the number of shallow tubewells has increased greatly since the last minor irrigation census of 1993–94. Another factor is the increasing number of well drillers, who made the STW installation more easy and cost effective. That diesel pumps are not always fixed permanently to tubewells also makes it possible to evade the SWID regulation of 200 m spacing between two shallow tubewells. However, drilling of a deep tubewell requires a large investment as well as

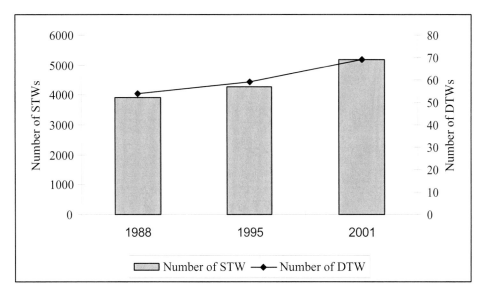

Figure 3.   Increase in number of shallow tubewells (STWs) and deep tubewells (DTWs) in Berhampore block of Murshidabad district, West Bengal.

自吸泵组 Self-Priming Pump Set

Figure 4.   Chinese version of the centrifugal pump fitted to a powerful Z170F diesel motor (3.5 to 4.4 hp).

difficulties in obtaining a permit (one km spacing between two DTWs). The deep tube-well technology is also more complex, requiring underground pipeline systems. Further, STW allows farmers to extract water using a diesel pump as and when required, where as DTWs and MDTWs are dependent on the uncertainty of electricity supply. Getting an electricity connection for a tubewell is not easy given the tedious procedure and various restrictions.

4.4   *Exploring well driller's knowledge for understanding local geo-hydrology*

The local well drillers have knowledge gathered over years of experience. The drilling business started some where around late 1960s to early 1970s during the green revolution which has also been called quite aptly, a 'tubewell revolution' (Repetto, 1994). Boom in well drilling business occurred between 1990 and 1995. However, their business has dropped to just 70 tubewells per year now compared to the 1970s when it was as high as 120 per year with a peak in about 1995.

The most common well drilling method used is 'water jet drilling'. The borehole is drilled by rotating a cutter or drill and the soil cuttings are continuously removed by circulation of drilling fluid (water). Drilling fluid is pumped down through the drill pipe and out through the ports or jets in the cutter and then flows upward in the annular space between the hole and the drill pipe, carrying the soil cuttings in suspension to the surface. At the surface, the fluid is channeled into a settling pit, where the cuttings drop out, and the water is re-circulated back down the hole. Water in one well was struck at a depth of

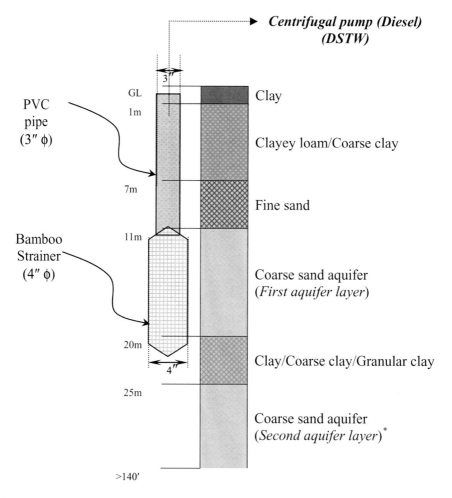

Figure 5.   Bore log information of the DSTW site at village Kaya and the cross section of low-cost bamboo filter section (local method of installation) of Diesel Shallow Tube Well (DSTW).
Ref: on-site Well Drillers boring on December 06, 2006.
Beyond 20 m the aquifer layer was plotted from the well driller's description.

11 m. The tubewell that was constructed had 3″ diameter PVC pipe followed by the bamboo strainer of about 4″ diameters. Cow dung was used to stabilize the coarse sand aquifer zone. On an average, one well driller constructs 200 tubewells in a year at a cost varying from INR 2000 to 4000 each. It takes a day and four laborers to construct one tubewell. The aquifer configuration is summarized in Fig. 5. The drillers report that the water table has declined by less than 2 m in the last 35 years.

5   SOCIO-ECONOMICS OF GROUNDWATER IRRIGATION

The analysis is based on data collected from 80 farm households who either owned or bought water from others. The main features of the agrarian structure of the study villages are small sized farm holdings with prevalence of fragmentation. Most households in the study area are marginal (less than 1 hectare of land) (Tables 4 and 5).

Cropping intensity in the study area has increased significantly with the growth of the groundwater infrastructure. Irrigation may also use water from *pukur* (pond), *khaal* (stream) and *bheel* (lake) but due to the non-perennial nature of these sources groundwater has emerged as a major source of irrigation (Table 5). The main irrigated crops in the study villages are *aman* paddy sown in the *kharif* (monsoon) season; wheat, rapeseeds and mustard in the *rabi* (winter) season; and *boro* paddy in the *summer* season. Besides these,

Table 3.   Distribution of households by landholding size in the study villages.

	Baninathpur		Kaya		Mankara		Paschim Gamini		All villages	
Landholding size	HH	%	HH	%	HH	%	HH	%	HH	%
Landless	96	19.43	0	0.00	182	62.75	560	76.29	838	46.74
Marginal farmers	346	70.04	220	80.00	74	25.51	103	14.03	743	41.44
Small farmers	36	7.28	51	18.54	26	8.96	59	8.03	172	9.59
Medium farmers	13	2.63	3	0.00	8	2.75	11	1.49	35	1.95
Large farmers	3	0.60	1	0.00	0	0	1	0.13	5	0.28
Total	494	100	275	100	290	100	734	100	1793	100

Source: Primary survey, HH = Household.

Table 4.   Extent of land fragmentation in the study area.

Particulars	Baninathpur	Kaya	Mankara	Paschim Gamini
Average size of landholding (ha)	0.59	0.36	0.71	0.73
Average size of plots (ha)	0.09	0.07	0.26	0.13
Average number of plots in ownership holding (numbers)	6.7	5.0	2.75	5.4
Largest number of plots owned by a HH in the village (numbers)	15	18	6	10

Source: Primary survey.

Table 5.   Cropping pattern and source of irrigation in the study villages.

Name of the major crop	Area under crop (ha)	Season	GW using Electric pump	GW using Diesel pump	Both	Average yield (kg/per ha)
Paddy (*Aman*)	98.4	Kharif	0	20.2	182.2	3337
Paddy (*Boro*)	23.5	Summer	4.0	0	50.6	5113
Wheat	14.6	Rabi	0	16.2	20.2	2065
Mustard	22.3	Rabi	0	18.2	37.7	1010
Jute	49.6	Summer	12.6	6.1	105.2	2150

Source: Primary survey.

vegetables like cabbage, cauliflower etc. and jute are the major irrigated commercial crops sown in *rabi* and summer seasons respectively.

The cropping pattern is highly dependent on the motive power of the water extraction mechanism i.e. diesel or electricity. In Baninathpur (*Dakshinpada*), where there is no electricity, people have mostly abandoned *boro* paddy in summer and diversified to low water consuming crops like jute. Villages like Kaya and Paschimgamini have a dependable electricity supply that allows more diversification in cropping systems like vegetables, sugarcanes and other commercial crops along with *boro* paddy. Mankara village has access to government deep tubewell but larger command areas under deep tubewells discourages cultivation of *boro* paddy due to unreliable water supply and farmers have diversified to low water consuming crops like jute in the summer season.

5.1   *Evolution of groundwater irrigation*

Before the 1970s farmers depended on small scale surface water bodies such as ponds and depression storages for supplemental and crop-saving irrigation. The groundwater development initiative came from the government in the form of deep tubewells and non-governmental organizations (NGOs). For example, the *Tagore Society* initiated groundwater irrigation in Kaya village back in 1976. They formed a co-operative for the purpose of operating and maintaining the STWs, and there are currently 12 STWs under the ownership of this co-operative.

During the mid-1970s, the West Bengal Comprehensive Area Development Corporation (WBCADC) was established with the help of World Bank funds to facilitate a rural development programme. One of the major activities of this corporation was to install medium deep tubewells and provide irrigation facilities to small and marginal farmers. But high costs of maintaining and operating these infrastructures[2] along with emphasis on participatory irrigation management led to the formation of water user's associations and since 1985 many tubewells have been handed over to the water user's association The WBCADC

---

[2]   The other reasons for transferring the tubewells were theft of equipment and farmers did not pay taxes on time. Many became defaulters that led WBCADC to disconnect 82 ESTW, 5 MDTW.

Table 6. History of irrigation in the study villages.

Name and type of intervention	Year	Village	Remarks
Tagore Society (an NGO) installed 12 STWS	1976	Kaya	These are now managed on an cooperative basis
WBCADC installed a number of medium duty tubewells	1970	Paschim Gamini, Baninthpur (North) and Mankara	These have been transferred to farmers organization in 1985
Irrigation department constructed deep tubewells	1970	Mankara	Still managed directly by the government
First diesel pump purchased by a private individual	1980	Baninathpur	Gopinath Mondal
First electric shallow tubewell owned by a private villager	1976	Kaya	A group of people

Source: Primary survey.

Table 7. Irrigated area, tubewell density and number of pumps in the study area.

Name of the villages	GWI area to NCA (%)	Tubewell density (per 100 ha of NCA)	Number of diesel pumps	Number of electric pumps
Baninathpur	88.12	25.5	44	2
Kaya	75.00	40.1	30	3
Mankara	96.89	2.8	18	3
Paschim Gamini	88.10	23.9	40	8

Source: Primary survey.

has now also transferred the medium duty tubewells to the farmers' organizations. The transfer of these tubewells was made in a stamped paper agreement with a leased out cost of Re 1/- per year. However, irrigation department's heavy duty tubewells (HDTW) are still maintained by the government.

5.2 *Significance of groundwater irrigation*

Table 7 shows the size and significance of groundwater irrigation in the study area. This indicates that more than 75% of the net cropped area is groundwater in all the villages. Further, tubewell density (per 100 ha of NCA) is high in Kaya where both cooperative wells and individual wells are providing water to agriculture, followed by Baninathpur and Paschim Gamini. Lowest tubewell density was found in Mankara where government tubewells supplies water to the majority of farmers so that individual farmers have not invested in tubewells. The numbers of diesel pump are high in Baninathpur where there is no electricity connection for agriculture purposes. However, in Mankara, though farmers have access to government operated electric deep tubewell, most also own Chinese diesel pumps as an insurance measure.

The majority of non-pump owners have access to irrigation through government infrastructures mostly prevalent in Mankara and Paschim Gamini. In these two villages, there is a water sharing mechanism through the institution of the 'water beneficiary committee'

Table 8.   Ownership and access to GW irrigation.

Particulars	All villages
Total number of pump owners	42.0
Total number of non-pump owners	38.0
Total members of beneficiary committee (numbers)	13.0
% of pump owners who sell water to others	33.3
% of pump owners who buy water from others	28.9
% of pump owners who sell as well as buy water	21.4
% of pump owners who neither buy nor sell water	19.1
Participation rate in private groundwater markets	80.9

Source: Primary survey.

formed by farmers with landholdings adjacent to a tubewell. WBCADC and Irrigation Department installed these tubewells. Farmers of the command area come together to a general consensus to use water in a sustainable manner. In the other villages where government infrastructure is lacking, private pump irrigation rental markets (for diesel pumps) or groundwater markets (for electric pumps) are the major water sharing institution. However, farmers with large land holdings participate less in water trading as opposed to small and marginal farmers who have surplus water to 'sell' after meeting their own requirements. These informal groundwater markets are ubiquitous in West Bengal, while the non-market based water sharing arrangements are quite common. (Mukherji 2007, Rawal 2002)

Participation in private groundwater markets is high in all the four villages at more than 80% (Table 8). Access to groundwater for irrigation through groundwater markets is a well-established arrangement in this region. Apart from informal water markets, groundwater is also sold by various agencies such as the government and WBCADC through formal mechanisms.

5.3   *Private groundwater markets*

Given the fragmentation of landholdings and increasing energy prices, the groundwater economy is dominated by informal groundwater market transactions. These markets are well developed in the region and have provided scope for non-pump owners to participate in groundwater irrigation. The size of water markets vary across the region and also over time depending on as motive power, cropping pattern and various institutional mechanisms (Mukherji 2007).

The buyer-seller ratio is an indicator of dependency of buyers on seller. This ratio is high for diesel pump owners (7.1 in Kharif, 18.6 in Rabi and 10.6 in summer respectively). Though diesel pumps service a larger number of water buyers, but it is electric pump owners who service the larger amount of land. A seller-buyer ratio of more than one indicates that buyer is dependent on more than one sellers in all the seasons. This may be due to land fragmentation (Table 9).

The water markets are prosperous in the region. In the study villages, the terms and conditions of payment may be either cash or kind. Cash payment is common for buyers who buy water from government or WBCADC tubewells and kind payment for operator (locally called as *agaldar)* for safeguarding the crops and distributing water. In the case of diesel pumps, both cash and kind payment in terms of labour days or crop sharing is

Table 9.   Water trading under different institutional mechanisms.

Particulars	Government/CADC			Private		
	Kharif	Rabi	Summer	Kharif	Rabi	Summer
Number of buyers	41	36	42	15	27	23
Area in ha	12.4	21.3	31.9	6.8	10.8	10.2
Hours of water bought/ha	338.8	79.8	377.9	104.9	82.0	147.2
Particulars	Electric			Diesel		
	Kharif	Rabi	Summer	Kharif	Rabi	Summer
Number of buyers	50	43	52	63	197	120
Number of sellers	6	6	6	14	14	13
Buyer-seller ratio	8.3	7.2	8.7	7.1	18.6	10.6
Area irrigated/WEM	21.5	21.5	23.2	14.3	22.9	16.4
Number of hours of water sold/WEM	410.5	309.8	492.0	71.0	139	164.3

Note: CADC = Comprehensive Area Development Corporation.
Source: Primary survey.

common. Payment in terms of kind transaction is also due to social relations among farmers. Table 10 shows various modes of payment and water prices for different crops. It shows that government owned and user association operated HDTW and MDTW sell water at cheaper rates than private diesel and electric tubewell owners. But then, among the private water sellers, electric tubewells sell water cheaper than their diesel counterparts.

5.4   *Co-operatives and water sharing mechanisms*

The water sharing mechanism is at sole discretion of the water user beneficiaries committees (WBCs). However the complete control of the WEMs was vested to WBCs only in the case of WBCADC tubewells. However, in case of minor irrigation department owned tubewells, the control is only partially transferred to the WBCs (Fig. 6). In villages like Kaya a group of farmers came forward and installed tubewells and formed themselves into user groups. These user groups can sell water to non-members at a higher price to cover the operation and maintenance costs of the structure. At present, the cost of water from WBCADC/Irrigation Department owned electric tubewells is INR 15/hour for members and INR 30/hour for non-members. Water sharing and pricing structures are decided by the cooperative members. However, the pricing structure and decision on crop cultivation are done in consultation with all the members at the beginning of each year, preferably before monsoon season. A schematic diagram showing characteristics of different informal institutions facilitated by government are shown (Fig. 8).

5.5   *Area and cropping pattern*

Table 11 shows that pump-owners bring more area into cultivation during non-rainy seasons, although the cropping intensity is less than that of non-pump owners (i.e., 174% for pump-owners and 197% for non-pump owners). This is due to the pump-owners cultivating a smaller area in *kharif* season. In contrast, non-pump owners maintain near constant cultivating practices to absorb the full capacity of land and water.

Table 10. Water rate in the study villages (2005–06).

| Crops | Private | | Government (HDTW) (Operated by water beneficiary committee) | CADC (MDTW) (Operated by water beneficiary committee) |
	Diesel	Electric		
Kaya	Paddy (Aman) @ INR 40/hr. Jute and mustard @ INR 40/hr to 50/hr.	Paddy (Aman) @ INR 3000/ha/season to 5250/ha/season. Paddy (Boro) @ INR 6000/ha/season to 7500/ha/season. Mustard @ INR 30/hr to 35/hr Vegetables @ INR 30/hr.	Not applicable	Not applicable
Baninathpur	Paddy (Aman) @ INR 40 to 45/hr + 110 kg/ha Paddy (Aman) @ INR 45/hr +190 kg/ha Wheat and Jute @ INR 40 to 45/hr Mustard @ INR 45/hr	Not applicable	Not applicable	Paddy (Aman) @ INR 3375/ha/season. Paddy (Boro) @ INR 3375/ha/season Mustard @ INR 3375/ha/season
Mankara	Paddy(Aman)@ INR 40 to 50/hr Paddy (Boro) @ INR 40 to 50/hr Wheat and Jute @ INR 40 to 45/hr Mustard @ INR 40 to 45/hr	Not applicable	Paddy (Aman) @ INR 1750/ha/season Paddy (Boro) @ INR 2250/ha/season. Wheat @ INR 1750/ha/season Mustard @ INR 450/ha/season Jute @ INR 750/ha/season	Not applicable
Paschim Gamini	Paddy (Aman) @ INR 40 to 45/hr + 200 kg/ha or labour Paddy (Boro) @ INR 50/hr + 200 kg/ha for tubewell operator Wheat and Jute @ INR 40 to 45/hr Mustard @ INR 40 to 45/hr	Not applicable	Paddy (Aman) @ INR 1250/ha/season plus INR 675/ha/season + 225 kg/ha/season for tubewell operator Paddy (Boro) @ INR 3000/ha/season + 225 kg/ha/season Jute @ INR 600/ha/season Wheat @ INR 525 + 90 kg/ha/season for tubewell operator	Paddy (Boro) @ INR 2250/ha/season + 190 kg/ha/season Wheat @ INR 1350/ha/season + INR 60/irrigation Jute @ INR 1500/ha/season + 60/irrigation

Source: Primary survey.

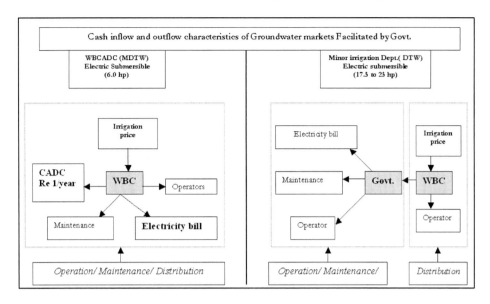

Figure 6.   Characteristics of groundwater market facilitated by the Government.

Table 11.   Area under groundwater irrigation in different seasons (ha).

Category of farmers	Kharif	Rabi	Summer	Annual	Net cultivated area (ha)	Cropping intensity (%)
Pump-owner	16.0	19.6	20.5	2.9	59.0	174
% of area	47.4	58.0	60.8	8.6	100	
Non-pump owner	9.3	7.8	8.5	1.7	27.3	197
% of net cultivated area	70.4	56.7	63.9	12.2	100	

Source: Primary survey.

It is the availability of groundwater irrigation that makes it possible to cultivate summer crops such as *boro* paddy, vegetables and jute in this region. Further, the yield rate per hectare is highest in the case of *boro* paddy (5000 kg/ha) for pump-owners and vegetables (14430 kg/ha) in the case of non-pump owners. The area under high yielding varieties (HYV) is more than 50 percent in all the cases (Table 12).

5.6   *Estimation of cost of groundwater irrigation*

In the sample farmers, there were 53 tubewells of which 51 were STWs and remaining 2 were MDTWs. The average depth of STWs is 20 m and MDTWs is 33.5 m. The historical investment has been amortized (apportioned) at nominal interest rate of 2% for the average of 15 years that ranged from INR. 206 to 13,513 per year. This amortized cost divided over the gross area irrigated (i.e., area irrigated in all crops over all seasons) gives the irrigation cost on per ha basis (Table 13).

The amortized cost per ha of net area irrigated is high (INR 10377.8) for diesel pumps due to the higher operation and maintenance cost for these pumps. The amortization

Table 12.   Area and yield of different crops.

	Pump owners			Non-pump owners		
Crop	Total area (ha)	Area under HYV (ha)	Yield (kg/ha)	Total area (ha)	Area under HYV (ha)	Yield (kg/ha)
Paddy (Aman)	14.0	10.4	3493	8.6	8.79	3184
Paddy (Boro)	6.4	4.4	4943	3.1	5.77	5286
Wheat	2.6	1.6	2164	1.3	1.8	1968
Jute	12.4	11.4	2156	11.78	4.7	2144
Mustard	6.6	3.9	1010	8.71	1.1	1010

Source: Primary survey, HYV = High Yielding Varieties of seeds.

Table 13.   Amortized cost (AMC) of well and pumps in the study area.

Particulars	Rupees per unit
Range of AMC/year	206 to 13513
Irrigation cost per ha of gross area irrigated	5062.8
AMC per ha of net irrigated area (NIA)	9026.7
Diesel	10377.8
Electric	7698.6
AMC per hour of irrigation	27.94
Diesel	43.00
Electric	17.65
AMC per m^3 of water extraction	0.73
Diesel	0.45
Electric	0.28

Source: Calculated from primary survey data.

cost per hour is high for diesel-operated pumps (INR 43.00) followed by electric pumps (INR 17.65). This is related to the number of hours of operation. The amortized cost per m^3 of water extracted from the WEM is less also high for diesel pumps (INR 0.45) as compared to electric pumps (INR 0.28).

5.7   *Groundwater irrigation and productivity*

The land productivity is a reflection of agricultural efficiency, which represents land quality, investment on inputs such as fertilizer, pesticides, and irrigation facilities. Groundwater productivity was calculated taking gross production of a particular crop divided by total quantity of groundwater applied to that crop. The higher land and groundwater productivity has been achieved by pump owners rather than non-owners for selected crops (Table 14).

The achievement of high land and water productivity is dependent on several factors. First, timeliness of water application to the crop will yield higher productivity. Second, use of excess water can lead to decline in crop productivity due to land and physiological crop characteristics (Table 14).

Table 14.   Cost and groundwater productivity by pump owners and non-owners.

Crops	Area (ha)	Groundwater applied of cultivation per ha(m³)	Cost of GW (INR/m³)	Cost of GW (INR/ha)	Average land productivity (kg/ha)	Average applied GW productivity (kg/m³)
Pump owner						
Paddy (Aman)	14.0	2720	0.40	2665	3554	0.53
Paddy (Boro)	6.4	8302	0.30	6096	4918	0.24
Wheat	2.6	854	0.74	3851	2356	1.12
Jute	12.1	632	0.74	1153	2139	1.37
Mustard	6.6	583	0.75	1087	1013	0.70
Non owner						
Paddy (Aman)	8.5	6291	0.17	2596	3206	0.21
Paddy (Boro)	3.1	14005	0.12	4246	5239	0.15
Wheat	1.3	975	0.84	2013	1966	0.82
Jute	4.8	870	1.08	2324	2144	1.00
Mustard	3.5	775	0.89	1709	1010	0.53

Source: Primary survey.

The higher water productivity can be observed for crops such as wheat and jute for both pump-owners and non-pump owners. This is because these crops have a low water requirement compared to other crops. The wheat crop requires irrigation at sensitive growth stages to enhance productivity. The owners of WEMs ensure timely irrigation whereas non-owners cannot do so and this is reflected by average land productivity figures (Table 14). In contrast, the non-owner of WEMs have higher land productivity (2121 kg/ac) in case of *boro* paddy in comparison to WEM owners (1991 Kg/ac). This contrasting picture is due to several factors such as soil type, timely application of water, timely application of necessary inputs such as pesticides and fertilizers.

5.8   *Cost and benefits of groundwater irrigation*

The costs and returns of some of the major crops are computed from primary data collected in 2005–06. The analysis has been carried out for pump-owners and non-pump owners to compare the cost and benefits of groundwater irrigation and the economic sustainability of these farms. The total cost (excluding the imputed costs of family labour), gross return, net return (gross return minus total cost) and return per rupee spent on inputs have been worked out using primary data.

The net return, for pump-owners, per hectare of vegetables is the highest followed by jute, wheat, *boro* paddy and *aman* paddy (Table 15). However, in case of vegetables, the amount of operating costs required to generate the net return is high and this limits the extent of its cultivation on small and marginal farms. The average net returns per rupee is the highest in vegetables (INR 1.78) followed by jute (INR 0.94), wheat (INR 0.61), mustard (INR 0.55) *aman* paddy (INR 0.31) and *boro* paddy (INR 0.30). The analysis indicates that given the small fragmentation of landholding size, jute and vegetable crops provide greater opportunities for a high return per rupee of cost. However, the jute crop cannot be grown extensively due to its heavy requirement and cost of labour. The average cost and net return for non-pump owners also indicates that the vegetables are the most profitable among all the crops. Their cost of cultivation is higher than that of pump-owners except in the case of *aman* paddy and mustard. This is because they have to pay additional

Table 15.   Average cost and returns of principal crops (2005–06).

Crops	Total cost of production (INR/ha)	Gross return (INR/ha)	Net return (INR/ha)	Net return per rupee of cost
Pump-owners				
Paddy (Aman) (N = 45)	14721	19310	4589	0.31
Paddy (Boro) (N = 26)	20664	26930	6266	0.30
Wheat (N = 25)	13375	21551	8176	0.61
Mustard (N = 32)	9939	15400	5461	0.55
Jute (N = 36)	13387	25950	12562	0.94
Vegetables (N = 40)	20674	36801	16127	1.78
Non-pump owers				
Paddy (Aman) (N = 39)	19639	22455	2816	0.14
Paddy (Boro) (N = 21)	25981	35968	9987	0.38
Wheat (N = 15)	13618	17592	3967	0.29
Mustard (N = 26)	11735	17540	5804	0.49
Jute (N = 32)	21812	30298	8486	0.39

Note: Total cost includes costs on purchased and own inputs at the current market prices during 2005–06 but does not include imputed cost of family labour.
Source: Primary survey.

amounts for purchased water. The net return per rupee of cost is also low compared to pump-owners.

The analysis reveals that the pump-owners earn a higher income than non-owners. Even under the assumption of lack of constraint on the availability of working capital, irrigation and marketing facilities, non-pump owners with 0.4 ha of land can generate a total net return of INR 12730 provided they grow vegetables. However, if they grow other crops such as paddy (*aman*), paddy (*boro*), wheat and jute, the net output will be less than for vegetables. But, capital requirement is high for vegetables and even for *boro* paddy and hence non-owners are less likely to cultivate these crops than pump owners.

5.9   *Coping mechanisms to meet increasing fuel costs*

The average depth of STWs in the study area has not changed over time (Table 16). Many of the tubewell owners had either re-constructed or relocated their tubewells due to age of the well or rusting of iron pipes. The average investment per well has declined due to low cost drilling popularized by the local well drillers.

Groundwater in this area occurs in a thick zone in the alluvium. The aquifer is made up of different grades of sand and gravel which occur to a depth of 90 m in the east and 140 to 150 m in the west of Bhagirathi river.

However, economic scarcity is an important issue in the study region as most of the respondents are poor marginal and small farmers who do not have enough capital to invest in tubewells and pumps. Farmers use different coping mechanisms to overcome economic scarcity. The major coping mechanisms include purchase of low cost Honda/Chinese pumps, use of rubber pipes as conveyance system, changing cropping pattern and mixing of diesel with kerosene, etc.

Table 16.   Morphological changes in groundwater structures (STWs).

Year	Range of depth (m)	Average depth (m)	Average investment per well (INR)
1966–1980	18–33 (6)	24	4800
1981–1990	12–23 (7)	16	2043
1991–1995	16–27 (7)	21	2264
1996–2000	12–40 (27)	19	2796
2001–2004	12–30 (5)	19	1860

Note: Figures in parenthesis denote number of tubewells. INR = Indian Rupees.
Source: Primary survey.

Table 17.   Brands of water lifting devices (pumps) in the study area.

	Ownership of pumps(%)	
Category of farmer	Chinese and lightweight kerosene operated Honda and similar pumps (3.5 HP)	Kirloskar and other traditional 5-HP pumps
Marginal (>0.5 ha)	52.0	48
Small (0.51 to 1.0 ha)	28.0	72
Medium (1.01 to 2 ha)	16.0	84
Large (>2 ha)	4.0	96
Total number of pumps	25	8
Year of purchase	2000–2001 onwards	Before 2000

Source: Primary survey.

6   FORMAL INSTITUTIONS INVOLVED IN GROUNDWATER MANAGEMENT

Since the 1990s, there have been changes in the water policy and the statutes that have been proposed and adopted at the union and the state level. These changes have been broadly in the area of regulation of groundwater, setting up of water regulatory authorities and the introduction of participatory irrigation management through the setting up of water users associations.

Therefore, groundwater management, in view of declining water table and its deteriorating quality, is a major concern for policy makers in West Bengal. Several Bills and Acts were enacted in the State Legislative Assembly to control and regulate the over development of groundwater resources in West Bengal. The most important one is 'West Bengal Groundwater Resources (Management, Control and Regulation) Bill, 2000' which came into effect from 15 September 2005. This requires that all tubewells constructed 15 September 2005 have to be registered. Issue of permit is essential to construct new tubewells. This *issue of permit* has replaced the *certification* process by the SWID. It was previously mandatory to obtain a 'no objection' certificate from the State Water Investigation Department (SWID) in order to obtain an electricity connection. The permit operates at three level (1) District Level Authority, (2) Corporate Level Authority (only for Kolkata) and (3) State Level Authority.

Table 18.   Electricity consumption pattern in agricultural use (constitutes mostly the groundwater extraction pumps) in the state.

Year	No. of consumers	Total consumption (MKWh)	% of consumers	Average rate per unit of consumption (Paise)
2005–06	106,954	894.568	6.07	150
2004–05	105,399	814.588	6.01	108
2003–04	102,887	770.000	5.97	98
2002–03	102,271	726.002	7.70	117

Source: WBSEB (2002–2006), Government of West Bengal.

Table 19.   Rates of the time of day.

Time of the Day	Rate (Paise)/unit of consumption	Remarks
6 AM to 5 PM	135	Normal rates
5 PM to 11 PM	475	Peak rates
11 PM to 6 AM	70	Off-peak rates
Average for 24 hours	153.95	

Source: WBSEB (2006), Government of West Bengal.

A better arrangement may be to bring together all the farmers within the 100 m circular periphery of a source and register the tubewell on a joint ownership with water sharing schedule apportioned to the area of the share holder under the command. The command area size would then be as high as 3.14 ha per STW (As per 2000–01 census it is 1.77 ha/ STW).

The groundwater irrigation structures are being facilitated by the West Bengal State Electricity Board (WBSEB). WBSEB was setup in 1956 with a mandate to provide electricity to different sectoral consumers. During that time rural electrification that includes the bulk of agrarian consumers was addressed only in a sporadic manner. Its goal is to electrify all the villages by 2007 and individual households by 2012. There is a consistent rise in consumers in this sector every year (Table 18). There is also a recent move to introduce 'time of the day' meters for every agricultural consumer by 31st March 2008 and this seems to be a challenging task (Table 19).

7   CONCLUSIONS AND POLICY IMPLICATIONS

The purpose of this study was to understand the issues relating to physical and economic access to groundwater irrigation in a cluster of villages of Murshidabad district typically representing the high water table tract of the state of West Bengal. Groundwater is relatively abundant and water tables are within 6–9 m of the ground surface in the study region and recharge in post monsoon season is adequate. However, not everybody has access to the groundwater resource. There remains an economic scarcity of groundwater in the region brought about by low rates of rural electrification and high diesel prices. Groundwater is a predominant source of irrigation. Private groundwater markets coupled with government owned deep tubewells and water user association operated shared tubewells are the main sources of irrigation for those who do not own any means of irrigation. The dynamics of water usage between water sellers and buyers is asymmetric, with pump owners harnessing more benefits than the water buyers in terms of quantity and timeliness of availability of irrigation water. The absence of electricity connection for irrigation purposes is another major constraint and this forces many farmers to depend on expensive diesel driven pumps.

A positive move is to energize all the WEMs by March 2008. This would result in less use of diesel pumps which are not cost effective for both water buyers and pump owners. However, introduction of TOD meters for every agricultural consumer owning a tubewell would have a mixed impact on the irrigation economy of the water sellers as well as buyers. The energized tubewell owners have to incur more cost per irrigation hour as compared to that in the flat tariff regime and the water buyers might have to pay more.

REFERENCES

Chambers, R. (1986). Irrigation against rural poverty. Paper for the INSA National Seminar on Water Management—The key to the Development of Agriculture, held at Indian National Science Academy, New Delhi, 27–29 January.

Chambers, R., N.C. Saxena and T. Shah (1989). To the hands of the poor: water and trees, New Delhi: Oxford and IBH Publishing Co. Pvt. Ltd.

Deb Roy A. and T. Shah (2003). 'Socio-ecology of Groundwater Irrigation in India', in Llamas, R&E. Custodio (eds) *Groundwater Intensive Use: Challenges and Opportunities*, Swets and Zetlinger Publishing Co., The Netherlands.

GOI (2001). *The 3rd Minor Irrigation Census of India* 2000–01, Government of India, New Delhi.

Hussain, I. and E. Biltonen (2001). Pro-poor irrigation intervention strategies in irrigated agriculture in Asia: Developing the project framework. In Managing water for the poor: proceedings of the regional workshop on pro-poor intervention strategies in irrigated agriculture in Asia, (ed.) Hussain Intizar and Eric Biltonen. Colombo, International Water Management Institute.

Kanhert, F. and Levine, G. (1989). Key findings, recommendations and summary. Groundwater irrigation and rural poor: options for development in the Gangetic basin. World Bank. Washington DC, USA.

Mukherji, A. (2004). 'Groundwater Market in Ganga-Meghna-Brahmaputra Basin: A Review of Theory and Evidence', *Economic and Political Weekly*, Volume 30 (31): 3514–3520.

Mukherji, A. (2007). 'The energy-irrigation nexus and its impact on groundwater markets in eastern Indo-Gangetic basin: Evidence from West Bengal, India', *Energy Policy*, Vol. 35 (12): 6413–6430.

Mukherji, A. (2008). Spatio-temporal analysis of markets for groundwater irrigation services in India, 1976–77 to 1997–98, *Hydrogeology Journal*, 16 (6): 1077–1087.

NSSO (1999). 54th round: *Cultivation practices in India, January 1998-June 1998*, Department of Statistics and Programme Implementation, GOI, August 1999.

Rawal, V. (2002). Non-market interventions in Water Sharing: Case studies from West Bengal, India. *Journal of Agrarian Change*. Vol. 2 (4): 545–569.

Repetto, R. (1994). *The "second India' revisited: population, poverty and environmental stress over two decades*, Washington D.C., World Resources Institute.

Shah, T. (1993). Groundwater markets and irrigation development: Political economy and practical policy. New Delhi: Oxford University Press.

SWID (1998). Groundwater resources in Murshidabad district, West Bengal, Report, State Water Investigation Department, Government of West Bengal.

Thornthwaite, C.W., Mather, J.R., (1957). Instructions and tables for computing the potential Evapotranspiration and the water balance, Drexel Inst. Of Tech. Publ. Climatology 10 (3): 183–311.

Vaidyanathan, A. (1996). Depletion of groundwater: some issues. *Indian Journal of Agricultural Economics*. Vol. 51 (1& 2): 184–192.

Vaidyanathan, A. (1999). Water resource management: institutions and irrigation development in India. New Delhi: Oxford University Press.

WBSEB (2002–2006). West Bengal State Electricity Board Year Books, 2002, 2003, 2004, 2005 and 2006, Government of West Bengal, Kolkata.

WBSEB (2006). West Bengal State Electricity Board Year Book 2006, Government of West Bengal, Kolkata.

INTERNET AND OTHER SOURCES

http://dw.iwmi.org/dataplatform/Home. aspx downloaded on 15th February 2007.

CHAPTER 8

The impact of shallow tubewells on irrigation water availability, access, crop productivity and farmers' income in the lower Gangetic Plain of Bangladesh

A. Zahid
Department of Geology, University of Dhaka, Dhaka, Bangladesh

M.A. Haque
Ground Water Hydrology, Bangladesh Water Development Board (BWDB), Dhaka, Bangladesh

M.S. Islam
Water Resources Planning Organization (WARPO), Dhaka, Bangladesh

M.A.F.M.R. Hasan
Department of Information and Communication Engineering, University of Rajshahi, Rajshahi, Bangladesh

ABSTRACT: Field surveys were carried out to understand the development of groundwater irrigation and its impacts on crop-economy and rural livelihoods in the Gangetic Plain of Bangladesh. The study covering five villages revealed that the aquifer characteristics and the surface water-groundwater relationship are favourable for the development of groundwater in the area. Here, like elsewhere in Bangladesh, evolution of minor irrigation led to the expansion of groundwater markets. However, the average cropping intensity in the area is lower than the country's average. Supply of energy in terms of electricity and diesel fuel plays an important role in the steady expansion of irrigated agriculture. The total cost for a shallow tube well run by diesel motive power is almost twice that for electric motive power. There are great prospects for the efficient use of the water resources, while the problems are associated with the energy-irrigation issue and inefficient management. The irrigation water market should be geared towards poverty alleviation through proper agricultural, credit, subsidies and energy support policies.

1 INTRODUCTION

Groundwater is the main source of irrigation in Bangladesh and is one of the key factors that has made the country self-sufficient in food production in recent times. Importance of groundwater irrigation increased with the introduction of high yield variety (HYV) seeds in the late 1960s in order to meet the food demand of a growing population. The Government of Bangladesh (GoB) started emphasizing groundwater irrigation in the mid-1970s, first with the publicly owned deep tubewell (DTW) projects, but soon shifted focus to privately owned shallow tubewells (STW)[1]. In Bangladesh, farmers with less than 0.2 ha of cultivable land comprise 66% of the population. Most of these farmers are poor and rarely benefit from the large water development schemes. About 90% of the total irrigated area of Bangladesh is under private sector led 'minor irrigation'[2]. Easy and cheap installation along with low maintenance cost encourages farmers to invest in STWs. These are either owned by individuals or a collective group of users, and in most cases, are installed without any financial support from the government. Bangladesh has a total cultivated land of 7.1 Mha of which 5.4 Mha is irrigated. Out of this 5.4 Mha of net irrigated area, 3.8 Mha is under minor irrigation (GoB, 2006).

Bangladesh has a total geographical area of 44,000 km^2 and covers a large part of the Bengal Basin that was formed during the Cretaceous period. It is the lower riparian country located in the floodplains of three major rivers, namely the Ganges (or Padma—as it is known in Bangladesh), the Brahmaputra and the Meghna and their tributaries, which form the largest active delta in the world. Generally, there is abundant water during the monsoon causing widespread floods and drainage problems in large areas, and little surface water during the dry season to meet irrigation and other requirements. In the southwestern part of the country, the delta has been classified as inactive but the major part in the south and southeast, including Matbarer Char of Shibchar upazilla under Madaripur district, where this study is conducted, is very active. Due to regular erosion and formation of land influenced by the river Ganges, agricultural development in this area is low compared to other more stable areas. However, there is great potential of agricultural expansion by applying and increasing suitable technologies to utilize water resources. The economy of Madaripur, as well as Bangladesh, depends on the proper harnessing of the water resources. The growth of groundwater irrigation in the country, especially supplied by private STWs purchased by individuals for cash or credit at unsubsidized prices led to the emergence of irrigation water markets. Here, farmers without wells benefit and attain access to irrigation through purchase of groundwater from farmers owning wells. This paper looks into the aspect of groundwater markets.

The dominant food crop of Bangladesh is rice. HYV seeds, application of fertilizer, and irrigation have increased yields, although these inputs have also raised the production costs. With the increasing use of irrigation, there has been a shift to *boro* rice, which is cultivated during the dry season from October to March. In 2006, the total production of *boro* in the country was 13.8 million metric tons (MMT), or 55% of the total rice production (GoB, 2006).

Sustaining the groundwater markets and a growing groundwater-based irrigation economy is critical for food security and for lifting millions of poor farmers out of poverty. Hence, understanding the facts of groundwater irrigation in contemporary Bangladesh and devising ways of supporting a sustainable development is crucial. The main purpose of this paper is to acquire an integrated understanding of the development of groundwater irrigation and its impacts on crop-economy and rural livelihoods in the lower Gangetic plain,

---

[1] Shallow tube wells are small irrigation wells having discharge capacities of 12–15 l/s with maximum depths of 40–60 m and well diameter of 100–150 mm. The discharge capacity of deep tube wells are about 50 l/s having greater depths (up to 100–120 m) and larger diameter compared to shallow wells.
[2] Minor irrigation is the private sector development of small scale shallow irrigation by individuals or collective group of farmers.

integrating assessments of the water resources, the energy-water issue, cropping pattern, socio-economic conditions and the role of existing policy and institutional frameworks.

2   METHODOLOGY AND OBJECTIVE OF THE STUDY

Technical and socio-economic field surveys were carried out in five representative villages of Matbarer Char Union under Shibchar Upazila in Madaripur district. The survey covered all the farmers who owned a well, and practically all of these farmers were selling water to neighbouring farmers. In total, 34 tubewell owners termed water sellers (WS) in the villages were included. In addition, 49 farmers buying water (WBs) were included. Finally, a number of well drillers, local officials and village elders were selected for focused group discussions. Questionnaires were developed and administered for gathering hydrogeological, agricultural, groundwater irrigation, and socio-economic data and information about the study villages of Naktikandi, Kharakandi, Natunkandi, Purankandi and Sikderkandi. The area was selected because of the relatively low agricultural growth and groundwater development in an attempt to understand the current constraints and options for future growth. The study covered 44 wells, out of which three (7%) were fitted with electric centrifugal pump and the rest (93%) were with diesel centrifugal pumps (Table 1). Most wells in the area have been installed during the last 10 years, which is somewhat later than in other parts of the country.

In addition, available lithologic logs and aquifer test data from the study area from the Bangladesh Water Development Board (BWDB) were used to determine the type and extent of aquifers and the properties of aquifer sediments. Hydrological data including rainfall, river discharge, river stage, and groundwater level of BWDB were analyzed to evaluate the availability of water resources and the surface water-groundwater interaction. Important chemical parameters of shallow groundwater samples procured during the project conducted by the Department of Public Health Engineering (DPHE) and the British Geological Survey (DPHE-BGS, 2001) were evaluated to inform on the groundwater quality in and around the study villages.

The specific objectives of the study were:

- Evaluation of the groundwater resource, with due consideration to groundwater—surface water interaction;
- Reconstruction of groundwater irrigation history in this region;
- Analysis of the impact of groundwater irrigation on crop economics and livelihoods;

Table 1.   Summary of the respondent count.

Water transaction code	Number of respondents					
	Naktikandi	Kharakandi	Purankandi	Natunkandi	Sikdarkandi	Total
WB	9	12	10	15	3	49
WS[a]	5	4	8	11	6	34
Total	14	16	18	26	9	83

[a] One well owner was not selling water at the time of the survey.
Source: Author's Survey, 2006–2007.

- Identification of the problems and prospects of groundwater market; and
- Assessment of the role of policies and programmes in supporting a groundwater-based agriculture.

3   STUDY AREA

The Madaripur district on the Shibchar flood plains is covered by a sequence of fluvio-deltaic deposits of Quaternary age. The alluvium is subdivided into the older

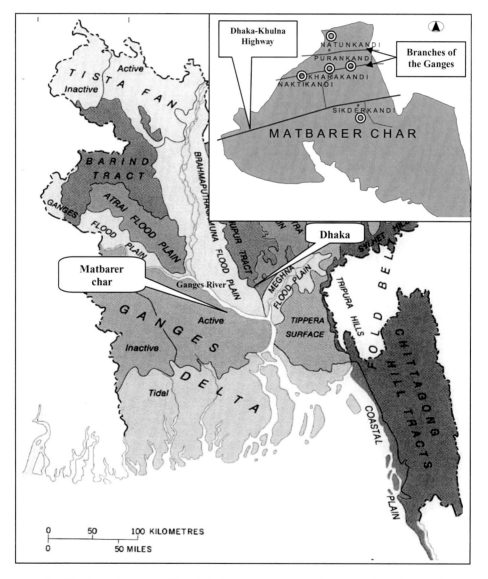

Figure 1.   Physiographic map of Bangladesh (Alam et al., 1990) and location of the study area (Matbarer Char).

alluvium (Pleistocene) and the newer alluvium (Holocene) (Acharyya, 2005). The elevation of the Shibchar flood plains ranges between 4 and 7 m above mean sea level (mamsl). Matbarer Char, which is the part of Madaripur district where the five study villages are located, is about 70 km south of Dhaka, the capital city of Bangladesh (Figure 1). The villages are separated from one another by a number of tributaries, and isolated channels associated with the Ganges. Numerous rivers, channels, tidal creeks, swamps and depressions are present in the area. Floodplains traditionally have been used for cultivation because of abundant water from rain, canals and the high groundwater table in the region. Although Matbarer Char is surrounded by the river Ganges and its shifting channels, shallow aquifers under Holocene sediments are the principal source of irrigation water in the *boro* season. In the critical dry month of March when the water demand is at its peak, the river flows reduce to a minimum, which is far below the total need of the paddy crop. Demographic details of the study villages are given in Table 2.

Madaripur receives an average annual rainfall of about 2000 mm. Rainfall is highest between June and October (Figure 2a). During this period, much of the study villages are submerged under water and no cultivation is possible. The average minimum temperature is 10 to 11°C (January) and the average maximum temperature ranges from 32 to 35°C, (May). The highest humidity is about 95% during October and the lowest is about 50% in January. Long-term rainfall data (since 1994) show that the annual rainfall varies between 1352 and 2697 mm, with the lowest in 2003 and the highest in 2004 (Figure 2b).

In 2004, 140 diesel and 150 electric operated STWs and 555 diesel and 400 electric operated Low Lift Pumps (LLPs) were used in Shibchar *Upazilla* during the *boro* season. The total irrigated land during this season was about 6275 ha, or 19% of the total area of the *Upazilla* (BADC 2005). Only about 10 Mm^3 out of a total renewable recharge of 93 Mm^3 was abstracted in 1995–96 (BMDA 2004). Though this figure has increased over the years, it indicates that substantial groundwater resources are available for further irrigation development in the future.

The crop year is divided into two main seasons—*rabi* (dry) and *kharif* (wet). The *kharif* season is further subdivided into *kharif*-I and *kharif*-II. The rabi season starts from mid-October and ends in mid-March, while *Kharif*-I season spreads from mid-March to mid-July and *Kharif*-II from mid-July to mid-October. Crop selection is greatly influenced by the seasons. Further, farmers select their crops based on land type, availability of irrigation water or residual soil moisture and on domestic consumption needs, the local market

Table 2. Demography of the study villages.

Village name	Area (Acres)	Population			Literacy rate % (7+ years)		
		Male	Female	Total	Male	Female	Total
Naktikandi	368	172	161	333	24.8	16.2	20.2
Kharakandi	330	525	503	1028	35.6	25.5	30.5
Purankandi	490	1700	1647	3347	42.0	20.0	35.6
Natunkandi	560	632	609	1241	35.6	20.3	28.2
Sikderkandi	480	435	375	810	26.9	12.3	19.7
Total	2228	3464	3295	6759	37.1	19.8	30.8

Source: Matbarer Char Union Council (2006).

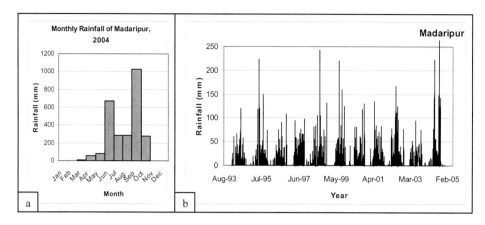

Figure 2.   Distribution of rainfall in Madaripur area (a): Annual rainfall pattern; (b): Daily rainfall record from August 1993 to February 2005.

Table 3.   Major crops cultivated under Shibchar Upazilla (2004–2005).

Crop name	Land cultivated (ha)		
	Rabi	Kharif-I	Kharif-II
Boro Rice (HYV)	6014	–	–
Boro Rice (Local)	1100	–	–
Wheat	2238	–	–
Onion and Garlic	3891	–	–
Mustard/Coriander	3277	–	–
Coriander/Vegetable	5222	–	–
Lentil	1255	–	–
Sugarcane	700	700	700
Jute/Aman/Aus Rice	–	13627	–
Ropa Aman Rice	–	–	5002
Pulses	–	–	5154
Others	211	4080	2049
Non-cultivated	–	5502	11004
Total agricultural land	23909	23909	23909

Source: Upazilla Agriculture Office, Shibchar (2006).

demand and their socio-economic condition. Rice, wheat, mustard, jute, lentil, spices, vegetables, sugarcane are the major crops cultivated in the area (Table 3).

In *kharif*-I, broadcast *aus* and *jute* were the main crops grown on high to medium high lands (to avoid damage from flooding) while broadcast *aman* was also sown in March/April on lower lands subjected to flooding. This was then harvested in November/December (Figure 3). In *kharif*-II, transplanted *aman* was the main crop grown under rainfed condition. Rabi crops like wheat, pulses, vegetable, and mustard were grown on residual moisture and sometimes with available irrigation water.

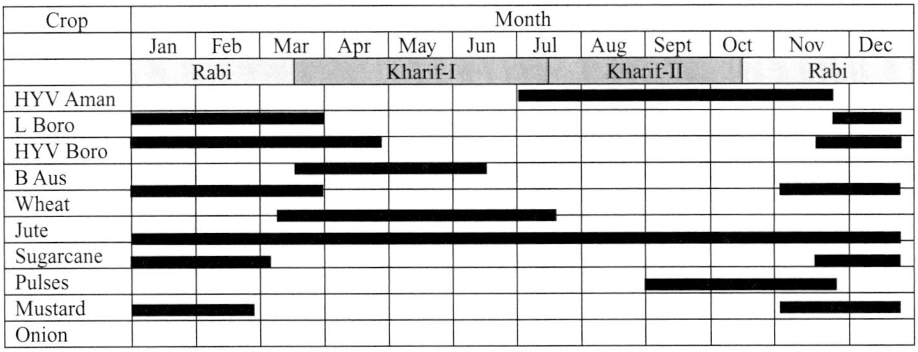

Crop	Month											
	Jan	Feb	Mar	Apr	May	Jun	Jul	Aug	Sept	Oct	Nov	Dec
	Rabi				Kharif-I			Kharif-II			Rabi	
HYV Aman												
L Boro												
HYV Boro												
B Aus												
Wheat												
Jute												
Sugarcane												
Pulses												
Mustard												
Onion												

Figure 3.   Crop calendar showing the growth period of major crops.
Source: Authors' survey, 2006–2007.

4   ASSESSMENT OF GROUNDWATER RESOURCES

4.1   *Aquifer and sediment properties*

In the study area, groundwater for irrigation purposes is abstracted from shallow depths (15–40 m). The aquifer sediments consist primarily of fluvio-deltaic deposits of the Ganges river and are deposited mainly under active meander environment. Based on local drillers' logs of shallow wells drilled at Matbarer Char and deep borelogs of BWDB conducted at Rajoir and Madaripur town, about 20 km southwest and 30 km south of Shibchar respectively, simplified lithological sections illustrate the sediment sequences and extent of the aquifers (Figure 4).

In Matbarer Char, the surface silty clay appears with a thickness of about 2–3 m. The sediments comprise brown silty clay followed by reduced grey clay (Figure 4). Below this aquitard, shallow tubewells are installed in the upper aquifer consisting of grey, fine and very fine sand with subordinate medium grade sand. The permeability and specific capacity of the aquifer sediments in the Shibchar area were estimated at 46 m/d and 6.42 l/s/m, respectively (Table 4). Transmissivity and storativity is between 675 and 1,105 m^2/day and 0.0016 and 0.0022 respectively. In Matbarer Char, the thin surface silty clay layer have a vertical hydraulic conductivity of between 0.005 and 0.008 m/day for wet soils and 0.005 and 0.040 for dry soils favours vertical percolation of rain and flood water into the shallow aquifer (MPO, 1987). These results reveal that the shallow aquifer is semi-confined to unconfined in nature and has much potential for groundwater development due to good recharge, storage and transmission properties.

However, depth, thickness and properties of different sediment units vary even within short distances depending on formation and depositional history. The deep bore logs show that the thickness of the upper clay and silty clay layer varies between 30 m at Rajoir and 5 to 7 m at Madaripur, and is characterized by high porosity and low permeability. In the Rajoir area, two deeper confining clay layers were encountered between 165–175 m and 207–212 m depths. A deep silty clay layer intercalated with silty sand layers between 253 m and 293 m separates the deeper aquifer from the upper aquifer system. In Madaripur town, a 28 m thick clay layer was encountered from 130 to 160 m below ground level. Sediments above the confining unit form a shallow aquifer predominately composed of very fine to

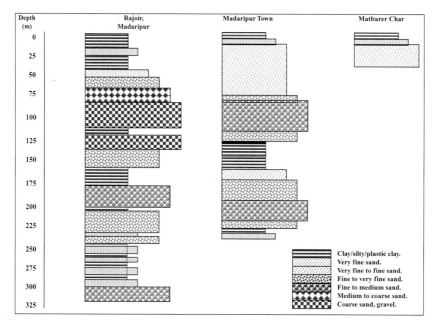

Figure 4.   Generalized lithologic sections in and around Matbarer Char (BWDB 1994).

Table 4.   Properties of aquifers in and around Shibchar Upazilla (BWDB 1994).

Area	Transmissivity (T) m²/d	Storativity (S)	Permeability (K) m/d	Sp. capacity of wells l/s/m
Shibchar, Madaripur	675–1,105	0.0016–0.0022	46	6.42
Sadarpur, Faridpur	1,500	0.01	20	14.55
Bhanga, Faridpur	1,100	0.15	25	35.12
Charbhadrasan, Faridpur	1,700	0.04	21	12.89
Sreenagar, Munshiganj	700	0.02	25	1.68
Bhagyakul, Munshiganj	1,055	–	48	10.04

fine sand. Sediments below the confining unit form the deep aquifer, which contains fine to medium sand (Figure 4).

4.2   *Groundwater—surface water interaction*

Recharge of the shallow aquifer is basically from rainfall in the wet season. Artificial sources, like return flow from irrigation and leakage from water distribution systems contribute additional recharge potential. The streambed sediments have relatively high permeability and good hydraulic connectivity with the underlying aquifers, and also promote recharge from the streams. In July, when river discharge is highest, river water levels and groundwater levels are at their maximum (Figure 5). River recession lowers the groundwater table, indicating a close connection between the two. Lowering of river flow during the dry season and increased irrigation pumping of groundwater near the rivers determines whether river flows would be gaining and loosing.

There is a clear difference in the study area between potential recharge and actual recharge. Actual recharge is the quantity of water that enters the aquifer system until it is full and can take no more into storage, while potential recharge is the amount of water that is available for recharge, above ground. The point of the 'aquifer full condition' is attained towards the end of June and excess water or rejected recharge moves over the land surface directly to the rivers. This shows a potential for further harnessing of water if capturing and storage facilities were available. During the peak irrigation season in March and April, the groundwater hydrograph reaches its minimum and the steepest rise in the hydrograph occurs immediately after the irrigation pumps are switched off (mid-April to mid-May). There is no tendency of long term decline in the groundwater level, again supporting the argument that the system with the present groundwater exploitation is in balance if not underutilised. During the monsoon, the hydrographs rise steadily until the levels are within 1–2 m of the surface at which point a dynamic equilibrium is established, primarily through exchange with the rivers (Ravenscroft, 2003). Groundwater pumping for irrigation from the shallow aquifer mainly in the dry season causes large seasonal water level fluctuations, and in the Matbarer Char area, water level ranges from the surface during the monsoon to about 6.3 m depth in the dry season.

4.3   *Groundwater recharge and irrigation abstraction*

Calculating the water balance by summing all the inputs and outputs to the system is important for groundwater assessment. The simple expression of this analysis is: INPUT-OUTPUT=±Δ STORAGE. However, consideration is needed of all the components within

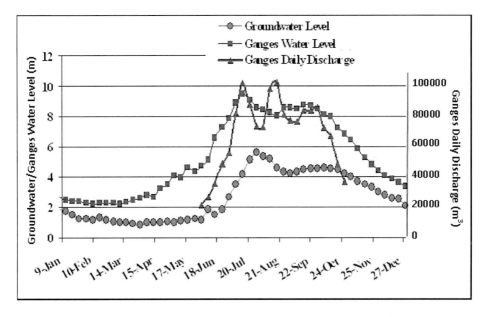

Figure 5.   Correlation of river discharge, river water stage and groundwater level for the year 2004.

the water budget, e.g. precipitation, evapo-transpiration, surface water interaction, lateral groundwater flux, infiltration from irrigation, and abstraction wells. Different models and various computer codes can be used for recharge estimation, however they are data intensive. The following water-table fluctuation method (Scanlon et al., 2002) is the simplest way for calculating actual recharge in unconfined aquifers.

$$R = S_y dh/dt = S_y \Delta h/\Delta t$$

where, S_y is specific yield, h is water table height, and t is time.

In Tables 5 and 6, the actual annual recharge and total irrigation abstraction in the five study villages are presented. For calculating the groundwater recharge, the specific yield of the aquifer was estimated by analyzing average grain size from lithologic logs. For calculating water abstraction, an average well discharge of 50 m^3/h was applied, based on field measurements. Lowering of the water level permits more recharge to groundwater. Compared to 3.04 Mm^3 of total irrigation abstraction by the five villages, actual recharge is about 4.73 Mm^3. Though large uncertainty in the calculations should be expected, the data imply that the resource is not stressed. Considering a domestic and drinking water use of 15 m^3/person/year, then the yearly total groundwater abstraction by the 6,800 inhabitants of the five villages for this purpose is about 0.1 Mm^3, or only 3% of the total recharge.

4.4   *Groundwater quality and impact of agriculture*

In Matbarer Char, the soil aeration conditions range from continuously reducing controlled by the high groundwater table to alternating reducing and oxidizing because of groundwater fluctuations. For cultivation of paddy, the soil is usually kept submerged, or at least saturated with water throughout the growing period of the crop and reduction of the top soil starts. Thus, conversion of original soils into paddy soils changes many properties. An iron-enriched subsurface horizon is formed rapidly, which is a diagnostic feature of this type of paddy soil (Zhang and Gong, 2003). Naturally occurring iron-containing minerals (biotite, chloride etc.) are common in Holocene alluvial sediments in the area (Sengupta et al., 2004) and due to weathering slowly transform to iron-oxyhydroxides (FeOOH) coatings. In floodplain areas, paddy soil containing orange to dark yellow-brown iron-oxyhydroxides bands, a few centimetres above the transition zone between the oxidizing conditions and reducing conditions, and have a generally high concentration of arsenic. Layers rich in iron-oxyhydroxides and manganese-oxides that occurs under the plough pan (Koenigs 1950) are observed in the topsoils formation at Matbarer Char. With better soil aeration, crystalline iron oxides and their ratio to total iron increase with time (Zhang and Gong, 2003).

Mobile arsenic in soils can bind to FeOOH (Norra et al., 2005) and release to solution when anoxic conditions allow reduction and dissolution (Bhattacharya et al., 1997, Acharyya et al., 1999, Harvey et al., 2002, Yan et al., 2000). Higher concentrations of reduced, dissolved iron and manganese in water samples from shallow depth in paddy soils from the study area support this enrichment (Table 7). The bands enriched with iron and manganese in the soil of the area are subordinate to the older cultivated areas. The iron content in the shallow groundwater (18–21 m) under Shibchar Upazilla ranges between 0.54 and 11.5 mg/l, all above the recommended WHO concentration (0.3 mg/l).

Table 5.   Groundwater recharge in the study villages.

Village name	Area (m²)	Specific yield	Groundwater Table (m)			Recharge (Mm³)
			Max.	Min.	Fluctuation	
Naktikandi	1,489,296	0.12	6.2	1.5	4.7	0.84
Kharakandi	1,335,510	0.12	6.2	1.5	4.7	0.75
Purankandi	1,983,030	0.12	6.2	1.5	4.7	1.12
Natunkandi	2,266,320	0.12	6.2	1.5	4.7	1.28
Sikderkandi	1,942,560	0.10	4.8	1.0	3.8	0.74
Total						4.73

Source: Authors' survey, 2006–2007.

Table 6.   Total withdrawal of groundwater for irrigation by the five villages in 2005–06.

Village name	Nos. of STW	Total operational days			Operational hours		Well discharge (m³/h)	Total abstraction (Mm³)
		Min.	Max.	Mean	Total	Mean		
Naktikandi	05	120	300	180	12,480	2,496	50	0.62
Kharakandi	04	120	240	157	9,180	2,295	50	0.45
Purankandi	08	120	150	127	13,245	1,655	50	0.66
Natunkandi	11	120	135	122	19,665	1,787	50	0.98
Sikderkandi	06	120	150	132	6,705	1,117	50	0.33
Total	34							3.04

Source: Authors' survey, 2006–2007.

The concentration of manganese in most samples exceeds the recommended WHO value of 0.1 mg/l, except in Panch Char (0.016 mg/l) area, and ranges between 0.88 and 1.84 mg/l in other sites. All shallow samples contain arsenic above WHO recommended concentration of 10 μg/l with concentrations up to 280 μg/l. Deep groundwater in the area are free from arsenic contamination (DPHE-BGS, 2001). The influence of high arsenic concentrations in irrigation water on crop content and potential human health impacts from consumption of the crops are not known, and research on this topic is required.

Increasing the intensity of crop production using chemical fertilizers and pesticides may affect groundwater in the area as the subsurface silty clay layer (1–2 m) is vulnerable to leaching of contaminants into the shallow aquifer. Fertilizers, such as urea, triple super phosphate (TSP) and marinate of potash (MP) and pesticides, such as insecticides, herbicides, fungicides, and rodenticides are used by farmers. High concentration of phosporous and sulfate in Matbarer Char and many other samples (Table 7) indicates the possibility of such contamination. While water quantity is not an issue in this region, water quality issues are quite pertinent.

Table 7.   Concentration (mg/l) of important water quality parameters of shallow groundwater samples in Shibchar Upazilla (DPHE-BGS, 2001).

Union	Depth (m)	As µg/l	Fe mg/l	Mn mg/l	P mg/l	SO mg/l
Matbarer Char	19	66.8	2.27	1.370	1.2	4.9
Sannyasir Char	18	82.4	1.17	0.959	0.5	81.3
Panch Char	21	11.3	0.54	0.016	<0.2	<0.2
Bayratala	20	86.7	1.82	1.840	0.6	<0.2
Banskandi	19	289.0	6.16	1.800	0.7	4.1
Umcdpur	20	74.6	11.50	0.887	0.8	129.0
Dattapara	21	62.4	0.58	1.350	<0.2	16.8
WHO Limit		0.01	0.3	0.1		

Source: Authors' survey, 2006–2007.

5   HISTORIC DEVELOPMENT AND SIGNIFICANCE OF GROUNDWATER IRRIGATION

In 1972, the Bangladesh Agricultural Development Corporation (BADC) initiated capital-intensive methods for DTW installation in Bangladesh and provided subsidized well components (pumps, drilling equipment, etc.) for rapid expansion of larger public ground-water irrigation schemes. BADC used to install DTWs by reverse circulation drilling methods using very costly power rigs. Subsidized projects provided cheap irrigation water to farmers. Privatization and expansion of minor irrigation and withdrawal of government subsidy in irrigation equipment led to rapid growth of farmer-financed STWs. The man-ually driven percussion method became popular for STW installation. Most of these STW were drilled to a depth of 20–45 m. The direct circulation rotary drilling method with donkey pumps is another low-cost technology for the installation of up to 300 m deep even larger diameter wells. STWs have increased significantly in numbers throughout the country from 133,800 in 1985 to 1,182,500 in 2006. In comparison, there were only 28,289 DTWs in operation in 2006 (Figure 6). The average discharge of shallow tubewells was found to be 50 m^3/hr. The average irrigated area (command area) of the diesel-operated STWs in the study area was 3.2 ha while it was only 0.8 to 1.2 ha for the electric-operated STWs.

The development of groundwater irrigation in the area started in late 1990s. Electric-ity operated shallow tubewells have been introduced very recently (only three electric WEMs present), and is limited by slow development of the electricity network in the area. Figures 7a and 7b show that, in spite of the increase in cost escalation of drilling, the development of groundwater has been evolving in the area over the years.

5.1   *Expansion of groundwater markets*

Informal groundwater markets in Bangladesh in general have contributed significantly to improving irrigation management and practices of irrigated rice production, tillage mecha-nization and farming integration and most importantly provided irrigation to the non-pump owners. An estimate shows that despite the increasing use of smaller capacity pumps in recent years, the command area per STW has remained at 0.34 ha in 1997/98, compared to 0.32 ha in 1982/83 (Mandal, 2000).

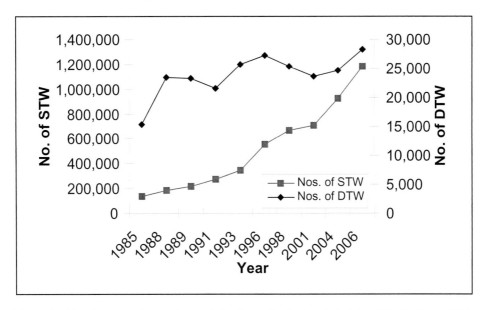

Figure 6.   Development of groundwater irrigation wells in Bangladesh (AST-DAE. 1991, ATIA. 1995, Mandal 2006).

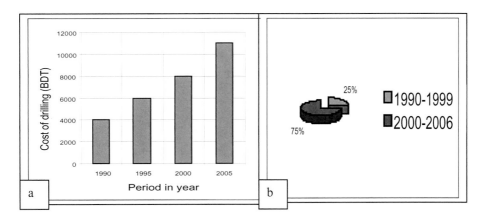

Figure 7.   (a) Cost escalation in Bangladesh Taka (BDT) (1 USD = 70 BDT as in May 2008) of drilling one STW (Cost includes installation (labour, supervising), well materials and pump for one well and (b) Percentage of numbers of installed tubewells before and after 2000.
Source: Authors' survey, 2006–2007.

The expansion of the groundwater market in Bangladesh has created new kinds of associated businesses in the rural areas, e.g. drillers and local workshops for manufacturing irrigation pumps, pipes, and spare parts, thereby creating alternative livelihoods. The market is dominated by a dynamic private sector involving about 1.7 million owners and managers of mechanized pumps, 0.76 million owners and operators of non-mechanized and traditional irrigation devices and a total of 7.6 million irrigator farmers (Mandal, 2000).

The existence of an irrigation market is also recognized in areas of South Asia such as Pakistan and India. The market varies in structure, conduct, coverage, public sector support and overall growth and efficiency (Palmer-Jones, 1997; Shah, 1993). Shah (1997) puts forward a tentative hypothesis regarding agrarian transformation through accumulation of machine capital, particularly pump capital and emergence of groundwater markets in much of South Asia. He proposes that *initial* conditions of low groundwater development preclude the poor and landless farmers from benefiting substantially from agriculture. This stage is also characterized by dependence of human and animal labour, and the nature of an asset lease market is crude and rudimentary. The *next* phase is characterized by accumulation of machine capital (particularly pump sets) by the rural elite. Water markets or lease market for other assets (except land) do not develop fully at this stage, because large farmers use all the water that their machines pump for self-cultivation. However, in the *later* stages, even the small and medium farmers start investing in machine capital in general and pump capital in particular, and water markets develop because they cannot use all water on their own small fields (Shah, 1997). An alternate hypothesis was that an average South Asian pump irrigation seller is more likely to be a small farmer with fragmented land holding and the chief motive in selling pump irrigation would be the need for making the investment in WEMs viable by improving the capacity utilization (Shah et al., 2006).

5.1.1   *The energy-irrigation issue*

Supply of energy in terms of electricity and diesel fuel for pumping plays an important role in the steady expansion of irrigated agriculture. Energy pricing and supply policies for agriculture have shaped the size and structure of groundwater irrigation economies that have emerged in different hydro-economic zones. In Bangladesh, about 90% of all the STWs are diesel powered. This figure is 93% for the study villages. In the rural areas, electricity is supplied through the Rural Electrification Programme (REP), under the overall direction of the Rural Electrification Board (REB). Power is supplied through Rural Electricity Co-operatives, each covering an average of five to seven Upazillas. Debroy and Shah (2003) have drawn attention to the energy-divide as a major feature of India's groundwater economy. They have suggested that, owing to different electricity pricing and supply policies pursued by electricity utilities, India faces an energy divide. Eastern states in India have suffered progressive rural de-electrification since the mid-1980s (Sharma 1989; Shah 2001). As a consequence, groundwater irrigation has come to be increasingly dependent upon more costly diesel fuel. Shah (2001) also argued that because of its diesel-dependence, groundwater-rich eastern India and Nepal Terai under-irrigate and in general are unable to take the full advantage of the groundwater resource potential. More recently Mukherji (2007) has analysed the energy-irrigation issue in the context of water abundant West Bengal, India and found that farmers are facing a severe 'energy-squeeze' due to escalating diesel prices and slow pace of rural electrification.

The same condition in a different form has been prevailing in four and partially one other of the study villages, Sikdarkandi, where the WEMs are run both by electricity and diesel. The high price of diesel hinders the full-fledged development of groundwater irrigation in the area. In Bangladesh, the diesel price has increased to from 14 to 32 BDT/liter from 1996 to 2006. As a result, some well owners decided to quit pumping groundwater as they were incurring losses. Many of the respondents were of the view that the farmers could enjoy a great deal of benefits if they could get electricity connection from their tubewells. Lack of electricity and the cost of diesel are squeezing the already meagre profit margins of

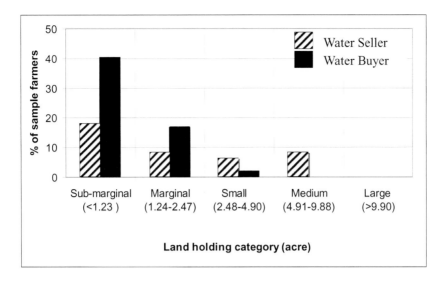

Figure 8. Distribution of water sellers and water buyers in different size classes of land holding. Source: Author's Survey, 2006–2007.

the farmers. Many of the farmers irrigate, even after incurring losses, just for subsistence. There is a two-way solution to get rid of this problem: either there should be diesel supply at subsidized rate or a secured power supply to run the irrigation pumps on demand.

5.1.2 *Access to groundwater irrigation under market conditions*

The small farmers share of STW ownership has increased significantly over the years in Bangladesh (Mandal, 2000). In the study area, sub-marginal and marginal farmers with small landholdings (land holding size <1 ha) are mostly water buyers (WBs), while the small and medium farmers (with landholding between 1.01 and 4.0 ha) mostly own the STWs and sell water from the same (Figure 8). This indicates that the groundwater market is characterized by larger well-owning farmers selling water to the smaller and marginal farmers. In this set-up, small farmers benefit by getting access to water which may not otherwise be accessible for them due to lack of entry into the market as an owner due to high capital investments. Outside the groundwater market, there are households who have few hectares of lands but do not cultivate irrigated crops, and instead earn a better living by doing business in towns. Yet, many others who have small plots of lands are not able to be a part of irrigation market and can not cultivate irrigated *boro* paddy. These people earn their living by cultivating *kharif* crops without irrigation, and by offering themselves in the labour market. However, these groups of villagers are very poor.

5.1.3 *Water charge and mode of water transaction*

The water charge for groundwater irrigation in Bangladesh generally evolved in response to local conditions and varies between regions, crops, soil type, topography and seasons. There are generally three modes of payment of purchased water: hourly rates (BDT/h), area-based rates (BDT/ha) and crop share, which is based on paying a share of the harvested crop to the water seller. In large parts of the country, a one-fourth crop share payment has evolved, mainly in the context of cash constraints by the water-buying farmers in the beginning of the irrigation season when they also need to pay for fertilizers, seedlings and labour. In the study villages, crop sharing is practiced and payment is done at the field immediately after harvesting. But in the Matbarer Char area, a one-third crop share is practiced. With the increasing price of fuel, maintenance and supervision cost (extra manpower is engaged for pump operation and water distribution) WSs are now demanding money payment in addition to one-third of the crop. The in-kind system is innovative because it shifts the burden of irrigation cost to the water sellers (or suppliers) for about four months and allows them to receive a crop share as water charge at harvest. It may appear a little higher than cash payment but it is not necessarily exploitative when one includes interest on operating capital committed by the water suppliers as well as premium for risk of crop failure or damages.

5.1.4   *Economics of groundwater abstraction*

The costs of groundwater abstraction for diesel-driven and electricity-driven STWs are given in Table 8 Costs have been indexed at current prices using wholesale price indices. The study shows that the total cost for a STW run by diesel motive power is almost twice that for electric motive power. This is why the farmers in the study villages are strongly in favour of electric motive power.

6   IMPACT OF GROUNDWATER IRRIGATION ON CROP ECONOMY

6.1   *Cropping intensity and yield of major crops*

The cropping intensity is denoted as the ratio of total cropped area under multiple crops within a year to net cultivated area, which is expressed in terms of percentage. Farmers typically grow one or two crops per year and leave their land fallow for the rest of the year. The highest cropping intensity was found to be 173% in Purankandi village while the lowest cropping intensity was 124% in Sikderkandi village. The average cropping intensity in the study area was 146%. The country has achieved an estimated cropping intensity of about 185% (NWMP, 2001). The reason for the lower-than-average cropping intensity could be due to the severe flooding and very active sedimentation processes going on in the area, impeding year round cropping.

The yield of different crops in the study area is presented in Table 9 The survey produced mixed evidence about an expected increase in yields of different crops over the last five years. Compared to the country's average, yield of major crops are less in the study area (Table 9). For irrigated *boro* rice, only about a quarter of the respondent farmers reported an increase, whereas 65% reported decreases and the remainder had no change in yield in the year 2006 compared to 5 years ago. For other crops, there were mixed responses. According to the farmers the main causes of yield decline/stagnation of HYV *boro* rice were intensive cropping, unbalanced fertilizer application with excessive use of nitrogenous fertilizer and/or no fertilizer use and over-mining of soil nutrients from crop removal. Irrigation water availability did not appear to be limiting crop yields.

6.2   *Groundwater productivity of boro rice*

The *boro* rice variety widely adopted in the study area is BR-29. The growing season is 120 to 130 days. The preferred planting method is transplanting of seedlings using wet

Table 8.   Comparison of average cost (capital and annual running) of groundwater abstraction by diesel-driven and electricity-driven STW in the study area.

Item	Diesel cost (BDT)	Electricity cost (BDT)
Well construction cost	1,000	900
Distribution line cost	566	566
Repair and maintenance cost	1,380	300
Supervision cost	11,364	8,000
Diesel/electricity cost[a]	34,976	15,000
Lubricant (Mobil) cost	1,351	–
Total cost per STW	50,637	24,766

[a] Assuming an average representative cropping rotation and pumping duration of the area.
Source: Author's Survey, 2006–2007.

Table 9.   Yield of major crops in the study area.

Crop Name	Average yield (kg/ha)	Countrywide average yield (kg/ha) (LGED 2005)
HYV Boro	4804 (for WB) 5152 (for WS)	4999
T Aman	916	963
Wheat	1778	1976
Mustard	366	371
Sugarcane	30035	34827
Onion	5691	6190
Jute	4804	4999

WS = Water seller, WB = Water buyer.
Source: Author's Survey, 2006–2007.

bed technique. After the land is prepared, 35–40 day old seedlings are transplanted in the field and then the irrigation water supply period commences. A 3-day rotation method is practiced for water distribution in the study area, meaning that each farmer gets water every three days. The WSs maintain the rotational system. On average, water is applied 35–40 times to the fields during the rice crop season. Most of the farmers are familiar with the rotational irrigation schedule and have accepted it as an equitable system for water sharing. The applied groundwater productivity and applied irrigation rate for water sellers and buyers are presented in Table 10. Water sellers consistently apply more water (average for all crops: 15,301 m³/ha) than that of the water buyers (10,468 m³/ha).

There is a common misunderstanding among the farmers that the greater the depth of water in the field, the better is the yield of irrigated rice. However, even under irrigated conditions, occasional drainage is necessary for aeration. A series of water management studies conducted by the Bangladesh Rice Research Institute (BRRI) indicated that a range of water depths ranging from soil saturation only to 10 cm standing water gave statistically insignificant differences in rice yields, provided other management practices were uniform (Islam,

Table 10.   Groundwater productivity and applied irrigation rate for Boro rice in the study area.

Village name	Water productivity (kg/m^3)		Applied irrigation rate (m^3/ha)	
	Water seller	Water buyer	Water seller	Water buyer
Naktikandi	0.27	0.36	19068	13190
Kharakandi	0.57	0.58	8998	8292
Purankandi	0.31	0.35	16463	13718
Natunkandi	0.42	0.46	12091	10374
Sikdererkandi	0.35	0.41	19891	6765
Average	0.38	0.43	15302	10468

Source: Author's Survey, 2006–2007.

1986, 1987). It was observed that the average groundwater productivities in the study area were 0.38 and 0.43 kg/m^3 for water sellers and water buyers, respectively. This indicates that water buyers are getting better yields per unit input of water. An explanation for this is that water buyers are paying more for their water and hence tend to use it more judiciously. The National Water Management Plan (NWMP) recommended total water irrigation for HYV Boro rice to be between 1,200 and 1,500 mm per season, depending on the soil condition (LGED 2005). The average water used by farmers (12,886 m^3/ha or 1,289 mm) was within the recommended limits, and hence not supporting an excessive irrigation.

6.3   *Crop economy and cost of* boro *rice cultivation*

Crop production inputs such as human labour, draft power and most seeds are generally self-supplied by the farmers in Bangladesh. With the shift to HYV technologies, however, there has been a growth in importance purchased inputs such as chemical fertilizers, pesticides, irrigation and power tillage services, which are marketed through private-sector traders.

The input use and costs of *boro* rice cultivation in the study villages are presented in Table 11. The cultivation costs were BDT.35,407 BDT 34,271 and BDT 27,839 per ha for the water buyer, water seller with diesel pump, and water seller with electric pump, respectively. The cultivation costs of the water buyers were slightly higher than that of the water sellers. It is seen that the cost of labour and irrigation accounts for approx. 80% of the total cultivation costs.

A gross income analysis of cultivation of major crops (*boro* rice and mustard) over one year for both water sellers and water buyers in the study area is presented in Table 12. The gross income for water buyers was estimated from crop yield per ha multiplied by its unit price (subtracting costs associated with crop sharing). The income from water selling makes the difference between water buyers' and water sellers' gross income per year. The gross income of water sellers is 32% higher than that of water buyers. This shows that the water market based irrigation economy is more profitable for the water sellers than the water buyers. However, presumably the water buyers are better off under the water market based situation as they have secured access to irrigation water. The farmers indicated that the profitability of *boro* rice had decreased over the years, mostly due to increases in fuel prices.

Table 11. Cost of cultivation of a Boro rice crop in the study area.

Line Items	Qty. (kg)/No	Rate (BDT/kg or BDT/day)	Total BDT/ha
Seed	20	30	1482
Seed bed preparation	2 (man-day)	125	618
Seed bed management	2 (man-day)	125	618
Land preparation (power tiller)	8	125	2470
Transplanting, weeding, harvesting	35 (man-day)	125	10806
Pesticide			741
Fertilizer			3458
Irrigation (Water Buyer)[a]			15215
Irrigation (Water Seller with diesel motive power)[b]			14079
Irrigation (Water Seller with electric motive power)[b]			7647
Total cost for water buyer			35407
Total cost for water seller with diesel motive power			34271
Total cost for water seller with electric motive power			27839

[a] Calculated from the market price of *boro* rice, assuming 1/3 crop sharing.
[b] Calculated from estimated running costs during the rice season.
Source: Author's Survey, 2006–2007.

Table 12. Gross income of cultivation of major crops over one year for water buyers and water sellers in the study area.

Income	WBs (BDT/ha)	WSs (BDT/ha)
Boro rice	52488	56289
Mustard	6946	6946
Water selling	–	15215
Gross income	59433	78450

Source: Authors' survey, 2006–2007.

7 POLICIES AND SUPPORT PROGRAMMES PERSPECTIVE ON GROUNDWATER AGRICULTURE

In Bangladesh, 75% of the country's population live and earn their livelihood in rural areas, mostly from farming. The GoB has identified agriculture and rural development as a priority sector for rapid poverty reduction. Various policy options like provision of subsidies on major agriculture inputs, price support policies, price stabilization measures, special credit programmes and favourable tax policies/tariffs have been pursued to favour or augment crop sector growths. In the government policy framework, intensification of major crops such as cereals, and diversification to high-value non-cereal crops such as vegetables, are highlighted as important objectives. Agriculture and the rural economy are recognized as the key drivers of a growth strategy. In the draft Poverty Reduction Strategy (PRS) of the Government of Bangladesh (GoB, 2005), accelerated production of high value crops, strengthening of agricultural research, expansion of irrigation with emphasis on efficient management of water resources, expansion of utilization of surface water and rational utilization of groundwater are emphasized. The National Water Policy (NWPo, 1999) also supports private development of groundwater irrigation for promoting continuous agricultural growth, with surface water development where feasible. NGOs are also involved and playing a more proactive role in transforming agriculture and the rural economy.

The GoB is also considering giving support to other broad-based areas for the development of agriculture and farmers in a sustainable manner. These include getting prices (of seed, fertilizers and diesel at peak periods) right, expanding output price support, providing credit for small farmers, subsidizing the installation and maintenance of electricity connections to irrigation pumps, maintenance of rural roads, processing and marketing of perishable high-value farm products, and support to agricultural education and research. But the farmers of the study area are yet to benefit from those supportive roles of the government. In most cases, people of the area, like other areas in Bangladesh, are unaware of those support policies.

8   CONCLUSION

Groundwater development in the study area is based on extraction of shallow groundwater using shallow tube wells. There are plenty of water resources in the area and groundwater development has made *boro* rice cultivation feasible by securing irrigation water throughout the cropping season. The study area is relatively poor with low cropping intensities and crop yields due to seasonal flooding, inefficient agricultural practices, and poor infrastructure. Small scale groundwater irrigation development during the last 8–10 years has been the single most important driving force behind the steady expansion of agricultural output in recent years. An emerging groundwater market is a win-win situation for the water sellers as well as water buyers by securing access to irrigation water for the water buyers and providing additional income to the sellers. The market is in the early stages and there is room for expansion in terms of numbers of farmers still to participate, further groundwater resources available for extraction, and potential income gains from shifting to higher valued crops. Major hindrances to further development of groundwater irrigation and water markets are high diesel prices, lack of electricity connection and better connection to markets. Groundwater in the region contains high levels of arsenic and impacts on crops and human health, via exposure through crop consumption, need to be assessed. Increasing expansion of crop production using chemical fertilizers and pesticides may affect groundwater quality in the longer term.

The profitability of rice cultivation in the study area is crucial to the sustainability of irrigated agriculture. Irrigated *boro* cultivation is still profitable. But, the profits have declined due to rises of input costs. Labour and irrigation costs together normally account for two thirds or more of the total cost of production. Major increases in the prices of these inputs can greatly depress *boro* rice profitability. Returns to individual pump owners, therefore, can be quite unstable from year to year. The demand for irrigation coverage fluctuates in response to the profitability of the crops, particularly of rice.

In the study area, groundwater irrigation can be geared towards poverty alleviation through proper agricultural, credit, subsidies and energy support policies. The provision of electricity for irrigation pumps should be given top priority as the cost of irrigation by electric pumps is about 50% lower than that for diesel-run pumps. There should be diesel supply at subsidized rates until and unless the electricity reaches the study villages. Furthermore, if the polices and guidelines with regards to groundwater irrigation management and agricultural practices are implemented in the area through knowledge sharing, technology transfer or other innovative ways of communication, the study area in particular and the country in general will attain food sufficiency and eventually find ways for poverty alleviation.

REFERENCES

Acharyya, S.K. (2005). Arsenic levels in groundwater from Quaternary alluvium in the Ganga plain and the Bengal Basin, Indian Subcontinent: in sights into influence of stratigraphy. International Association for Gondwana Research, Japan. *Gondwana Res* 8 (1): 55–66.

Acharyya, S.K., Chakraborty, P., Lahiri, S., Raymahashay, B.C., Guha, S., Bhowmik, A. (1999). Arsenic poisoning in the Ganges delta. *Nature* 401: 545.

Alam, M.K., Hasan, A.K.M.S., Khan, M.R., Whitney, J.W. (1990). Geological map of Bangladesh. Geological Survey of Bangladesh, Dhaka.

AST-DAE. (1991). 1990–91 Census of Minor Irrigation in Bangladesh. Agricultural Sector Team and Department of Agricultural Extension, Ministry of Agriculture.

ATIA. (1995), 1993–94. Census of Minor Irrigation in Bangladesh. Main Report: National and Regional Summaries. Department of Agricultural Extension. Project BGD/89/039.

BADC. (2005). Survey report on Irrigation Equipment and Irrigated Area in Boro 2004 Season. Bangladesh Agricultural Development Corporation. Dhaka, Bangladesh.

Bhattacharya, P., Chatterjee, D., Jacks, G. (1997). Occurrence of arsenic contaminated groundwater in alluvial aquifers from the delta plains, eastern India: options for safe drinking water supply. Water Resour Dev 13: 79–92.

BMDA. (2004). Data Bank: Groundwater and Surface Water Resources Bangladesh. Barind Integrated Area Development Project (BIADP). Barind Multipurpose Development Authority, Ministry of Agriculture, Government of the People's Republic of Bangladesh.

BWDB, (1994). Report on the compilation of aquifer test analysis results as on June, 1993, Bangladesh Water Development Board Water Supply Paper no. 534.

Debroy, A., Shah, T., (2003). Socio-ecology of groundwater irrigation in India: an overview of issues and evidence. In: Liamas R Custodio E (eds) Intensive use of groundwater: challenges and opportunities, Swets, Lisse, The Netherlands.

DPHE-BGS. (2001). Arsenic contamination of groundwater in Bangladesh. British Geological Survey Technical Report WC/00/19, Volume 4: Data compilation.

GoB. (2006). Economic Review. 2006. Ministry of Finance, Government of the People's Republic of Bangladesh, Dhaka.

GoB. (2005). National Strategy for Accelerated Poverty Reduction. Planning Commission, Government of the People's Republic of Bangladesh. Revised as on January 12, 2005.

Harvey, C.H., Swartz, C.H., Badruzzaman, A.B.M., Keon-Blute, N., Yu, W., Ali, M.A., Jay, J., Beckie, R., Niedan, V., Brabander, D., Oates, P.M., Ashfaque, K.N., Islam, S., Hemond, H.F., Ahmed, M.F. (2002). Arsenic mobility and groundwater extraction in Bangladesh. Science 298: 1602–1606.

Islam, A.J.M.A. (1986). Review of agronomic research and its future strategy. Adv. In Agronomic Res. in Bangladesh. Bangladesh Soc. of Agron. BARI. Joydebpur. Vol 1.

Islam, A.J.M.A. (1987). Farmers perception of soil fertility and crop productivity trend—a study conducted by AST/CIDA with direction of the Ministry of Agriculture, Govt. of Bangladesh.

Koenigs, F.F.R. (1950). A Sawah profile near Bogor (Java). Institute for Soil Research, Bogor, Indonesia. Aric Res St Bulletin no. 105.

LGED (2005). Agriculture production hand book, Local Government Engineering Department, Government of Bangladesh. April, 2005.

Mandal M.A.S. (2000). Dynamics of Irrigation Market in Bangladesh. In: Mandal M.A.S. (ed) Changing Rural Economy of Bangladesh. Bangladesh Economic Association, Dhaka. pp. 118–128.

Mandal M.A.S. (2006). Groundwater Irrigation Issues and Research Experience in Bangladesh. Presentated at the International workshop on Groundwater Governance in Asia. 12–14 November, 2006, IIT Roorkee, India.

Matbarer Char Union Council (2006). Personal communication.

Mukherji, A. (2007). 'The energy-irrigation nexus and its implications for groundwater markets in eastern Indo-Gangetic basin: Evidence from West Bengal, India', *Energy Policy*, Volume 35 (12): 6413–6430.

MPO. (1987). The Groundwater Resources and its Availability for Development. Technical Report no. 5. Master Plan Organization. Ministry of Irrigation, Water Development and Flood Control, Govt. of Bangladesh and Associates.

Norra, S., Berner, ZA., Agarwala, P., Wagner, F., Chandrasekharam, D., Stuben, D. (2005). Impact of irrigation with arsenic rich groundwater on soil and crops: A geochemical case study in West Bengal Delta Plain, India. Applied Geochemistry 20:1890–1906.

NWMP. (2001). National Water Management Plan, Water Resources Planning Organization, Government of Bangladesh.

NWPo. (1999). National Water Policy 1999, Ministry of Water Resources, Dhaka, Bangladesh.

Palmer-Jones, R.W. (1997). Groundwater management in South Asia: What Role for the Market? Paper presented at 18th European Conference on Irrigation and Drainage, Oxford.

Ravenscroft, P. (2003). Overview of the Hydrogeology of Bangladesh. In. Groundwater Resources Development in Bangladesh: Background to the Arsenic Crisis, Agricultural Potential and the Environment. Editors: A Atiq Rahman and Peter Revenscroft. The University Press Limited, Bangladesh. pp. 43–86.

Scanlon, B.R., Healy, R.W., Cook, P.G. (2002). Choosing appropriate techniques for quantifying groundwater recharge. Hydrogeology Journal (2002) 10:18–39 DOI 10.1007/s10040-0010176-2.

Sengupta, S., Mukherjee P.K., Pal, T., Shome, S. Nature and origin of arsenic carriers in shallow aquifer sediments of Bengal Delta, India. Env. Geology (2004) 45: 1071–1081.

Shah, T. (1993). Groundwater Markets and Irrigation Development: Political Economy and Practical Policy, OUP, Mumbai, India.

Shah, T. (1997). Pump irrigation and equity: machine reform and agrarian transformation in water abundant Eastern India. Policy School, Working Paper 6. The Policy School Project, Anand, India.

Shah, T. (2001). Wells and welfare in the Ganga Basin: public policy and private initiative in eastern Uttar Pradesh, India. IWMI research report 54. IWMI, Colombo, Srilanka, 43 p.

Shah, T., Singh, O.P., Mukherji, A. (2006). Some aspects of Sout Asia's groundwater irrigation economy: analyses from a survey in India, Pakistan, Nepal Terai and Bangladesh. Hydrogeology Journal 14: 286–309.

Sharma, I. (1989). Underdevelopment, inefficiency and inequity: case of groundwater management in Bihar. Paper presented in workshop on Efficiency and Equity in Groundwater Use and Management, February 1989, Institute of Rural Management, Anand, India.

Upazilla Agriculture Office, Shibchar (2006) Personal communication.

Yan, X.-P., Kerrich, R., Hendry, M.J. (2000). Distribution of arsenic (3), arsenic (5) and total inorganic arsenic in pore waters from a thick till and clay rich aquitard sequence, Saskatchewan, Canada. Geochim Cosmochim Acta 62: 2637–2648.

Zhang, G.L., Gong, Z.T. (2003). Pedogenic evolution of paddy soils in different soil landscapes. Geoderma 115: 15–29.

CHAPTER 9

Reaching the poor: Effectiveness of the current shallow tubewell policy in Nepal

D.R. Kansakar
Department of Irrigation, Government of Nepal, Lalitpur, Nepal

D.R. Pant
International Water Management Institute (IWMI), Nepal Office, Lalitpur, Nepal

J.P. Chaudhary
Nepal Television, Kathmandu, Nepal

ABSTRACT: Expanding groundwater irrigation in Terai is important for Nepal to meet its increasing food requirements and for poverty alleviation. The current government policy, since 2000, aims to help small and poor farmers in investing in group-based Shallow Tubewells (STWs), by providing various support services, instead of a direct cash subsidy. This paper evaluates the outcomes in the two sites among many, where this policy has been implemented through a government project. Poverty alleviation and STW irrigation expansion, the two main goals of the new policy, are difficult to be achieved simultaneously. Although small and poor farmers now have a higher rate of access to STWs, it still fails to benefit the poorest such as tenant farmers and those without legal land entitlement. The process is further slowed down due to small landholding sizes and the time consumed in organizing small farmers. The paper suggests that promoting individual STWs through rural electrification could be a more effective policy for a faster groundwater development and poverty alleviation.

1 INTRODUCTION

Annual growth in food production has remained poor in Nepal, although the agriculture sector has been given a high government priority in all the development plans since 1956. In 1957, the total food grain production of Nepal was 2.84 million tons and that increased to only 7.24 million tons in 2002. This increase was barely sufficient to meet the food requirements of the population of the country, which was growing annually at the rate of 2.27% over the period of 1991–2001. The slow agricultural growth was due to the domination of rain-fed agriculture. In 2001, irrigation was available in only 33% of the total net cultivated land (WECS, 2002). Only 3.3 million tons of food grains were grown in the irrigated lands, and this was less than half of the country's total cereal crops production. The agricultural productivity in general is low in Nepal, in comparison to that in other countries in Asia. For example, the average food grain productivity increased from 1.8 ton/ha in 1957 to only 2.0 ton/ha in 2002. The productivity of paddy was 1.9 ton/ha in 1957, and 2.5 ton/ha in 2002. This is very low when compared with China's productivity of paddy at 7.8 tons/ha (IWMI, 1999). High productivity in China was possible because of its facilitating policy environment, efficient water management practices, and appropriate technology. It is believed that Nepal can achieve a productivity level of 4.0 ton/ha in paddy even under the existing level of agricultural technology (NPC, 2005). Similar potentials for increases exist in other principal cereal crops like wheat and maize.

In the past, food production had increased due to expansion in the cultivated land area, mainly in the Terai and the Inner Terai valleys (the foothills and the plain area south of the Himalayas). The cultivated area expanded from 1.24 million hectare in 1957 to 2.64 million hectares in 2001. But, the scope for further expansion in cultivated area is limited. Therefore, food production has to increase by means of a rise in the productivity level. Agricultural productivity has to increase also for alleviating poverty, because 80% of the active labour force, and 35% of the population are living below the poverty line (NPC, 2002; NPC, 2005).

Irrigation, including year-round irrigation, is important for increasing cropping intensity, which in turn is essential for increasing crop productivity. The average cropping intensity is 183% for the country, although it is slightly higher in irrigated areas (up to 220%) (CBS 2004). The Agriculture Perspective Plan (APROSC and John Miller Associates 1995) and the National Water Plan (WECS 2005) of the Government of Nepal have emphasized year-round irrigation development for agricultural growth in order to achieve the country's economic development targets. Groundwater irrigation development, particularly through shallow tubewells (STWs), is stressed in the Terai, where the available annual groundwater recharge remains under-utilized.

2   EVOLUTION OF SHALLOW TUBEWELL POLICY IN NEPAL

The recent policy emphasis on STW development supports and reinforces previous initiatives to promote STW technology in Nepal. With the early advent of STW technology in the mid-1970s, the Nepalese government implemented different policy measures for its promotion in the Terai region. From 1980 to 2000 the government provided direct subsidy on the capital cost of the STW. The subsidy rates varied with time, from a low of 30% to a high of 85% of the total cost of a STW system. The subsidy rate was higher for STWs owned by groups of farmers rather than for individual wells, in an attempt to rationalize the investments and encourage group ownerships.

The Agricultural Development Bank of Nepal (ADBN), the main channel for distributing government subsidy, installed 49,034 STWs between 1980 and 2000. Although it has reported irrigation development in 168,832 ha of land, the actual area irrigated by these wells may be less, as an ADBN well was found to irrigate only 2.3 ha, on average (Koirala, 1998; APROSC and John Mellor Associates, 1995), as against the expected command area of 3.4 to 4.0 ha (GDC, 1994).

In the earlier ADBN programme, the STW subsidy was tied up with the Bank's loan program, for which 0.68 ha of land was required as loan collateral. There was no institutional effort in organizing the farmers group. As a result, small farmers who had smaller plots could not draw this government subsidy. In Terai, a large section of the farmers (44% of the landholdings) owns less than 0.5 ha land (CBS, 2004). Therefore, the subsidy was utilized primarily by the large and medium farmers. Despite the lower subsidies, the majority of the benefiting farmers (97%) preferred individual wells (Koirala, 1998). However, the tubewell utilization rates were lower in individual wells (178 to 200 hrs/annum), com-pared to group-owned wells (269 hrs/annum), because the latter commanded slightly larger areas (Shrestha & Uprety, 1995; Koirala, 1998; and GDC, 1994).

For economic reasons, the government agencies, mainly the Department of Irrigation (DOI), the Groundwater Resources Development Project (GWRDP), and the Department of Agriculture (DOA), focused on group-owned STW development. These organizations mobilized farmers to organize themselves into groups, for joint operation, maintenance, and management of the STWs. Under various projects, more than 10,000, mostly group-owned STWs, have been installed by these agencies, irrigating a total of 38,000 ha (*DOI internal reports, 2006*). However, the effectiveness of the group approach has been questionable, because of the reported collapse of the STW groups few years after the completion of the projects (*personal communication with DOI/GWRDP officials, 2006*). Furthermore, the group approach did not favour the smaller farmers in terms of offering joint collateral when adding smaller pieces of land.

In general, growth in STW installation has been slow in Nepal, when compared to those in neighbouring countries, like Bangladesh and India (Shah et al., 2003). Interestingly, the policy itself has been a main constraint, because the annual budgetary resources allocated by the government was always limited (Koirala, 1998; GDC, HTS and East Consults, 1997). Again, the benefits of this policy could not reach the small and poor farmers, who needed the government subsidy most. Nearly half of the Terai landholders were not eligible for subsidy programme, because their landholding size was insufficient for the collateral.

In July 2000, the Nepalese government stopped the direct subsidy policy, and initiated a new policy of providing various complementary incentives and services that facilitated the farmers in installing STWs on their own and helped them improve their agricultural practices towards obtaining higher return.

2.1   *The Community Groundwater Irrigation Sector Project*

The Community Groundwater Irrigation Sector Project (CGISP) was designed specifically for implementing the new government policy of 'no direct subsidy' on STWs. The project was started in 1998, but the STW installation activity began in 2000, after the enforcement of the new policy. The project target was to install 15,000 STWs—13,500 group-owned and 1,500 individually owned wells—by 2007 and develop irrigation facility in 60,000 ha area in the 12 Terai districts of the Eastern and the Central Development Regions of Nepal (Figure 1). Within these districts, the project was implemented only in about 300 Village Development Committees (VDCs), where sub-surface geology allowed the use of manual drilling. Village Development Committee (VDC) is the smallest political and administrative unit. Each VDC is divided further into nine wards. Each VDC was considered as a sub-project area for project implementation, and the project was implemented on the basis of the demand from the community in each sub-project area. By July 2006 only 7,000 STWs had been installed under the project. This project was scheduled to end in July 2007.

The CGISP project was targeted specifically at the 'small' and the 'marginal' farm-ers, who own less than 1 ha of land. Partner NGOs were hired to mobilize and organize these target farmer communities into tubewell user groups. As in the earlier government programmes, this project also promoted STW systems consisting of a four-inch diameter tubewell fitted with a diesel or electric centrifugal pump. The well depth depended on the local hydrogeological conditions, but the depth of STWs in the study areas were generally less than 37.5 m. The cost of a STW system, including drilling and pump installation, ranged from Nepali Rs. (NPR) 41,000 to 65,000 (1 USD ~ 66.02 NPR as in May 2008).

For each well, a Water User Group (WUG) was formed among the landholding members, whose fields together constituted a contiguous plot of at least 4 ha. A group must have at least 5 members, among whom the office bearers are elected for managing the STW and the group. The project provided loan to each group for purchasing a STW. Local Participating Financing Institutes (PFIs) provided the loan, on a group-guarantee basis without requiring land as loan collateral. However, land collateral was required when a farmer wished to install an individual well. But, the project faced difficulty in organizing groups that met the joint command area size requirement. Therefore, the project reduced the minimum command area size to 3.5 ha and later on to 2.5 ha. The project (i.e. PFIs) did not provide loans for STWs to groups whose command area was smaller than the requirement, but all other benefits, such as training and agricultural extension services, were provided. The PFIs also provided loans for purchasing agricultural inputs to all the farmers, large and small.

As the electricity supply network was scarce in the project districts at the time of the project preparation, the project was designed for diesel pump operated STWs only. However, the project later financed electric pumps in its sub-projects where an electricity supply already existed. Although technically, a 2 to 3 HP pump is sufficient for operating the STW and the target command area, the project encouraged farmers to buy 5 HP or smaller pumps, because the farmers were accustomed to using much higher capacity pumps. The project was supposed to educate the farmers on pump selections, but the ultimate decision on pump selection was left to the farmers. As an incentive to attract more farmers in STW irrigation, and to ease the access to market centres, the project also financed the improvement of farm-to-market roads in the villages, on a cost-sharing basis. There was a ceiling of US $ 50,000 for this activity in each sub-project area. The community contributed 10% of the cost, in terms of labour, and the project financed the remaining cost. Furthermore, the project also contributed $50 in cash as a matching grant to the tubewell maintenance fund of each group. This was to help the WUGs in sustaining their STWs.

3   THE OBJECTIVES OF THE STUDY

The present study has been carried out to assess the effectiveness of the current STW policy in expanding the STW irrigation and in reaching the small and marginal farmers in Nepal Terai—the stated goals of the government policy. This study has attempted to analyze the strengths and the weaknesses in this policy. The study was carried out in the two sub-project areas (VDCs) of the CGISP in Eastern Terai region, namely the Dangihat VDC in Morang District, and the Arjundhara VDC in Jhapa District (Figure 1).

These study sites were selected because of their distinct STW development patterns, in spite of their geographical and hydrogeological similarities and comparable infrastructure development status. The CGISP project was launched early in 1998 in Dangihat, but only 22 STWs have been installed against the target of a minimum 50 wells in each sub-project area. A large number (72) of STWs have been installed in Arjundhara in a two-year period between 2004 and 2006. All these STWs have electric pumps, whereas there were only six electric pumps used in Dangihat, although electricity was available in both areas. The similarities between the two areas include their physical settings; both

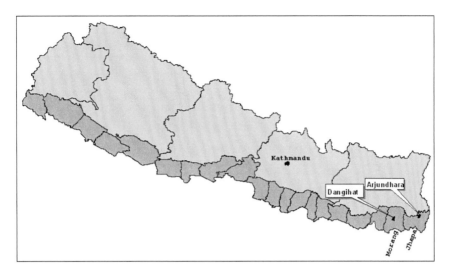

Figure 1. Map of Nepal showing the study areas. It shows the five development regions and the twenty districts in Terai region (darker grey). The CGISP program is implemented twelve eastern most districts of Nepal Terai.

the areas are located in the northern part of the main Terai plain, and consists of hard sub-surface geologic formations that are difficult to drill through by manual methods. Consequently, only a few STWs existed in these areas before the implementation of the CGISP project. Both areas have good development infrastructures like schools, health posts, electricity distribution network, village road network, and links to the national highway.

4   THE STUDY AREAS

The Dangihat VDC lies in the northern part of the Morang district. It has a total surface area of 23,886 ha, out of which the cultivated area is only 1,096 ha. There are 3,054 households and a population of 8,475. Landholding size distribution is uneven among these households (Figure 2), and a few farmers hold large tracts of land. There are 1,405 households that have no landholding; these households depend on the 'large' farmers in the role of tenant farmers or sharecroppers, or as simply farm labourers (Karki and Basnet, 2004). Except for ward numbers 4 and 5, where groundwater is the only option available for irrigation, seasonal monsoon irrigation is available in most parts of the VDC from the nearby streams. The CGISP programme was launched mainly in those areas, where surface irrigation facilities were not available.

Situated in the northern part of the Jhapa district, the Arjundhara VDC has a total area of 154.66 km². Out of a total 7,596 ha of cultivated land, seasonal monsoon irrigation was available in 6,750 ha. There are 4,025 households and a population of 16,178 in total. Distribution of land among these households is shown in Figure 2. It may be noted that there are fewer landless people in Arjundhara although the percentage of large landholders

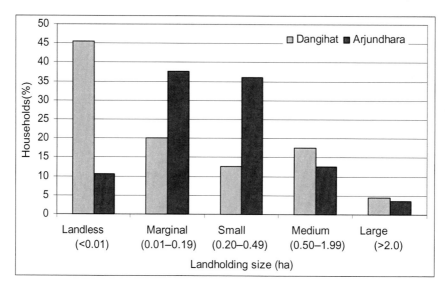

Figure 2.   Landholding size patterns in the study area (in hectares).
Source: Field survey.

Table 1.   Composition and distribution of the study sample by water transaction types.

Category	Danghihat	Arjundhara	Total
STW groups (part of CGISP) [water seller]	19 (19) [7]	29 (29) [0]	48 (48) [7]
Members of CGISP STW groups	101	136	237
Individual well owners (part of CGISP) [water seller]	12 (0) [7]	11 (10) [5]	33 (10) [12]
Exclusively rain-fed cultivators	3	5	8
Water buyers (from CGISP wells)	15 (9)	15 (15)	30 (24)

Source: Field survey.

is similar to that in Dangihat area. For the farmers here, STWs are important for winter and spring irrigation. They could also supplement the surface irrigation in summer crops.

5   METHODOLOGY

The present study is based on a structured household questionnaire survey among the STW irrigators and the rain-fed cultivators. It collected the basic social, agricultural, economic, and technological data from the respondents. Separate unstructured questionnaires were developed and used in the survey of the CGISP beneficiaries, including those farmers, who were not well owners but bought water from the CGISP well. These questionnaires were used to collect data on (i) the number of beneficiary households in a well-sharing group, (ii) the land area of each group member and the total irrigated area by each well, (iii) the reasons for lack of STW installation before the project; (iv) the project

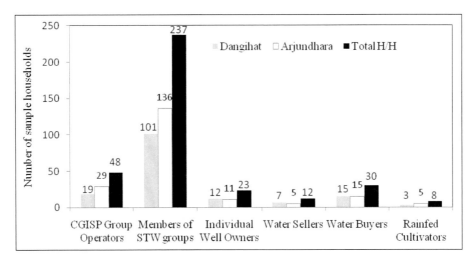

Figure 3.   Composition and distribution of the study samples (households) by their water transaction status.
Source: Field survey.

facilities which they valued the most; and (v) the issues involved in the installation, operation and management of the STWs in the groups. These issues included (a) mechanism for pump operation and maintenance; (b) water distribution among the members; (c) repayment of loan; and (d) their preference in the type of STW ownership.

The survey samples were selected randomly from (i) the CGISP tubewell group members, (ii) water buyers, both from CGISP and private wells (iii) private/individual STW operators, and (iv) rain-fed cultivators. The composition of the sample and their distribution in the two study sites are given in Table 1 and are shown in Figure 3. Furthermore, the views and experiences of the partner NGOs, the Participating Financial Institution (PFIs), and the project officials that were expressed during the field survey have been used in data analysis and their interpretation.

6   RESULTS

6.1   *Effectiveness of the STW policy*

6.1.1   *Group ownership*
No well-sharing groups existed besides the CGISP programme beneficiaries in the study areas. Therefore, it may be derived that the group concept among the farmers was solely a product of the project policy, because it was required that the farmers form a group in order to obtain a collateral-free loan. It was found that the private operators shared their wells in the form of water sale or renting of their pumping set to their neighbouring farmers. In Arjundhara, 14% of all CGISP wells surveyed were owned by individual farmers, showing that individual ownership was popular among the farmers, in spite of the requirement of land collateral for a loan. But in Dangihat, no individual well was constructed under the

CGISP, perhaps because the farmers were poorer or were not willing to accept a loan under the land collateral condition. Individual ownership is still preferred here, as the survey showed that 7 out of the 20 water buyers and rain-fed cultivators interviewed preferred to have their individual wells in the future. The current group-based STW policy has not been suitable for all farmers in the Terai. This finding is in agreement with the previous experiences from the ADBN programme, and points to the possible need to re-assess the feasibility of the group concept if rapid growth of STW is to be encouraged in the future.

6.1.2   *Group formation process*

Since the individual landholding sizes are generally small in Nepal Terai, individual owner-ship in a standard sized STW is not profitable for the farmers due to the high costs involved in installation and operation. This fact was understood by 43% of the respondents who saw the advantage in operating STWs in groups. In Dangihat, 63% had appreciated the group approach in the current policy. In contrast, only 20% respondents in Arjundhara were appreciative of this approach, perhaps because the farmers here were economically better off than those in the Dangihat area.

With small landholding sizes, policy required numerous farmers to form a STW group that met the criterion of design command area size. Social mobilization and group organization activities consumed much of the project implementation time (*Personal communications with NGO, 2006 and project staff, 2006*). Different facilities and the supporting programmes in the project have been the main incentives for farmers to organize into the groups.

A contiguous plot of 4 ha or even 3.5 ha land required by the project for a STW group involved many households (between 3 and 11 members in the study sample). With numerous members, social, financial, and personality dynamics caused complexities in maintaining cohesion in the group. Therefore, proper functioning of such groups for a longer period of time is questionable, as the observations in the earlier projects have shown.

6.2   *Access of the poor farmers to STWs*

6.2.1   *Group size and households reached*

Data show that a total of 237 households now have access to year-round irrigation from the 48 CGISP STWs in the study sample. On average, a group consisted of 4.9 members (i.e. farmer households). The average size of a group was slightly larger (5.2 members per group) in Dangihat area, compared to Arjundhara (4.9). The median CGISP group has 5 members, which was the project's target size. The largest group had 11 members, while the smallest one had only 3 members. The extremes in the group sizes were in Arjundhara area.

6.2.2   *Landholding size*

According to landholding sizes, 55.7% of the group members were found to be 'medium' farmers, while the 'marginal' and 'small' farmers constituted 13.1% and 27.8% respectively (Table 2). The largest landholding size was 3.22 ha, and the smallest was 0.07 ha; the median holding size was 0.61 ha. The present data show that access of the marginal farmers in a STW is still low (13.1%), considering their relative numbers in Nepal Terai (CBS, 2004).

Table 2.   Categorization of project beneficiary households by their landholding size.

		Number of group members in percentage ($n = 237$)		
Farmer class*	Landholding size	Dangihat, Morang	Arjundhara, Jhapa	Total
Marginal	0.01–<0.1 ha	12.9	13.2	13.1
Small	0.1–<0.5 ha	35.6	22.1	27.8
Medium	0.5–2.0 ha	50.5	59.6	55.7
Large	>2.0 ha	1.0	5.1	3.4

Source: Field survey.

Table 3.   Distribution of command area among the CGISP group members.

		Group STW irrigated area in hectares (and as % to total) ($n = 237$)		
Farmer class	Landholding size	Dangihat, Morang	Arjundhara, Jhapa	Total
Marginal	<0.1 ha	4.39 (2.99)	4.57 (1.7)	8.95 (2.2)
Small	0.1–<0.5 ha	31.17 (21.27	25.88 (9.7)	57.05 (13.8)
Medium	0.5–2.0 ha	105.49 (71.98)	189.94 (71.17)	295.43 (71.5)
Large	>2.0 ha	5.5 (3.75)	46.48 (17.42)	51.98 (12.6)
Total		146.55 (100)	266.87 (100)	413.42 (100)

Source: Field survey.

6.2.3   *Irrigated area of the group members*

The 48 group STWs in the study sample were found to irrigate 167.4 ha area in total, i.e. an average of 3.5 ha by a well (Table 3 and Figure 4). Data shows that 71.5% of the irrigated area belonged to medium farmers, who constituted 55.7% of the members. 40.9% of the group members (marginal and small farmers) had only 16.0% of land within the command area. Irrigated area expansion will be low if the policy focuses primarily on marginal and small farmers. However, significant reductions in poverty can be achieved by such a policy.

6.2.4   *Land size of the individual STW operators*

Among the 23 individual STWs operators surveyed, 12 were from Dangihat area and the remaining 11 were from Arjundhara area. All 12 wells in Dangihat were installed earlier, with the subsidy under the ADBN programme. 10 out of the 11 wells in Arjundhara were installed with the loan from the CGISP programme. These well owners had deposited land collateral for the loan. The remaining one well was financed privately, because its owner, a marginal farmer, was ineligible for a loan, because he did not legally own the land. Overall, 59.3% of the private well operators were found to be large farmers, and 33.3% were medium farmers. In Arjundhara, the number of large farmers was much higher (87%); and 10 of 11 private operators were medium farmers.

These results indicates that (i) even the large farmers needed credit for STW installation, and that land collateral was not the issue for these farmers; (ii) large and medium farmers generally preferred individual wells, whether there is subsidy or no subsidy; and that

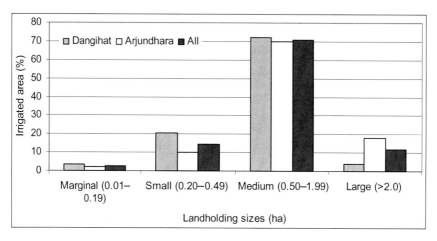

Figure 4.   Distribution of command area among the CGISP group members.
Source: Field survey.

(iii) in practice, the policy has excluded those farmers, who do not have land entitlements. Hence, the new credit policy is still not effective enough to reach the core of the poor farmers in Terai.

6.2.5   *Credit facility*
Prior to the CGISP, lack of funds had hindered STW installation for 42% of all the respondents, and 63% did not have either the sufficient land, or did not like to deposit their land for loan collateral. Relaxation in the land collateral requirement for loan, with the CGISP, had enabled these farmers to install STWs. By mobilizing the micro-credit institutions (PFIs) instead of regular commercial or development banks, the policy had achieved wider outreach to benefit the rural community.

Farmers used the loan facility for STW acquisition only, not for support to agricultural inputs. The reason for this is not known at present, but the usage of high yield variety seeds, chemical fertilizers and other inputs were found to be low among all the respondents in general.

6.2.6   *Loan recovery*
The present data show that smaller cooperative type micro-financing PFIs had high loan recovery rate (highly represented in Arjundhara), compared to the larger commercial or development banks (only represented in Dangihat). Among all the borrowers, 75% paid their loans back on time, all from the Dangihat area.

Group conflict was the cause for loan-default in 42 cases, while another 33 respondents were not paying because of a 'false rumour' that the government would cancel their loans. Interestingly, one respondent reported loan default because the bank staff did not come to the village for collecting money, as the small PFIs were doing. Thus, it may be concluded that a farmer-friendly credit service is necessary for the success of credit-based development programmes.

6.3   Agricultural practices and farm economics

6.3.1   Cropping intensity

Earlier, the cropping patterns in the study areas used to be generally, paddy-wheat, paddy-wheat-maize, paddy-maize-mustard, and paddy-maize. With the availability of assured year-round irrigation from the STW, farmers had changed their cropping patterns and their cropping intensities also had increased (Table 4). In Dangihat, where the sole source of irrigation water was groundwater, the cropping intensity was generally higher, and the main crop was banana, grown by 63% of the respondents. In Arjundhara, such shifts are just beginning to take place, but it should be recalled that the project was implemented much later here.

6.3.2   Crop diversification

Present data shows that 42% of the farmers were growing exclusively cash crops, whereas 32% grew cash crops along with the cereal crops. Only 26% grew just the traditional cereal crops. Shift to high value crops was higher in Dangihat, where nearly 49% farmers had switched over completely to cash crops (mainly banana and some other fruits), while another 19% have introduced banana in parts of their fields. In Arjundhara, 70% farmers have introduced higher value crops (e.g. sugarcane, tea, beetle nuts, and vegetable crops) in their cropping pattern. Only about 28% were growing the cereal crops exclusively (Table 5). According to the respondents, this positive change enhancing the farmers' income from farming (see next section) was partly attributed to the agricultural training and extension programmes undertaken as a part of the project.

6.3.3   Agricultural productivity and farmers' profitability

Table 6 shows that cereal and oilseed crops were not profitable; some farmers were found to be at loss from these crops, particularly in Dangihat. These farmers did not use adequate irrigation, nor did they use improved seeds and chemical fertilizers. However, the high value crops, particularly banana, were profitable. In Arjundhara, the newly introduced vegetable and annual cash crops such as sugarcane, tea, beetle nuts etc., have been profitable for the STW irrigators.

Table 7 shows that the profit level in Dangihat increased with the availability of STW irrigation. In spite of the higher price the water buyers had to pay for STW irrigation, they could make higher profit compared to the rain-fed cultivators.

Table 4.   Cropping intensity among the STW irrigators in the study areas.

STW operator type	Cropping intensity (Average in %)	
	Dangihat	Arjundhara
CGISP group members	273.6	224.4
Private operators	232.0	221.5
Water sellers	274.6	230.0
Water buyers	277.8	233.0

Source: Field survey.

Table 5.   Crop diversification among the STW irrigators in the study areas.

Crops grown	STW irrigators (in %)	
	Dangihat	Arjundhara
(Traditional) cereal crops only	32.4	28.3
Cereal plus annual cash crops	18.9	35.0
Annual cash crops only	48.7	1.7
Cereal plus vegetable crops	0.0	35.0

Source: Field survey.

Table 6.   Net return from different crops among the STW irrigators.

Crop	Average net return (NPR/ha)	
	Dangihat	Arjundhara
Paddy	1,356	3,986
Wheat	(−) 506	1,968
Maize	1573	3,345
Mustard	336	880
Banana/Other fruits	20,758	10,453
Vegetable	–	9,581
Sugarcane	–	31,070
Beans	–	2,982
Beetle nuts	–	5,518

Source: Field survey.

Table 7.   The average net profit from banana cultivation, in Dangihat, Morang.

Water transaction type	Average net profit from banana crop (NPR./ha)
Rain-fed cultivators	21,779
All types of STW irrigators	34,438
CGISP group members	35,914
Private operators	35,895
Water sellers	33,923
Water buyers	29,172

Source: Field survey.

6.3.4   *Cost of STW operation*

The main cost in STW irrigation is the cost of energy consumed in pump operation. The equipment repair and maintenance are the other costs. There are two types of pumps used in the study areas—diesel engine and electric motor operated centrifugal pumps. Diesel

Table 8.   Operation cost of the centrifugal pumps used in the study areas.

Pump capacity	Average pumping cost (Rs./Hour)[a]	
	Electric centrifugal pumps	Diesel centrifugal pumps
5 HP		64.27
7 HP		87.42
8 HP		85.04
10 HP		82.93
2 HP (1.5 kwh)	17.95	
3 HP (2.2 kwh)	29.53	
5 HP (3.7 kwh)	37.40	
7 HP (5.5 kwh)	23.85	.

[a] Costs include repair and maintenance costs.
Source: Field survey.

centrifugal pumps were commonly used in the past, because of the lack of electricity distribution system. With the expansion of electricity supply networks, electric centrifugal pumps are getting more popular, mainly because electricity is much cheaper than diesel fuel. Again, electric pumps are less expensive to buy and easier to maintain than diesel pumps. It was more expensive to operate a diesel pump than an electric pump. Again, high capacity pumps consumed more energy (electricity or diesel fuel), and therefore were more costly to operate (Table 8). With numerous high capacity pumps in use, many farmers in the study area were found spending more on pumping than was necessary.

6.4   Limitations of the current STW policy

6.4.1   Group formation

During implementation, the project had to face several limitations in the group approach to STW development. It was seldom that small farmers had their lands situated together. A design command area of 4 ha or even 3.5 ha involved several farmers, as the individual plot size was small. Organizing a group among many members was time-consuming, if not difficult. Because of the practical difficulties in forming groups, the CGISP could install only 7,000 wells up until 2006, against a target of 15,000 wells at the end of the project in 2007 (CGISP 2006).

6.4.2   Loan policy

Legal ownership in land within the command area is mandatory in the policy for a farmer to obtain STW loan. This condition has restricted STW installation among those farmers, who operated on the trust-owned lands and public lands. Similarly, the farmers operating under different forms of rental arrangements, such as renting for a fixed amount of money, or for a fixed quantity of produce or a fixed share in produce, or in exchange for service or for mortgage, were also not eligible for the loan facility. There were also farmers who farmed on newly encroached lands that had not been registered. Since STW loan was not

Figure 5.   A 4-inch diameter shallow tubewell fitted with a 8HP diesel pump, in Dangihat.

Figure 6.   A less expensive Chinese electric pump (2 HP), fitted on a 2-inch diameter shallow tubewell in Arjundhara, Jhapa.

provided for such lands, large areas in Terai have been left out and numerous farmers, who depended on such lands, have been deprived of STW irrigation development.

6.4.3   *STW technology*

The average discharge of a 4″ diameter well (Figure 5) was 46.4 m³/hr in Dangihat, and 75.4 m³/hr in Arjundhara area. Such well discharges are good enough to irrigate larger than 4 ha area (GDC, 1994). Similar well discharges had been designed to irrigate 6 ha in the earlier Community Shallow Tubewell Irrigation Project, and even 10 ha in Nepal Irrigation Sector Project. In the latter case, buried pipe system was also provided for efficient water conveyance. The CGISP designated a 4 ha plot, in order to justify the capital and the operation costs of a STW. As the policy was targeted to small farmers, the group approach was inevitable. But, this approach has slowed down the progress in STW installation in the project.

These days, shallow tubewell technology has become more divisible due to the inno-vations in small electric pumps and low cost construction material. It is now possible to construct smaller sized and less expensive STWs, which could be affordable even to the indi-vidual small and marginal farmers (Figure 6). STW numbers could grow rapidly if the policy is to promote such less expensive individual wells among the small and marginal farmers.

6.4.4   *Selection of the pumping set*

There is a common misconception among Terai farmers that 'high capacity pumps yield high well discharge'. In Dangihat, diesel pumps were more common (81.2%) than electric pumps. Fifty six percent of these pumps have higher than 5 HP capacities (i.e. 7, 8 and 10 HP). Even among the CGISP wells, 42% of the pumps were more than 5 HP. As a result, operation cost was high in these STWs (Table 8). In Arjundhara area, on the contrary, lower capacity pumps were more common. There were 16.1% pumps of 2 HP and 32.2% of 3 HP capacities. The 5 HP pumps constituted 48.5% of the total pumps in use. Only one pump was of 7 HP capacity. All the pumps were here run by electric motors.

The effect of pump capacity on well discharge and pumping cost in Dangihat and Arujundhara areas are shown in Figures 7 and 8, respectively. These figures show that the cost of pumping went up as the pump capacity increased, but there were no outstanding changes in the well discharges. The results show that electric pumps were 5 to 6 times cheaper to run than the diesel pumps.

The present study has shown that the farmers in Dangihat had to bear the burden of high pumping costs, because of their use of high capacity diesel pumps. Only a few farmers used electric pumps here—4 new STW owners and 2 old ones, who replaced their diesel pumps. In Arjundhara, pumping cost was found to be generally much less, because all the pumps were run with electricity.

6.4.5   *Ranking of CGISP support components*

In the present survey, it was found that 63% of the respondents in Dangihat favoured the group concept, but only 26.3 favoured group-based loan arrangement, because of the con-flict in the groups. In Arjundhara, the majority of the respondents (80%) were keener on the farm-to-market road improvement component than they were on group ownership provision (20%). Their second attraction was agricultural training and extension programme. The project has not achieved the goal of empowering farmers in procuring STW construction

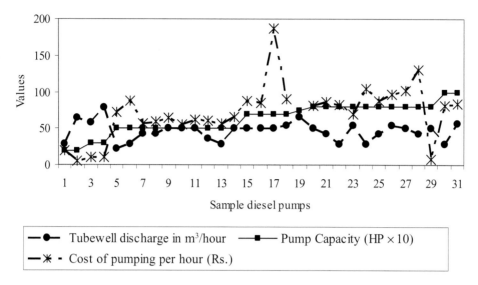

Figure 7.   Tubewell discharges and pump operation costs for diesel pumps of different capacities in Dangihat Area, Morang.
Note: The pump capacity figure is multiplied by 10, for enhanced graphic representation.

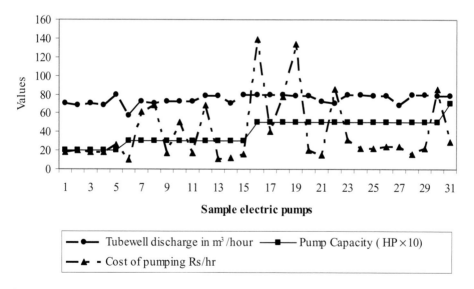

Figure 8.   Tubewell discharge and pump operation costs for electric pumps of different capacities in Arjundhara Area, Jhapa.
Note: The pump capacity figure is multiplied by 10, for enhanced graphic representation.

Table 9. Ranking of the facilities provided in CGISP by the beneficiaries (%).

	Rank I			Rank II			Rank III			Rank IV			Rank V		
	D	A	All	D	A	All	D	A	All	D	A	All	D	A	All
Provision of group STW	63.2	20	40.8	36.8	–	24.5	–	80	14.3	–	–	8.2	–	–	15.2
Provision of collateral-free group loan	26.3	–	16.3	52.6	–	32.7	21.1	–	20.4	–	100	20.4	–	–	10.2
Agricultural extension and training	5.3	–	10.2	–	80	14.3	15.8	20	26.5	73.7	–	36.7	5.3	–	12.3
Improvement of farm-to-market road programme	5.3	80	20.4	5.3	20	16.3	68.4	–	32.7	15.8	–	20.4	–	–	10.2
Independency in purchasing STW material and pumping set	–	–	14.3	–	–	10.2	–	–	8.2	5.3	–	12.2	94.7	100	55.1

D = Dangihat (n = 19); A = Arjundhara (n = 30); All = Dangihat and Arjundhara Together (n = 49).

material and pumping equipment and the services from the private sector, because the NGO and the PFI staff were found to have made these decisions. The respondents have ranked this policy provision at the lowest (Table 9).

7   CONCLUSIONS

Compared to the earlier policy of "subsidy and credit" package, which was enjoyed mostly by the large and medium farmers, the current policy of "no subsidy, but the incentive package of liberal loan facility and other support programmes" has benefited large numbers of marginal and small farmers (40.9% of the total beneficiaries). The key to this achievement was the new credit policy that has relaxed the condition of land collateral. Yet, two conditions have been identified in the policy that has restricted a wider participation of the poor and small farmers. These are, (i) all group members must legally hold land within the command area of the STW; and (ii) that only formal groups that meets all the conditions will be eligible for collateral free credit facility. The first condition has put restriction to developing STWs in the rented lands, public lands, trust-owned lands, and newly encroached and unregistered lands. Similarly, the second condition has excluded those farmers who could not form a group for various reasons. Promotion of individual wells would have solved the latter situation, while relaxation of land ownership for the loan facility would have addressed the first one. From the present study, it may also be concluded that the credit facility is required even for the large farmers, and that land collateral is not an issue for this group of the farmers.

From the large number of STWs installed within a short time period in Arjundhara area, it may be concluded that STWs numbers may grow rapidly if the electricity supply network is expanded in the rural Terai. Again, promotion of lower capacity electric pumps (2 HP) can further help in STW growth, even among the smaller farmers, because it can reduce pumping costs.

Improved agricultural practices and diversification to high value crops leads not only to increased cropping intensity, but also yields high profit to the farmers. All these improvements are possible only when STW program is integrated with effective and adequate agricultural training and extension supports, as it has been the case with the CGISP. Similarly, improvement in farmers' access to market centres is also important for agricultural development. It is important that the government policies for agriculture development, including STW irrigation development, need to be integrative of all these aspects.

ACKNOWLEDGEMENTS

The authors would like to acknowledge the financial support from the International Water Management Institute for carrying out the study. They are thankful to Mr. H.P. Khatiwada, Association Organizer, for conducting the field survey, and to Mr. N. Khatri, Office-Chief, GWRDP/Biratnagar, for the logistic and staff support during the field studies. Thanks are due to Mr. M. Giri, The Sahara Nepal, Jhapa, for providing data on CGISP farmers. The authors are highly indebted to Dr. K.G. Villholth, Senior Researcher and Project Leader, GGA Project, IWMI, for her encouragement in carrying out this study, and for her valuable suggestions in improving the manuscript.

REFERENCES

APROSC (Agricultural Projects Services Centre) and John Mellor Associated, Inc., (1995). *Nepal Agricultural Perspective* Plan (*Final Report*). National Planning Commission, His Majesty's Government of Nepal and Asian Development Bank.

CBS (Central Bureau of Statistics), (2004). *National Sample Census of Agriculture, Nepal, 2001/02.* His Majesty's Government of Nepal/National Planning Commission/Central Bureau of Statistics, Kathmandu, Nepal.

CGISP (Community Groundwater Irrigation Sector Project), (2006). *Annual Progress Report of Community Groundwater Irrigation Sector Project,* Community Groundwater Irrigation Sector Project/Groundwater Resources Development Project/Department of Irrigation/Government of Nepal, Lalitpur, Nepal.

DOI (Department of Irrigation), (2006). *DOI Internal Reports,* Department of Irrigation, Government of Nepal, Lalitpur, Nepal.

GDC (Groundwater Development Consultants Ltd) (1994). *Reassessment of the Groundwater Development Strategy for Irrigation in the Terai.* Vol. 1, Main Report, His Majesty's Government/Department of Irrigation/Groundwater Resources Development Project, Nepal.

GDC, HTS and East Consult (Groundwater Development Consultants Ltd., Hunting Technical Services Ltd. and East Consult (P) Ltd), (1997). *Community Groundwater Irrigation Sector Project (TA 2589–NEP).* Asian Development Bank/His Majesty's Government of Nepal/Department of Irrigation, Lalitpur, Nepal.

IWMI (International Water Management Institute), (1999). IWMI *Annual Report 1999.* International Water Management Institute, Colombo, Sri Lanka.

Karki, L. and Basnet, B.B., (2004). *Land Ownership and Livelihood: A report on Dangihat VDC, Morang.* Buddha Air (P) Ltd. (in Nepali).

Koirala, G.P., (1998). *Clogs in Shallow Groundwater Use.* Research Report Series No. 39, Winrock International, Kathmandu, Nepal.

NPC (National Planning Commission), (2002). *Tenth Plan (2002–2007).* National Planning Commission/His Majesty's Government of Nepal, Kathmandu, Nepal.

NPC (National Planning Commission), (2005). *National Living Standards Survey 2003–2004.* His Majesty's Government of Nepal/National Planning Commission/ Central Bureau of Statistics, Kathmandu, Nepal.

Shah, T., A. Deb Roy, A.S. Qureshi, and J. Wang, (2003). Sustaining Asia's Groundwater Boom: An overview of Issues and Evidence. *Natural Resources Forum, 27, 130–141.*

Shrestha, J.L. and Uprety, L.P., (1995). *Study of Group Shallow Tubewell Irrigation.* 1995. Poverty Alleviation Project in Mid and Far Western Terai. FAO/UN.

WECS (Water and Energy Commission Secretariat), (2002). *Study on the Optimum Use of Tubewells for Irrigation in the Terai Districts of Eastern and Central Development Regions,* IRDS, Kathmandu, Nepal, January 2000.

WECS (Water and Energy Commission Secretariat), (2005). *National Water Plan* (2002–2027). His Majesty's Government of Nepal/Water and Energy Commission/Water and Energy Commission Secretariat, Kathmandu, Nepal.

CHAPTER 10

Agricultural groundwater issues in North China: A case study from Zhengzhou Municipal Area

R. Sun & Y. Liu
Department of Hydrogeology, China University of Geosciences, Wuhan, Hubei, China

Y. Qian
Hydrology Bureau, Yellow River Conservancy Commission, China

K.G. Villholth
Geological Survey of Denmark and Greenland (GEUS), Copenhagen, Denmark

ABSTRACT: Zhengzhou Municipal Area is a water-short region in north China. Irrigated agriculture mainly relies on groundwater, but development of groundwater has caused a series of environmental problems, such as groundwater table decline and groundwater pollution. The objective of this paper is to present the groundwater resources conditions, its socio-economic impacts and the policy-institutional options in Zhengzhou. Based on field surveys of 22 villages in Xinmi County and Xingyang County in Zhengzhou, the results indicate that water shortage has impacted the development of agriculture and the living standard of the farmers. Deep tubewells are replacing open dugwells and shallow tubewells in most villages, which has increased the cost of irrigation and had a negative impact on the income of the farmers. The ownership of tubewells was mainly collective. The high cost of drilling deep tubewells and lack of demand for buying water has limited the development of individual tubewells and water markets. Advanced water saving technologies have been adopted by only a few villages and water use efficiency in irrigation is low. Farmers have little incentive to adopt advanced water saving technologies because of high cost, small and fragmented landholdings, and low profitability of farming. Specific and practical laws, regulations and policies on groundwater management are lacking.

1 INTRODUCTION

Groundwater has played an increasingly important role in agriculture, domestic and industrial water supply in North China. With the rapid development of the economy and society, water demand has been rising sharply. Water shortage has become a major factor impeding further development of agriculture and the improvement of living standards for the farmers (Li, 2002).

The use of groundwater for irrigated agriculture in China has a very long history right from ancient times. Agricultural irrigation has reduced poverty in China by directly boosting yields and giving farmers food security (Pei, 2003). In the North China Plain, groundwater is the most important source for industrial, agricultural and domestic water use. In recent years, over-exploitation and poor management of groundwater has resulted in several environmental problems, such as groundwater table decline, land subsidence and groundwater pollution (Xia, 2002).

Zhengzhou Municipal Area (called Zhengzhou, Capital of Henan province) is located in the lower reaches of the Yellow River Basin (Fig. 1). Zhengzhou is a water-scarce region in north China and water resources per capita are less than 230 m^3/year (Wang, 2001).

Development of the economy of this area mainly depends on groundwater. According to the comprehensive evaluation report of water resources and its development in Zhengzhou made by the Water Resources Bureau of Zhengzhou (WRBZ, 2006), groundwater use accounted for nearly 92% of the total water use and 90% of the agricultural water use in 2000.

Development of groundwater and primary dependence on it for irrigation makes ground-water governance a challenge in Zhengzhou region. Perhaps for this reason, governance of water resources has up till now focused on surface water and the effort on managing groundwater has been minimal. Although groundwater governance is starting to receive attention from the government, it is so far mainly restricted to urban areas.

Traditionally, managers of groundwater resources have been mostly hydrogeologists, who may have limited knowledge on the socio-economic aspects of groundwater use. Similarly, the few social scientists who work on groundwater issues have limited under-standing of the science of hydrogeology, so a holistic understanding of inter-disciplinary perspectives is needed. For better governance, an understanding that integrates groundwater resource conditions, socio-economic perspectives, and policy and institutional perspectives is necessary.

This study was undertaken to integrate groundwater resource conditions, with socio-economic and policy-institutional perspectives on groundwater governance in Zhengzhou to elevate the current understanding and to promote sustainable development of groundwater resources in the future.

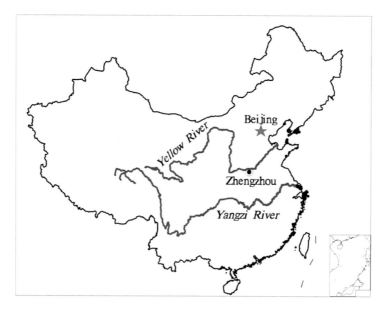

Figure 1.   The location of Zhengzhou in China.

2   METHODOLOGY

The study was based on analysis of existing secondary data on the Zhengzhou area as well as a field survey conducted in December 2006 and January 2007 among local farmers and officers charged with groundwater management to reveal the intricacies of groundwater use and management.

The secondary data collection consisted of a search from reports, news and literature from the Internet to gather information on weather, population, social and economic information, and water usage and management of groundwater. Data on hydrogeology were collected from the local officers in charge of water resources including the Water Resources Bureau of Zhengzhou (WRBZ) and those of Xinmi and Xingyang. The policy and institutional perspective was analyzed based on semi-structured interviews with government officials in the Water Resources bureaus of Zhengzhou, Xinmi and Xingyang.

Prior to the field survey, the Assistant Section Chief of WRBZ was interviewed and material relating to groundwater was collected from the local officers in charge of ground-water resource management, agricultural irrigation, and water conservation. Based on a good preliminary understanding of the hydrogeological and socio-economic conditions, two representative counties were selected as survey regions, Xinmi and Xingyang Counties (994 km^2 and 980 km^2, respectively) (Fig. 2).

For both of the counties, the main irrigation water source is groundwater. Xinmi County is mountainous while Xingyang County is plain terrain. For each county, 10 villages were subsequently selected randomly for the actual survey. For every village, one village leader and two tubewell managers (all managers are also farmers) were selected randomly to undergo a questionnaire survey about agricultural groundwater irrigation.

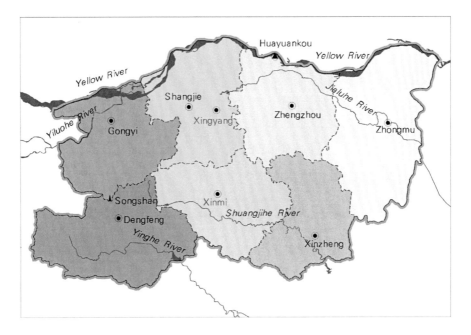

Figure 2.   The location of Xinmi and Xingyang County.

During December 2006, five semi-structured interviews with government officials from the Water Resources Management office and the office of Irrigation, Drainage and Rural Water Supply of WRBZ, Water Resources Bureau of Xinmi, and Water Resources Bureau of Xingyang were carried out to obtain information on local groundwater use and management. Detailed economic and social data were collected from the Statistical Yearbook of Zhengzhou, Xinmi and Xingyang.

The groundwater resource was estimated from groundwater budget calculation. The groundwater development was analyzed based on the data derived from the local officers and interviewees. Through analysis of the village leader questionnaires, the cropping patterns, use patterns of groundwater for irrigation and the overall socio-economic role that groundwater plays in agriculture in Zhengzhou was assessed. Based on the tubewell managers questionnaire, information on use of groundwater, groundwater irrigation technologies, water saving irrigation technologies and their adaptation and application, well depths and changes and the type and management of tubewells were obtained. Through interview with government officials, the laws and regulations and their application were evaluated. Finally, based on an inter-disciplinary analysis, an integrated understanding of the complex groundwater governance system was synthesized.

3   GENERAL DESCRIPTION OF THE STUDY AREA

The total area of Zhengzhou is $7,446$ km^2, of which 13.6% ($1,010$ km^2) is urban. According to Zhengzhou Statistical Yearbook, in 2005, the total population in Zhengzhou was 7.2×10^6, of which the rural population accounted for 62.8% (BSZ, 2006). In Xinmi County and Xingyang County, the total population is 0.8×10^6 and 0.6×10^6, of which the rural population accounts for 82% and 84%, respectively. The total GDP of Zhengzhou is 21.5×10^9 US $, of which only 4.4% derives from agriculture. The total GDP in Xinmi City is 2.1×10^9 US $, of which only 3.3% is from agriculture. The total GDP in Xingyang City is 1.9×10^9 US $, of which only 8.7% is from agriculture. Thus the agricultural economy, despite involving a majority of the population and a high fraction of the water use (Table 2), accounts for a low fraction of the total GDP in Zhengzhou.

As a whole, the elevation of the study area decreases from southwest to northeast. The highest peak is the Shaoshi Apex of Songshan Mountain with an elevation of 1494 m. The elevation is less than 200 m in the eastern plain. This area has continental monsoon climate. The average annual temperature is about 14.3°C . The mean annual precipitation of Zhengzhou is 641 mm and ranges from 751 mm in the south to 547 mm in the north, 65.0% to 67.2% of which is concentrated between June and September (Fig. 3). The mean annual evaporation is 1016 mm in Zhengzhou.

There are 124 rivers in the whole Zhengzhou area. The Yellow River is the largest (running across the northern border of the area) with a mean annual runoff of 44.4×10^9 m^3 at Huayuankou hydrologic station (Fig. 2). There are many reservoirs, of which 13 reservoirs are of medium size with storage of more than 10×10^6 m^3. Most surface water is of poor grade quality (grade V or below V in Chinese terminology, of which grade I is best and grade V is only applicable to agriculture irrigation). The surface waters have been polluted badly because of direct discharge of industrial and domestic wastewater, making groundwater even more important as source of water.

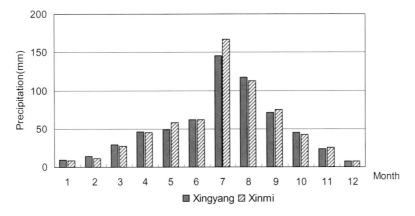

Figure 3.   Average precipitation per month in Xinmi and Xingyang County (1956–2000).

4   RESOURCE PERSPECTIVE

4.1   *Hydro-geologic conditions*

In Zhengzhou, rock outcrops from Archaeozoic to Neogene age dominate in the south and west mountain area (approximately 50% of the area) and Quaternary unconsolidated materials cover the north and eastern plains (Fig. 4).

There are three groundwater provinces, karstic water in Carbonate systems, water in fractured systems in hard rock and clastic rock, and groundwater within the Quaternary unconsolidated material. Fractured rock systems generally have a deep groundwater table and are relatively difficult to exploit. Highly porous Quaternary aquifers are widely distributed in the eastern plain.

The porous Quaternary aquifers are the most important in Zhengzhou. They can be divided into four more or less overlying sub-aquifers: shallow aquifer (unconfined aquifer less than 60 m deep), middle aquifer (confined aquifer 60 to 300 m deep), deep aquifer (confined aquifer 300 to 800 m deep) and super-deep aquifer (confined aquifer deeper than 800 m) (Li et al., 2005). The shallow aquifers have abundant groundwater because they can be easily recharged by precipitation, irrigation, rivers, and reservoirs. The confined aquifers, on the other hand, have abundant groundwater because of their huge extent and volume.

The tubewells for agriculture are mostly located in the eastern plain. Because of the superposition of porous aquifers, the tubewells almost always have several screens that are located in different sand or gravel layers.

4.2   *Groundwater resource*

A groundwater budget calculation for the Zhengzhou area shows that during 1980 to 2004, the mean annual groundwater recharge was $933.3 \times 10^6 \text{m}^3/\text{y}$ and the total discharge $1099 \times 10^6 \text{m}^3/\text{y}$ (Table 1). The groundwater budget calculation shows that ΔS, the difference between recharge and discharge, is a deficit which means that the groundwater discharge is higher than groundwater recharge, i.e. groundwater is in a state of overexploitation.

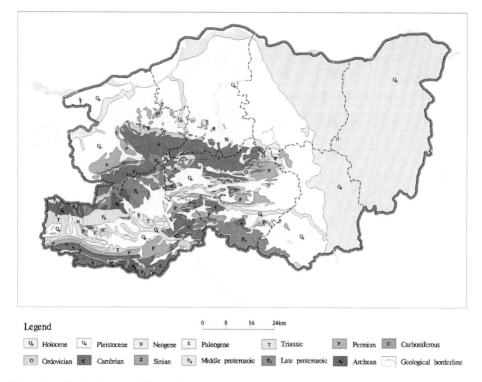

Legend

	0 8 16 24km

Q_h Holocene	Q_p Pleistocene	N Neogene	E Paleogene	T Triassic	P Permian	C Carboniferous
O Ordovician	$\in$ Cambrian	Z Sinian	P_t Middle proterozoic	P_t Late proterozoic	A Archean	Geological borderline

Figure 4.   Geologic map of Zhengzhou (See colour plate section).

Table 1.   Groundwater balance in the Zhengzhou Area (1980–2004).

Area (km^2)	Recharge (10^{6}m^3/a)						Discharge (10^{6}m^3/a)					ΔS
	R_{rain}	R_{riv}	R_{res}	R_{irr}	R_{lat}	Total	D_{eva}	D_{riv}	D_{lat}	D_{exp}	Total	
7446.2	620.5	115.8	22.2	113.2	61.6	933.3	76.2	49.4	10.6	962.8	1099.0	−165.7

Note: R_{rain} = recharge from rainfall, R_{riv} = recharge from rivers; R_{res} = recharge from reservoirs, R_{irr} = recharge from irrigation, R_{lat} = subsurface boundary flank inflow, D_{eva} = evaporation discharge, D_{riv} = groundwater discharge to river, D_{lat} = subsurface boundary flank outflow, D_{exp} = exploitation of groundwater, Δs is the change in groundwater storage.

In 2004, the total groundwater exploitation is 962.8×10^6m^3, which account for 103.2% of the mean annual recharge.

4.3   Development of groundwater resources

According to the comprehensive evaluation report of water resources and its development in Zhengzhou (WRBZ, 2006), groundwater contributed 62% of the total water use in the Zhengzhou area in 2000, and accounted for 92% and 74% of the total water use in Xingyang County and Xinmi County, respectively (Table 2).

Table 2. Water use in Zhengzhou in 2000.

Region	Total water use			Agricultural water use			
	Total (10^6m^3/y)	Groundwater (10^6m^3/y)	Groundwater ratio %	Total (10^6m^3/y)	Agriculture ratio %	Groundwater (10^6m^3/y)	Ground water ratio %
Xingyang	143.6	131.7	91.7	81.5	56.8	73.4	90.1
Xinmi	84.6	62.3	73.6	38.0	45.0	23.5	61.7
Zhengzhou	1448.8	890.9	61.5	797.9	55.1	542.7	68.0

Groundwater exploitation has taken place since the 1960s and the exploitation has accelerated since the 1980s. According to statistical data from WRBZ there were a total of 42,763 tubewells for various purposes in urban and rural areas in Zhengzhou in 2000, which means that there are on average 5.7 wells per km^2. Of the total groundwater use in Zhengzhou, about 43.7% was from shallow aquifers and 56.3% from middle and deeper aquifers. However, 93.8% in Xingyang County and 94% in Xinmi County of groundwater use was from the middle and shallow aquifers in 2000.

4.4 *Problems caused by overuse of groundwater*

4.4.1 *Decline of groundwater table*

In recent years, groundwater has been overexploited resulting in a decline in the overall groundwater table elevation. According to the comprehensive evaluation report of water resources and its development in Zhengzhou (WRBZ, 2006), there were 14 major cones of depression in the area in 2005, of which the two largest are in Zhengzhou City with an area of 153.7 km^2 in the shallow aquifer and 146.2 km^2 in the middle and deep aquifer (affecting nearly 30% of the city area). Table 3 shows the area of the cones of depression in the shallow and deep aquifers in Zhengzhou city and the two study areas. Large-scale drawdown of the groundwater table has led to significant land subsidence in the central areas of Zhengzhou city.

The major cones of depression are mainly in the urban areas. According to the survey, the regional groundwater table in rural areas is also declining and many dug wells and shallow tubewells in the countryside are being abandoned.

4.4.2 *Groundwater pollution*

Groundwater has been polluted by industrial wastes, fertilizers and pesticides in many parts of the Zhengzhou area. Polluted water has been threatening the safety of drinking water. For example, for the Shuangjihe River in Xinmi County, monitoring data in 2003 showed that chemical oxygen demand, COD, is 203 mg/l and ammonia nitrogen is 6.53 mg/l, well above the permissible limits. Because industrial wastewater has been directly discharged into the Shuangji River, the water has been badly polluted and fish death has occurred. The adjacent groundwater has also been polluted as the river recharges groundwater.

According to the report on drinking water safety in rural areas of the Zhengzhou Municipality (WRBZ, 2005), 1.5 million farmers, or 38% of all farmers, do not have access to safe water. In order to improve the health of people in certain areas with low quality water, the government initiated a project on ensuring safe drinking water in rural areas of Zhengzhou in 2005. In this project, 63 new public deep tubewells were drilled in 2006 with government funds to supply groundwater of potable quality.

Table 3.   Occurrence and extent of cones of groundwater depression.

Region	Type of aquifer	Area of cone (km^2)	Equivalent radius (km)
Zhengzhou	Shallow	153.2	7.0
	Deep	146.2	6.8
Xingyang	Shallow	28	3.0
	Shallow	26.2	2.9
	Deep	11.3	1.9
Xinmi	Karst	62.7	4.5
		21.4	2.6
		30	3.1

Survey results also showed that farmers have no clear understanding of groundwater resources and its behaviour. For example, when 43 tubewell managers were asked the question "Where does groundwater come from?", most of them answered "I do not know".

5   SOCIAL AND ECONOMIC PERSPECTIVE

5.1   *Income of farmers*

In recent years, the income of farmers in Zhengzhou has increased, from a mean annual income per capita of farmers of 408 US $ in 2001 to 617 US $ in 2004. But agriculture is not always the main income of farmers. According to the survey in 22 villages, the percentage of crop income to the total income of farmer families has decreased in most villages, on average over the villages from 58% to 47% over the period 1995 to 2005 (Fig. 5).

From the survey among the 22 village leaders, it was found that the mean income per capita does not increase with landholding size (Fig. 6). This is because crop cultivation is not the main source of income for farmers now. They have other income sources, such as a business in the local village or town, local factories, or work in another province or city. According to the current statistics, nearly 0.13 million of farmers in 2004 and nearly 0.14 million of farmers in 2005 left agriculture for urban jobs.

5.2   *Development of groundwater irrigation*

In 2000, the use of irrigation water in Zhengzhou was $6.59 \times 10^8 m^3$, of which groundwater use accounted for 68% (Table 2). As an example, Xingyang County will here be used to analyze the development history of groundwater in Zhengzhou.

Groundwater is the most important water source. More than 90% of the total water use is from groundwater in Xingyang County (Table 2). During 1993 to 1997, groundwater exploitation was partly related to the available precipitation. The year 1997 was a dry year (low precipitation of 293.4 mm). Large amounts of groundwater was pumped for industry and irrigation which resulted in the groundwater table decreasing rapidly and many tubewells being abandoned (WRBX, 1999; WRBX, 2004).

From 1993 to 1997, groundwater exploitation increased rapidly. In the same period, agricultural water use proportion decreased and industrial water use increased to the point that in 1997, industrial water use was higher than agricultural use. Since 1997, industrial

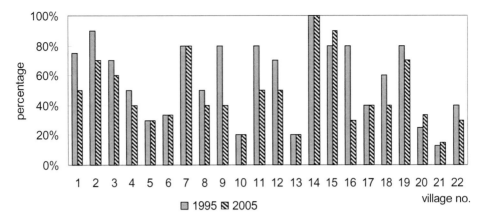

Figure 5.   The average ratio of cropping income of farmers to their total income.

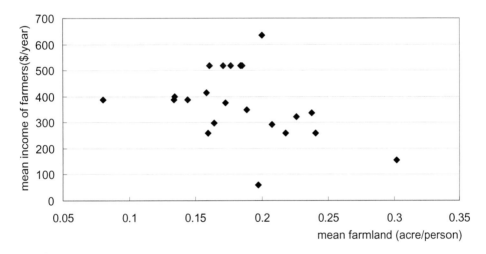

Figure 6.   Income and farmland of farmers in Zhengzhou, in 2005.

groundwater use decreased due to significant water saving measures in the industrial sector, until it again became lower than agricultural use (Fig. 7).

5.3   *Cropping pattern, cost and income*

According to Zhengzhou statistical yearbook, arable area in Zhengzhou was 332×10^3 ha of which 54.6% was irrigated in 2005. The total crop area was 526×10^3 ha. The main grain crops include wheat, maize and soybean (Table 4). Here the crops are mainly sold in the local market from where part of it is transported to other provinces or exported.

The crop net income for the last years was analyzed based on farmer's questionnaire. We found that farmers achieved high crop productivity due to favourable weather conditions. The crop input mainly includes irrigation, seed, fertilizer, pesticides and others (Table 5).

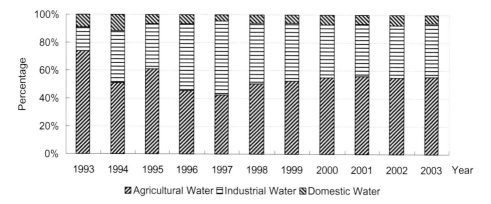

Figure 7.   Groundwater's share between various sectors in Xingyang County.

Table 4.   Cropping patterns and productivity in Zhengzhou in 2005.

Region	Arable area (10^3 ha)	Irrigation area (10^3 ha)	Total crop area (10^3 ha)	Wheat		Maize		Soybean	
				Area (10^3 ha)	Productivity (kg/ha)	Area (10^3 ha)	Productivity (kg/ha)	Area (10^3 ha)	Productivity (kg/ha)
Zhengzhou	330.7	180.4	516.8	177.0	4085.9	135.6	4719.4	15.1	1923.9
Xingyang	47.2	27.3	73.2	29.0	4699.4	20.8	5243.8	1.1	1378.8
Xinmi	47.0	16.0	66.4	28.1	3341.2	23.3	3926.6	2.1	1575.4

We find that cropping costs account for 43% of the gross cropping income. The net income of cropping is relatively low. Farmers are engaging in other economic activities to survive and they have little incentive to improve their farming activities.

5.4   Development, ownership and management of tubewells

5.4.1   Development of tubewells
In the 1980s, in Xinmi County nearly all irrigation wells were shallow wells, 20–30 m deep with a standing water level 10–20 m down. At present, excluding areas near to both banks of the Zhenhe River, which is a perennial river with good water quality, irrigation wells are 40 m deep. In other areas, groundwater quality is not good and nearly all the irrigation wells drilled before 1980 have been abandoned. New tubewells are 80–100 m deep and many of them are deeper than 100 m.

In Xingyang County, from 1993 to 2003, the percentage of deep tubewells (>60 m deep) increased and the percentage of shallow tubewells (<60 m) decreased (Fig. 8). Especially in 1997, low precipitation and development of groundwater made a large amount of shallow tubewells dry and obsolete. From 1998 to 2003, the total amount of tubewells increased, because of deep tubewell installation.

5.4.2   Investment in tubewells
According to the survey, in recent years, the capital cost of a new deep tubewell (nearly 100~120 m deep) was on average 5,700 US $. The electricity cost of running the tubewell in

Table 5. Average Crop cost and income per ha in Xingyang and Xinmi County in 2006.

Crop	Productivity (kg)	Crop price (US $/kg)	Gross income (US $)	Cropping cost Irrigation (US $)	Seed (US $)	Fertilizer (US $)	Pesticide (US $)	Others (US $)	Total (US $)	Net income (US $)
Wheat	6200.2	0.2	1075.2	59.0	53.1	210.9	32.6	112.4	469.3	604.7
Maize	6876.0	0.1	1003.8	44.5	60.3	190.4	30.4	104.5	432.3	571.6

Note: Income does not includes the part of the wheat and maize eaten by farmers themselves.

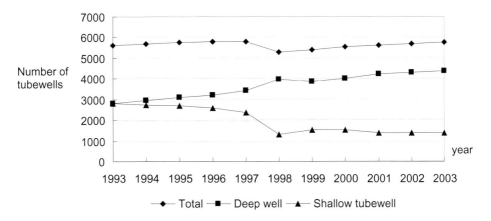

Figure 8. Changing patterns in type of wells.

2005 was on average 740 US $/yr, and the average maintenance cost of running the tubewell was 210 US $/yr. Comparing these costs and also comparing with the average annual net income of each individual farmer from cropping (1100 US $/yr/ha for two crops, Table 6) this is a very large investment.

According to the survey of 43 tubewells in Xinmi and Xingyang County, the investors in tubewells include: government, village or group, groups of farmers and individual farmers. In order to help farmers escape poverty, the government has invested in new tubewells during recent years, depending on the strategy and fiscal capacity of local government. For example, the industry of Xinmi is better developed than the one of Xinyang. So, in most villages in Xinmi County, many tubewells are funded partly or fully by the upper government through special projects. In contrast, in Xingyang County, there are few tubewells funded by the upper government.

5.4.3 *Ownership of tubewells*
There are three kinds of ownership of tubewells, collective, shareholding and individual ownership. It mainly depends on the investor of the tubewell. The investment for a collective tubewell is by the government or all the farmers in the collective, normally an entire village. The investment for a shareholding tubewell is by a smaller group of farmers and for an

Table 6.   Ownership types of tubewells in Xinmi and Xingyang County.

| | Ownership type | | |
	Collective	Shareholding	Individual
Investor	Government or all farmers in collective	A smaller group of farmers	Individual farmer
Management	Some farmer in collective	Shareholder in turn	Individual
Maintenance	All farmers in collective	All the shareholders	Individual
User	All farmers in collective	Shareholder, other farmers	Individual, other farmers

individual tubewell is by the individual farmer. The difference among the three ownerships of tubewells is summarized in Table 6.

The survey results in Xinmi and Xingyang County showed that nearly all tubewells were collective. In 2005, there were 535 tubewells in the 22 villages surveyed in the two counties and collective tubewells accounted for 95.9%.

Why do farmers opt for having collective rather than individual wells? The interviews and analysis revealed the following reasons:

1. Deep groundwater tables and high investment cost of tubewell installment. Investing collectively in a well is more economically feasible to most farmers.
2. Small and fragmented landholdings: In China, the land belongs to the state and the farmer only has the right of use. The land is distributed to every family according to land area and numbers of persons in the village. There is not much variability in the size of landholdings between families in the two counties. Arable land per person is less than 0.06 ha in most villages of Xinmi and Xingyang, according to the survey. That is to say, one family of five persons can have about 0.3 ha arable land. On the other hand, the land of one family is often divided into small parcels scattered in different places because they have to share different quality land with other farmers. This condition limits the farmers' interest in individual tubewells. The collective system, involving use from multiple wells by each farmer, ensures that he has more equitable and adequate access to water for his land fragments.

These are also the reasons why no groundwater market exists in Zhengzhou. The capital cost of water extraction mechanisms is excessive for individual farmers. Moreover, people cannot benefit from selling water because the wells and the water are not individually owned.

5.4.4   *Management mode of tubewell*
In Zhengzhou, management of tubewells includes pumping operation, irrigating farmers' land, collecting electricity charge for pumping from farmers, and maintenance of tubewells.

5.4.4.1   Collective tubewells
Nearly all the tubewells are installed by the farmers themselves. The farmers can be assigned by village leaders or selected by other farmers to become tubewell managers. One manager can manage one tubewell or multiple tubewells. According to the survey, 43 managers

manage 109 tubewells in Xinyang and Xinmi County. Of these, 99 collective tubewells are managed by 37 farmers.

An agricultural groundwater fee, payable to the electricity boards is charged by the tube-well managers according to the electric power consumption. In Zhengzhou, farmers need not pay a water resources fee. They only pay the electric charge for pumping. In 2005, the electricity tariff for irrigation was 0.087 US $/kwh in Xinmi County and 0.078 US $/kwh in Xingyang County. The tariff for irrigation is higher than for domestic use and lower than for industrial use. Generally, the tubewell manager will determine irrigation order priority among the farmers, mainly according to the order of receiving the farmer's request during irrigation, then the tubewell manager will note the electric power consumption of every family.

In addition, an operation and maintenance fee for collective tubewells is paid collectively and equally by the farmers to the tubewell managers. There are two kinds of payment, cash and farmland. Payment of cash is most common. Through overall discussion of all the villagers, tubewell managers decide on an additional charge for their efforts, over and above the electric charge. During the survey, the additional charge was only 0.0025 to 0.004 US $/kwh. For the payment in farmland, the mangers can get a small piece of farmland from the village, which is nearly 0.06 ha. As the benefit and compensation from managing tubewells are very low and the management is troublesome, most of farmers are not willing to manage tubewells. Now, some managers are village leaders and they are volunteers. Some managers are poor or old farmers and they get marginal benefit from managing tubewells.

5.4.4.2   Shareholding tubewells and individual tubewells

Shareholding tubewells are managed by the shareholders in turn. According to the survey, 10 shareholding and individual tubewells are managed by 6 farmers. The shareholding tubewells or individual tubewells can also irrigate other farmers land.

For shareholding tube wells, the additional charge for management is relatively higher than collective tubewells. But it is less than 0.052 US $/kWh. First, the income from the additional charge is used to maintain the tubewell. Then the surplus income is paid equally to every shareholder.

5.5   *Coping strategies in face of groundwater depletion*

As discussed in Section 4.4, farmers have no clear understanding of groundwater and its behaviour. About the coping strategies in face of groundwater depletion, all of the 17 managers said they would drill new tubewells. In addition, 11 managers said they would change the pump to a more powerful one, and only 2 managers said they would adopt water saving irrigation. Adjusting cropping pattern was not one of the reported options.

5.5.1   *Drilling deeper wells and abandoning shallow wells*

According to statistics in Xingyang County, nearly 100 wells were abandoned every year during 1993–2003. Especially in 1997, when precipitation was the lowest during 1993–2003 (293.4 mm) and a large amount of groundwater was pumped, 773 wells were abandoned in Xingyang because of groundwater table decline. Moreover, with the continued decline of the groundwater table, even more dug wells have been abandoned. Between 1993 and 2003, the numbers of new tubewells are higher than the number of abandoned tubewells per year.

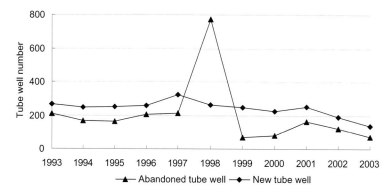

Figure 9.   Number of new tubewells and abandoned tubewells per year in Xingyang County.

But farmers are increasingly reluctant to replace abandoned wells because of increasing costs, although now the rate of increase of new tubewells is declining (Fig. 9).

5.5.2   *Adoption of water saving technologies*

According to our survey, traditional surface irrigation technologies have been widely adopted. The methods include land leveling, furrow irrigation, flood irrigation and border irrigation. These are basic irrigation methods with relatively low cost, ease of use and adoption for the farmers, but water use efficiency is very low.

Since the early 1990s, all levels of government realized the importance of water saving and paid attention to extend advanced water saving and irrigation technologies. With the support of the government, some modern water saving and irrigation technologies were introduced into Zhengzhou area.

Most of the villages have adopted the conservation tillage (or protective tilling) method since 1990. This method improves soil fertility and plant water availability, water productivity, and reduces water surface evaporation losses. In order to reduce water seepage losses and improve conveyance and distribution efficiency, underground pipe distribution systems instead of earth canals were built in most villages of Xinmi County with the financial support of the upper government. In Xingyang County, some of the villages also apply the pipe conveyance technology (surface pipe and underground pipe), but financial resources mainly come from raising funds by the collective in the village rather than the upper government.

To promote the more advanced technologies, a few demonstration zones were set up from 2000, for example to demonstrate rainwater harvesting in hilly areas in Xinmi and Xingyang, and sprinkler and drip irrigation were also demonstrated in some areas. In these demonstration zones, the main crop is a cash crop rather than grain crops and farmers receive a higher income. Progress has been achieved in implementing water saving technologies, but there are still many problems or challenges:

1. High cost of advanced irrigation technologies compared to farming income: sprinkler irrigation and drip irrigation and plastic film irrigation are the most efficient technologies, but these technologies are expensive. According to surveys, the cost of surface pipe and underground pipe systems are 58 - 78 US $/ha and sprinkler irrigation needs 1575–1960 US $/ha. In general, the farmers get a net income of about 1100 US

$/ha/year from cropping. In Zhengzhou, there is almost no water market and farmers cannot get much benefit from water saving.

2. Small and scattered farmland: arable land per capita is less than 0.06 ha in most villages of Xinmi and Xingyang. This condition limits the adoption of some water saving techniques, such as sprinkler irrigation. Scattered farmland also limits the incentive of farmers to change cropping patterns.

3. Lack of government attention: local governments do not pay much attention to the extension of advanced water saving technologies. Adoption of new technologies needs much money, but agriculture accounts for very low GDP.

Lack of public funds to extend modern irrigation technology: modern irrigation technologies are often adopted in demonstration zones with limited area. It requires extended funds to promote the application of new advanced water saving technologies in larger areas. Lack of government attention results in lack of a public fund.

5.5.3 *Raising price of water*
Farmers would adopt new technologies if the government paid the capital cost and some researchers think the water price is to low for sustainable irrigation in rural areas. Raising the water price may be an effective approach to resolving water scarcity problems. Raising the water price would force farmers to reduce water use and move to adopt water saving technology.

Farmers only pay the electric power for pumping, which accounts for nearly 10% of the total crop cost. That implies that the water price (cost of pumping) is already expensive for them. If cost of irrigation water rises, it will have a negative impact on farmers, as their net income would decrease. If the water price goes up, most respondents answered that they would not adopt more expensive water saving technologies if they pay for the technology themselves because they do not have enough money and it would not guarantee them increased income.

6 POLICY AND INSTITUTIONAL PERSPECTIVES

6.1 *Groundwater institutions*

Before 1998, groundwater management and surface water management belonged to different public entities. In 1998, groundwater and surface water management were united and now belong to the Ministry of Water Resources. The Water Resources Bureau of Zhengzhou (WRBZ) is responsible for overall water resources management in Zhengzhou Municipal City. Its subordinate water supply and water-saving office is charged with groundwater management.

6.2 *Laws and regulations on groundwater*

6.2.1 *Laws issued by the chinese government*
In China, groundwater belongs to the nation. Until now, there is no specific law pertaining to groundwater. Laws involving groundwater aspects are generally abstract. Laws and regulations formed by Chinese government with reference to groundwater are as follows:

1. Water law of the People's Republic of China
2. Environmental protection law of the People's Republic of China

3.   Law of the People's Republic of China on the prevention and control of water pollution
4.   Flood control law of the People's Republic of China
5.   Law of the People's Republic of China on water and soil conservation
6.   The administration of water abstract licensing and collection of water resources charges.

6.2.2   *Regulations or policies issued by Zhengzhou municipal government*
Since 1994, 14 regulations or policy documents have been issued by the Zhengzhou government. Of these, six laws are related to agricultural water use. But these six laws only mention water saving irrigation and pay no special attention to groundwater governance except the regulation on water abstract licensing.

6.2.3   *Implementation of water abstraction licensing*
On 1 August 1993, the Chinese government issued a regulation on the implementation of water abstraction licencing, according to which, groundwater users should apply water abstraction licensing when they want to abstract groundwater. But the regulation emphasized that users of groundwater for irrigation and those who are coping with a drought situation need not apply. On 21 February 2006, the Chinese government issued a new regulation: "the administration of water abstraction licensing and collection of water resources charges". The new regulation only relaxes the condition for obtaining a permit under drought situation. The difference between the two regulations showed that farmers now need to apply for license when they drill new tubewells for agriculture. But it is not being enforced for irrigation tubewells. In Zhengzhou city none of the irrigation tubewells have water abstraction licensing. The government officials told us that groundwater users for irrigation are not required to apply the water abstraction license and they seemed unaware of the new regulation.

6.3   *Vision and strategy of government*

The semi-structured interviews with the groundwater officials indicated that they have a good understanding about groundwater governance. They all agree that the previous emphasis of groundwater governance has been on urban groundwater users and not on the rural. For the sustainable groundwater governance, they think there are mainly three difficulties: insufficient funds, lack of attention of government and limited public awareness.

In general, groundwater governance has previously focused primarily on urban areas. This may be a reflection of larger interest and priority from the policy makers to promote economically viable cities. Drawdowns of groundwater is more pronounced in cities but declines are also evident in most rural areas. For example, water abstraction licensing regulation has been strictly enforced in Zhengzhou City, where drilling of new tubewells is strictly prohibited since recent years. In contrast, there is no limitation for drilling new tubewells for irrigation.

With lack of government attention, the funds invested by the government is limited. In Xingyang, groundwater monitoring is minimal with less than one monitoring well per town. Moreover, the monitoring equipment is ageing. According to Guo Xiaoguang (the Section Chief of Water Resources Management Office, Water Resources Bureau of Xinmi, personal communication), since 1980, not a single water resources investigation has been carried out. But sustainable groundwater governance depends on a clear understanding of the hydrogeological conditions and this is lacking.

Government seems to recognize the importance of groundwater governance. But because groundwater governance is very complex, it will need a long time to achieve good groundwater governance.

7   CONCLUSION: INTEGRATING THE THREE PERSPECTIVES

This paper has analyzed and discussed the three aspects related to groundwater management in Zhengzhou: the groundwater resources, the socio-economic and the policy and institutional perspective. Overall conclusions are the following:

1. Importance of groundwater: groundwater is a very important water resource for agriculture, domestic and industrial water supply in Zhengzhou area. It accounted for 91% and 81% of the total water use in two study regions, Xingyang City and Xinmi County, respectively.
2. Positive impacts of groundwater development on agriculture: the use of groundwater for irrigated agriculture in Zhengzhou has a long history. The groundwater has accounted for 68% of the total agriculture water use in recent years. Irrigation development increases the productivity of crops and the income of farmers.
3. Environmental, health, and geotechnical problems generated by over-exploitation and degradation of groundwater: groundwater is playing an increasingly important role, but, over-exploitation of groundwater has resulted in decline of the groundwater table and groundwater pollution. Drawdown is more pronounced in cities due to concentrated use, and pollution problems tend to affect the rural population more due to predominant shallow drinking water wells. These problems have negated the positive impacts on development of agriculture and improvement of living standard of the farmers. For example, the need for abandoning shallow wells and drilling deeper and deeper wells for irrigation has increased the cost of irrigated agriculture and water supplies.
4. The net income of farmers from irrigated agriculture is low: furthermore, the fraction of their total income from agriculture is declining, which is giving farmers little incentive to optimize irrigation water use. Farmers are diversifying their income and alternative income sources from urban jobs are becoming important coping mechanisms.
5. The ownership of most tubewells is collective and there are few private tubewells in the survey region. High costs of drilling, small and scattered farmland are the main reason for not having many privately owned shareholding or individual tubewells. The cost of drilling a tubewell is becoming increasingly prohibitive for the farmers due to deeper groundwater table and low incomes of most farmers.
6. There is almost no groundwater market in the survey areas: this is because the wells and the water are not individually owned, water use is inexpensive, and access to wells and groundwater is equitable. Farmers need not pay a water resources fee, only the electricity charge for pumping. Farmers cannot benefit from selling water, so it limits the development of individual tubewells and water markets.
7. Lack of adoption of advanced water saving technology: advanced water saving technologies is too costly for the farmers to afford. Advanced water saving technologies can

save much water but the farmers cannot benefit much from them due to low profitability of farming. Water saving is a strategic policy and in the public good, so government should play a more important role in extension of modern irrigation technologies and associated cash cropping.

8. Specific and practical laws, regulations and policies on groundwater management are lacking. Institutional roles are not clearly defined and separated.
9. Limited public awareness on groundwater governance.

It is clear from the study, though focusing on rural use of groundwater, that the question of sustainable groundwater management in an area like Zhengzhou, is a composite and complex problem of managing and balancing water resources use in the urban as well as rural areas. Whether there will be relief for groundwater from less agricultural use in the future is not clear. Low profitability in agriculture gives little incentive in saving water in irrigation, but it is important to optimize the limited resources available and there is a major challenge in getting such a message across to policy makers. There is a long way to go to realize sustainable management of groundwater in agriculture in Zhengzhou.

ACKNOWLEDGEMENT

The authors gratefully acknowledges Dr. Menggui Jin, Dr. Aditi Mukherji, Ms. Marcia Macomber for supervision of the paper preparation, and Dr. Jinxia Wang of CCAP for suggestions on the field questionnaire, Dr. Han Qiankun in WRBZ for his guidance during the field survey and providing data, and the International Water Management Institute for the funds that supported this research.

REFERENCES

BSZ (Bureau of Statistics of Zhengzhou City). 2006. Zhengzhou Statistical Yearbook 2006, China Statistics Press, Beijing. (in Chinese)
Li, G., Wang, X., Guo Y., 2005. Concentrated groundwater development status of middle aquifer in Zhengzhou Municipal Area. Yellow River, 27 (5): 44–46. (in Chinese)
Li, X. 2002. Pressure of water shortage to agriculture in the north and northwest regions of China. Arid Land Geography, 25 (4): 290–295. (in Chinese)
Pei, Y. 2003. Irrigation development strategy of China for early 21st Century. Technology of Water Resources, 23 (3): 1–5. (in Chinese)
Wang, H. 2001. Problems and policy of water resources in Zhengzhou Municipal area. China Water Resources, 5: 62. (in Chinese)
WRBX (Water Resources Bureau of Xingyang). 1999. Xingyang Water Resource Statistical Yearbook (1993–1998). Xingyang Statistics Press, Xingyang. (in Chinese)
WRBX (Water Resources Bureau of Xingyang). 2004. Xingyang Water Resource Statistical Yearbook (1999–2003). Xingyang Statistics Press, Xingyang. (in Chinese)
WRBZ (Water Resources Bureau of Zhengzhou). 2006. Comprehensive evaluation report of water resources and its development in Zhengzhou. (in Chinese)
WRBZ (Water Resources Bureau of Zhengzhou). 2005. Report on drinking water safety in rural area of the Zhengzhou Municipality. (in Chinese)
Xia, J. 2002. A perspective on hydrological base of water security problem and its application in North China. Progress in Geography, 21 (6): 51–526. (in Chinese)

CHAPTER 11

Groundwater use and its management: Policy and institutional options in rural areas of North China

J. Cao
Center for Chinese Agricultural Policy, Chinese Academy of Sciences, Beijing, China

X. Cheng
Yellow River Conservancy Commission, Xinxiang City, Henan Province, China

X. Li
China Geological Survey, Baoding, Hebei Province, China

ABSTRACT: The overall goal of the paper is to identify the characteristics of groundwater use and explore effective management strategies that can better meet the challenges of sustainable groundwater use in rural China. The data used in this paper were collected from a field survey in 20 randomly selected villages in Hebei Province of north China in 2006. Results show that with increasing water scarcity, groundwater tables are declining by 0.8 m per year. Farmers have adopted various strategies in order to cope with declining groundwater tables. These include construction of deeper tubewells, adoption of water saving technologies, change in ownership pattern from collective to individual ownership and finally, the development of informal groundwater markets. The paper shows that the regulatory capacity of the government to control groundwater exploitation is weak. Finally, this paper suggests that comprehensive policies to control groundwater use in rural area are needed.

1 INTRODUCTION

Water scarcity is one of the key problems affecting agricultural production in northern China—an area that covers 40% of the nation's cultivated area and houses almost half of the population (Lohmar, et al., 2003; Yang and Zehnder, 2001; Yang, et al., 2003). Water scarcity in northern China has arisen both because of limited water supply and increasing water demand. Water availability in northern China is only around 300 m^3 per capita, which is less than one seventh of the national average and far lower than the world average. (Ministry of Water Resources, 2002). At the same time the demand from rapidly growing industrial sector and an increasingly wealthy urban population is beginning to compete with the agricultural water demand (Crook, 2000; Wang, et al., 2005).

Over the last 20 years or so, groundwater has become the most important source of irrigation in northern China. Before the early 1980s, most of the increase in water availability in northern China came from the expansion of surface water systems, many constructed during the 1950s and the 1960s. China's leaders invested in building reservoirs, constructing new canal networks and increasing the utilization of the region's lakes and rivers.

By the end of the 1970s, much of the available surface water was already being utilized in most basins in northern China and the area irrigated by surface water more than doubled in the north China Plain (Nickum, 1988). However, due to poor management and lack of investment, especially in maintenance, many surface water canal systems began deteriorating in the 1980s (Lohmar et al., 2003). Since then groundwater has become the major source of irrigation. In north China, the share of groundwater irrigated areas increased from 5% in the 1950s to 68% of the total irrigated area in the 2000 s (Wang et al., 2007a).

Now the question arises that with further increase in population and the development of industry, can we meet the water demand of agriculture through additional groundwater extraction in northern China? The answer is "no", because the groundwater resource is not infinite and some environmental problems have already occurred as a result of over-exploitation of groundwater. According to China's Ministry of Water Resources (2002), between 1958 and 1998 groundwater levels in the Hai River Basin fell by up to 50 m in some shallow aquifers and by more than 95 m in some deep aquifers. The central government in China is developing the mega project "water transfer from south to north" to alleviate the crisis of water shortage in northern China. In fact, with the increase in water demand there is a need for supply augmentation measures, but also effective management systems and relevant policies geared towards demand management. The effective management and feasible policies should be based on comprehensive research on the characteristics of local groundwater resources, groundwater use, current policies, institutions and their effect. However, studies and reports on these aspects are few. Systematic analysis on groundwater use cannot be found except for those carried out by the Centre for Chinese Agricultural Policy, Chinese Academy of Sciences (Wang, et al., 2007, Wang, et al., 2006). In order to collect detailed information on groundwater use, Baoding City in Hebei province was selected as a sample study area.

The overall goal of this paper is to identify various drivers of groundwater use in rural northern China. To meet this overall goal, three specific objectives were pursued. The first is to describe the history of local groundwater use and emergence of current problems. The second to understand the evolution of tubewells ownership, management, adoption of water saving technologies and other responses made by farmers in view of declining groundwater tables. The third is to examine the institutions and regulations of groundwater management and their effectiveness.

2   DATA

The data used for the study come from both primary and secondary sources (see Figure 1). For primary sources, local officials were interviewed, and field surveys conducted using pre-designed questionnaires for village leaders and farmers who manage tubewells. The purpose of the interviews with local officials was to understand the general characteristics of groundwater use, policies and local environment in the study area. The secondary data is mainly from local statistical bureaus. The secondary data can help towards understanding the history of the groundwater development.

The respondents in our field survey come from 20 randomly selected villages in two counties in Baoding City, Hebei Province. After discussion with the local officials, two sample counties were selected (Mancheng county and Qingyuan county). After selecting the sample counties ten villages were selected in these two counties. In each village, three

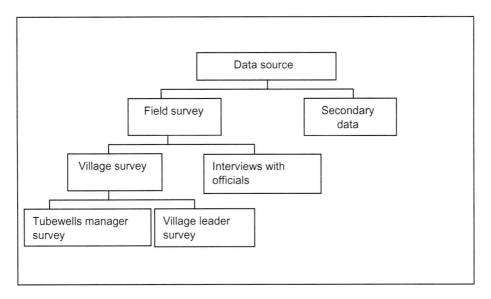

Figure 1.   Sources of data.

farmers who have tubewells were randomly selected. Finally, the total samples size was 20 villages and 60 farmers. The survey collected data for two years, 1995 and 2005. A survey of village leaders was conducted in order to understand groundwater status at the village level. Questionnaires for village leaders involved information on eight aspects in the specific year 1995 and 2005. These were the village socio-economic situation, changes in cropping patterns, local groundwater resource utilization, groundwater quality and security, the change in number of tubewells, the change in tubewells ownership pattern and management structure and finally adoption of water saving technologies. In addition to the village level survey, we also conducted household level surveys with questionnaires for tubewell owners or managers. The purpose of tubewell manager's survey (or farmer survey) is to understand perspectives of the farmers and their behavior *vis-à-vis* groundwater. The questionnaires for tubewell owners involved 8 aspects on groundwater use in 1995 and 2005: the technical characteristic of the tubewells and pump, the security and allocation of groundwater, water price and water market for selling water, regulations for groundwater use, the investment on tubewells and pumps, water volume consumed by different crops, income of farmers and their cost of agriculture production.

3   LOCATION AND BASIC CHARACTERISTICS OF THE STUDY AREA

3.1   *Location, topographical and climatic characteristics*

Baoding City is located in the central southern part of Hebei province. Mancheng and Qingyuan are the two study counties out of the total 25 counties managed by Baoding municipal government. The elevation of Baoding City decreases from west to east. Taihang Mountain is located in the western part of Baoding and has an elevation of about 1000 m. The eastern part of Baoding City is mainly a plain area with an elevation from 10 m to 30 m.

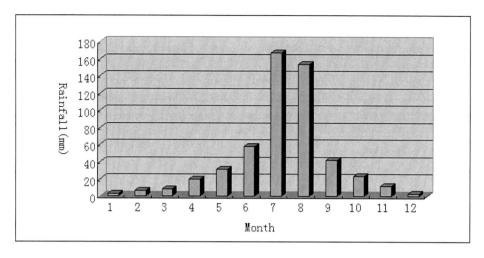

Figure 2.   Monthly rainfall distribution in the study area.

The climate of Baoding City is of the continental monsoon type. The annual temperature is 13.8°C, the average yearly sunshine duration is 2374.4 hours and the average yearly rainfall is 550.2 mm. Rainfall is distributed unevenly within the year with 61 percent of rainfall concentrated in July and August (Figure 2).

3.2   *Hydrological characteristics*

Daqing River, a tributary of Haihe river is the main river within Baoding City. Daqing river and its sub tributaries drain the entire basin. The southern part of Hebei plain is an area covered by lacustrine and riverine sediments which belongs to Huabei fractured depression which is surrounded by deep faults. Quaternary aquifers are porous layers and can be subdivided into four groups (I, II, III and IV) from top to bottom according to local hydrogeology. The fresh water is mainly located in the first and the second aquifer and they are the main aquifers. The general thickness of the southern part of Hebei plain is 70 m to 80 m, and a depth of 150 m to 220 m can be found to the north east of Shijiazhuang city and east of Baoding city. The aquifers change from coarse to fine, from one to more layers and the water quality deteriorates from south to north.

There are many sources of groundwater recharge. Infiltration from precipitation, surface water body, well irrigation and lateral seepage from mountain areas are the recharge sources for shallow groundwater in the south part of Hebei plain. Precipitation is the main source of groundwater recharge and it accounts for about 75 percent of the yearly average recharge. The main sources of discharge for shallow groundwater include artificial abstraction, unconfined groundwater evaporation and seepage to other layers. Of these, the artificial discharge is the main way which accounts for more than 80 percent of discharge. Seepage from deep karst, lateral flow from flood deposit fan, vertical infiltration or seepage are the main sources of recharge for deep ground water.

Groundwater flow follows the topography and generally flows from west to east. Before 1960, when groundwater abstraction was small, the shallow groundwater flow direction was in its natural state without being influenced by human activities. The gradient from the mountain front to the sea was 2 to 0.1%. After 1960, the annual increase in groundwater abstraction had serious impacts on groundwater flow and several large groundwater depressions were formed. The shallow groundwater flow direction changed and the groundwater gradient changed to 2.29 to 0.05%.

3.3 *Socio-economic characteristics*

According to the Hebei Economic Yearbook (2006), the total population of the province is 68.51 million people. Within Hebei province, Baoding is the most densely populated city (10.73 million) among 11 cities from Hebei. The per capital land availability in Baoding (0.07 ha) is lower than the average of Heibei province (0.09 ha). The agricultural GDP is 3.43 billions yuan ($\sim 4.4 \times 10^8$ US\$) in 2005 and is third highest in Hebei province. The per capital agriculture GDP (3200 yuan, ~ 416 US\$) is, however, lower than the whole province due to high population concentration in Baoding.

Since most of land area of Hebei province lies within the plain and has a good irrigation network, the share of irrigated area to net cropped area is high. The total irrigated land area is 651 461 ha and the share of irrigated land to net cropped area is 86%, which is higher than the average for the whole province (76%). In the sample villages, the share of irrigated land is still higher, at 98%. Since surface water is scarce, most of irrigated land depends on groundwater and it accounted for 87% of the net irrigated area.

The cropping pattern in this region is fairly simple. The gross sown area is 1181000 ha, and the cropping intensity is 1.55%. Wheat and maize are the main grain crops and are grown over an area of 848 200 ha. The yield of wheat and maize are 4839 kg/ha and 4459 kg/ha respectively. Cotton, oil crops, fruit tree and vegetable are the main cash crops with a planted area of 322 900 ha. The yield of cotton, oil crops and vegetable are 1006 kg/ha, 2732 kg/ha and 57404 kg/ha respectively.

In the samples villages (20 villages), the socio-economic situation is similar to Baoding city. Based on the survey data, the total population of these 20 villages was 52942 and land area was 4200 ha. The per capita land availability is 0.08 ha which is higher than that of Baoding city. The irrigated area is 4095 ha accounting for 98% of the farmland and 100% of irrigation was from groundwater. The gross sown area is 7203 ha and the cropping intensity reaches 1.71. Wheat, maize are the main crops, the others are fruit trees and cash crops such as vegetable.

4 DEVELOPMENT OF GROUNDWATER RESOURCES AND EMERGING PROBLEMS

Groundwater is the most important source of water in Baoding City and agriculture is the major water user. According to the Bureau of Hebei Water Resources (2000), the annual water availability (including groundwater and surface water) in Baoding City is about 3.1 billion m^3, of which groundwater accounts for 2.3 billion m^3 (74%). However, the annual water use is about 3.3 billion m^3, which indicates that water use especially groundwater resources use is more than the water resource available to Baoding City. Among all the sectors, agriculture is the main water user and 80% water is used for irrigation while the remaining 20% is shared by the domestic and the industrial sectors equally.

4.1   *Development of groundwater resources in northern China*

The development of groundwater resources in northern China can be divided into four stages since 1949. The first is from 1949 to the end of the 1950s. The major characteristic of this stage was the under-development of groundwater resources. During this stage, due to low productivity, water demand either from agriculture or from industry and other sectors was small and it was not necessary to develop groundwater resources. There were few wells, most of the wells were made of stones and bricks, and watermills were the main tool for groundwater extraction.

The second stage is from the end of the 1950s to the end of the 1960s. During this stage, groundwater began to be developed, but the scale of development was still limited. Agricultural production in rural China began to increase rapidly. In order to promote the growth of agricultural production, local governments started to encourage the use of groundwater but the focus was on using shallow groundwater resources.

In the third stage, groundwater was exploited at a large scale and its development was very rapid. This stage was from the early 1970s to the end of the 1970s. The year 1972 was a serious drought year. Faced with drought, both central and local government encouraged local collectives and farmers to construct tubewells. By the early 1980s, there were 86 424 tubewells in Baoding City. This was a growth of 128% from the previous decade.

Since the early 1980s, groundwater development has entered into the current stage of over-exploitation. With continuous increase in agricultural production, growth in industry and increase in population, the total water demand from all the sectors has gone up significantly. Due to the limited supply of surface water, groundwater resources have to be over-exploited to meet the increasing demand. The annual over-exploited volume for groundwater in Baoding City is about 200 million m^3, which is about 8% of the total water groundwater use.

4.2   *Tubewell expansion in the study areas*

According to the Report on Water Exploitation in Mancheng (1949–1988) and Qingyuan (1949–90), groundwater development in the study areas is at a similar stage to that elsewhere in northern China. In Qingyuan County, tubewell construction began in 1953 and by the end of the 1950s there were 1142 tubewells in the county. From the end of 1950s to the end of 1960s, the number of tubewells in Qingyuan County increased by 235%. With the support of local government, the number of tubewells in the county grew rapidly and reached 7787 units, an increase of 103% since the early 1970s. By the end of 2005, there were 13696 tubewells some 12 times the number of tubewells in 1960. Tubewell expansion in Mancheng County is similar to that in Qingyuan County. The total number of tubewells in Mancheng County is lower than that in Qingyuan county (Figure 3).

Tubewell expansion shown in the official data is largely supported by the information from the survey. The survey data show that the farmers started digging tubewells in 1953 and from then on the number of tubewells has been increasing, reaching 731 by the mid-1980s. The numbers of tubewells and the technological configuration of the tubewell changed between 1995and 2005 (Table 1). The table shows that within a span of 10 years, 312 additional tubewells were constructed, an increase of about 35% from the mid 1980s. Deep tubewells became more important than shallow tubewells. In 2005, 58% of tubewells were deep tubewells, while the share of deep tubewell was only 42% in 1995.

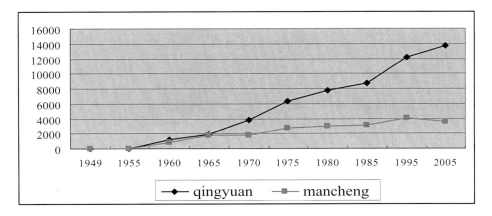

Figure 3.   Expansion in number of tubewells in Qingyuan and Mancheng County of Baoding City, Hebei Province.

Table 1.   The change of tubewells number and type.

			Deep tubewells		Shallow tubewells	
	Year	Number of total tubewells	Number	% to total	Number	% to total
Total	1995	902	228	25	674	75
	2005	1214	709	58	505	42
10 villages in	1995	570	210	37	360	63
Qingyuan county	2005	980	642	66	338	34
10 villages in	1995	332	18	5	314	95
Mancheng county	2005	234	67	29	167	71

Data source: Primary survey conducted in 20 villages in Baoding district in December 2006.

Table 2.   Average depth of groundwater at Baoding city (m).

	Well depth	1975	1980	1985	1990	1995	2000
Shallow water	<80	5.2	4.3	12.54	11.98	15.95	19.85
Deep water	>90	6.06	5.07	15.2	15	17	16.27

Source: Xu (2002).

4.3   *Decline of the water table and other environmental problems*

With increase in groundwater abstraction, the water table in the study area has declined gradually over time (Table 2). Based on the data from observation wells in Baoding, the shallow water table was 5.2 m below ground level (bgl) in 1975 but in 2000 the water table had declined to 20.0 m bgl. During the last 25 years the water table declined by 14.65 m or about by 0.6 m per year. From 1975 to 1985, the deep water table declined by 9.1 m or about 0.9 m per year. After 1985, the rate of decline in the deep groundwater table has slowed down and is currently around 0.05 m per year.

The village leaders' survey also reflects the same trend. In 10 villages of Mancheng County, the average of water table in 1995 was 26.8 m. But in 2005 the average water table was 35.6 m, showing an average decline of 0.9 m per year. During the same period, in Qingyuan county, the water table also declined by 0.9 m every year.

In some areas of intensive groundwater exploitation, cones of depression have been formed. Two such large cones of groundwater depression appeared in Yimuquan district and the Baoding city. The cone of depression in Yimuquan is located around the Yimu Spring water source area in north western Baoding city and had developed in 1967. The groundwater table depth at the funnel centre has increased from 4.4 m in 1967 to 33.6 m in 2001. The cone of groundwater depression in Baoding city was discovered in 1974. The groundwater depth at the centre of the cone of depression increased from 13.5 m in 1974 to 28.42 m in 2001. However, no such cones of depression have been found in rural areas.

5   STRATEGIES OF THE FARMERS TO COPE WITH GROUNDWATER DEPLETION

The farmers in north China have adopted several strategies to cope with depleting ground-water levels. The major strategies involve privatization of collective owned tubewells, conversion of shallow tubewells into deep tubewells, emergence of informal groundwater markets and adoption of various kinds of water saving technologies.

5.1   *Change in tubewell ownership from collectives to private and its effects*

5.1.1   *Evolution of tubewell ownership*
In a pattern similar to that found throughout the Hai River Basin (Wang, et al., 2006), the patterns of tubewell ownership in our sample areas also have changed over time, with ownership shifting from collective to private. From 1995 to 2005, the share of collective tubewells declined from 82% to 49% (Table 3, column 5). At the same time, the share of private tubewells increased from 18% to 51% (column 6).

Within the private sector, shareholding in tubewells is the major type of tubewell owner-ship (Table 3 and Table 4). In 1995, 98% of private tubewells were shareholding tubewells (Table 4). In 2005, the share of shareholding tubewells only declined by 2%. This result is somewhat contrary to Wang et al (2006). She found that share of shareholding tubewells had declined while that of individually owned private tubewells had gone up. However, in our study region, shareholding tubewell is still the pre-dominant form of tubewell ownership.

5.1.2   *Change in pattern of tubewell ownership and its relation with land and water scarcity*
As found by other researches in China (Wang, et al., 2006), the change of tubewell own-ership has close relationship with increasing water scarcity. Based on the survey data the percentage of private tubewells to total tubewells is 18% in villages where the groundwater table is less than 28 m (table 5), but it is 30% in villages where the groundwater table is between 28 m to 50 m. It shows that when water becomes scarce, private ownership tends to increase. A possible reason for this could be that private ownership gives better control over groundwater than access to collective tubewells. The problem of accumulating enough

Table 3.   Changes in pattern of tubewells ownership, 1995–2005.

			No. tubewells		Share of tubewells (%)	
	Year	No. tubewells	Collective	Private	Collective	Private
Total	1995	45	37	8	82	18
	2005	61	30	31	49	51
Villages in Qingyuan	1995	57	46	11	81	19
	2005	98	43	55	44	56
Villages in Mancheng	1995	33	28	5	85	15
	2005	23	16	7	70	30

Source: Primary survey conducted in 20 villages in Baoding district in December 2006.

Table 4.   The composition of private ownership tubewell in study villages (1995–2005).

		The private ownership tubewells			
		No. of tubewells		Share of tubewells (%)	
Year	Number of total private tubewells	Shareholding	Individual	Shareholding	Individual
1995	161	158	3	98	2
2005	620	597	23	96	4

Source: Primary survey conducted in 20 villages in Baoding district in December 2006.

Table 5.   The relationship between water table and tubewell ownership.

		Ownership (%)	
Depth to water table bgl	Sample size	Collective	Private
12 m–27 m	21	82	18
28 m–50 m	19	70	30

Source: Primary survey conducted in sample villages in 1995 and 2005.

capital to invest in a tubewell is overcome by the institutional mechanism of shareholding tubewells.

5.1.3 *Effects of change in tubewell ownership on cropping patterns*

The survey showed that the change of tubewell ownership possibly induces changes in cropping patterns, which is consistent with the finding by Wang, et al. (2006). The data (table 6) show that between 1995 and 2005 in Qingyuan county villages, when the share of private tubewells increased from 19% to 56%, the share of net sown area under wheat declined by 34% (columns 1 and 4, rows 1–2) and area under maize declined from 44% to 34%, while the area devoted to vegetables increased from 3% to 24% (column 8, rows 1–2).

In Mancheng county the change in ownership of tubewells resulted in similar shift in cropping patterns, but here instead of vegetables, the area under fruit crops increased from 19% in 1995 to 28% in 2005. This is because being partly a hilly county, fruit crops are the main cash crop in this county. Although the data show a fairly strong relationship between tubewell change and cropping patterns, the relationship between tubewell change and yields is less clear. It is true that the descriptive data show that yields increase over time as private tubewell change increases. But this certainly cannot be attributed to privatization of tubewells alone.

5.2   *Change of tubewell management regimes: from 'pure' collective to shareholding management*

With the change of tubewell ownership from collective to private, the management regimes of tubewell also have changed from collective to private (Table 7). The data show that the share of the different management regimes is similar to the share of the different tubewell ownership pattern. In 1995, the share of the collective management regime was 83% but this share declined to 49% in 2005. At the same time, the share of private management regime increased from 17% to 51%.

Table 6.   Relationship between the change in ownership of tubewell, cropping patterns and yields in Qingyuan and Mancheng counties (1995–2005).

	Total number of wells and tubewells	% of private wells and tubewells to total	Percentage of sown area under				Crop yield (kg/ha)	
			Wheat	Maize	Fruit	Vegetable	Wheat	Maize
Qingyuan								
1995	570	19	45	44	0	3	5513	5888
2005	980	56	35	34	0	24	6150	6900
Mancheng								
1995	332	15	36	36	19	0	5250	6585
2005	234	30	30	34	28	0	5355	6518

Source: Primary survey conducted in 20 villages in Baoding district in December 2006.

Table 7.   The share of different tubewell management regimes in the study area.

	Collective management (%)		Private management (%)	
	Pure collective management	Contracting management	Shareholding management	Individual management
1995	97	3	95	5
2005	86	14	96	4

Source: Primary survey conducted in 20 villages in Baoding district in December 2006.

Within the collective management regime, there are two sub-regimes. The first sub-regime can be called pure collective management. In this case, the tubewell is entirely managed by village leaders. The second pattern is called contracting management. Contracting management implies that the village leader rents out the collective tubewell to individual farmers, and then those farmers are responsible for managing those tubewells. The data show that the share of pure collective managed tubewells was 97% in 1995, and only 3% of tubewells were rented out to individual farmers. This increased to 14% in 2005 and the share of pure collective managed tubewells declined to 86%. Within the private management, there are also two regimes, shareholding and individual management. Table 8 shows that shareholding management is the main management regime which accounted for 95% in 1995, the share of individual management regime is only 5%. In 2005 there are no significant change on the share of shareholding management and individual management. This result indicates that shareholding management is still the preferred mode of management for private shareholding tubewells.

5.3 *Change of pump technology and ownership*

Pumps are an integral part of water extraction mechanisms. With decline in groundwater levels, pump technology had to be upgraded at least four times in the study area (Fig. 4). At the early stage of groundwater exploitation the watermill was the main tool for groundwater extraction. By the end of 1970s, water mills were completely replaced by impeller pumps. However, by the mid 1980s, most impeller pumps became outdated due to lowering of water table. As a result borehole submersible pumps were introduced. At present, over 90% of pumps are submersible pumps.

With the change of tubewells ownership, the ownership of pumps also changed. The definition on pumps ownership is the same as tubewell ownership. If the village community

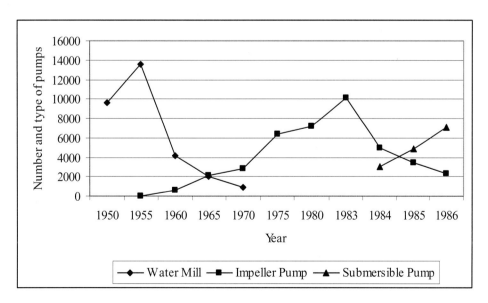

Figure 4.   Evolution of pumping technology in the study area.

buys the pumps it is a collective, if the pumps are purchased by several individuals pooling in their resources together, it is called shareholding and if only one farmer invests in the pump it is individual ownership. The pumps used with collective tubewells can be divided into all three ownership types (collective, shareholding and individual), though collective pumps are primarily used in collective tubewells. For shareholding tubewells, pumps could be either owned by the same set of individuals who own the shareholding tubewell or owned individually. In 90% of the cases, pumps used on shareholding tubewells too belonged to the same shareholding group. For individually owned private tubewell, the pump was found to be individually owned.

5.4   *Adopting water saving technologies*

During the survey of village leaders and water managers, it was found that there are many types of water savings technologies being used in northern China. The water saving technologies can be divided into three groups: traditional, household-based, and community-based (Table 8). Advanced water saving technologies (such as drip, intermittent irrigation, and chemical-based sprays) have so far very low levels of adoption.

In the study village, various types of water saving technologies have been adopted. The adoption rate of traditional technologies is higher than other two kinds of technologies. Based on our survey, the share of area sown with level fields and border/furrow irrigation technologies exceeded 70%. Except for traditional technologies, the underground pipe technology (community-based technologies) is also adopted over a large land area. Household-based water saving technologies have expanded rapidly during the last 10 years and this finding is similar to what Blanke et al (2006) found in their study. It is also seen that adoption of community based technologies has remained stagnant during the last 10 years, possibly due to stagnation in investment by the local governments. Increased

Table 8.   Adoption of water saving technologies in the study area.

Technologies	1995 Share of sown area with technology (%)	2005 Share of sown area with technology (%)	Initial year adoption of
Traditional technologies			
Border/Furrow irrigation	88	87	Before 1960s
Level fields	73	74	Before 1960s
Household-based technologies			
Plastic sheeting	1	8	1982
Drought resistant varieties	1	5	1988
Retaining stubble and zero or low tillage	0	15	1997
Surface pipe	1	11	1985
Community-based technologies			
Underground pipe	27	27	1987
Lined canal	5	4	1984

Data source: Primary survey conducted in 20 villages in Boading district in December 2006.

adoption of household technologies for water saving shows that there is a demand for and willingness to adopt water saving measures, but investment may be an important constraint.

5.5   *Emergence of groundwater markets*

Evolution and development of groundwater markets has been of great interest to researchers. In the early 1990s Pant (1991) found that 86% of the households in eastern Uttar Pradesh, India purchased water for irrigation; in central and western Uttar Pradesh, 65% of farm households purchased water. More recently studies in Pakistan by Strosser and Meinzen-Dick (1994) and Meinzen-Dick (1996) have found groundwater markets to be pervasive. In northern China, in response to the demand for water in an environment increasingly dominated by private wells, following a pattern similar to that observed in South Asia (Shah, 1993), groundwater markets have begun to emerge in recent years as a way for many farmers in rural China to gain access to groundwater (Zhang et al, 2008). But in the sample villages, a groundwater market was only found in one village. The usual practice in that village was that a water buyer will pay only a marginally higher fee to the tubewell owner based on their electricity consumption. Thus, the tubewell owner provides the irrigation service (by letting the buyer use the tubewell) and does not sell the water. It is expected that prevalence of groundwater markets would increase in the coming years in view of increasing de-collectivization and privatization of tubewells.

Water price could be an effective tool for balancing demand and supply of water and for improving water use efficiency. In essence, however, in rural China extracting groundwater is free; the only payment made by the agricultural water users is the cost of energy (electricity or diesel). As a result, the current water pricing policy in the agricultural sector of China (as opposed to the industrial and residential sectors) has not been effective in providing water users with incentives to save water (Wang, 1997). The data show that the average cost of water extraction (mainly electricity cost) was US$0.07 in 1995 and this nominally increased to US$0.08 in 2005. The total cost of pumping for growing a crop of maize is US$52 per ha and accounts for 14% of total cost of production. The result suggests the cost of groundwater extraction is relatively low and, therefore, farmers have little incentive to save water. China's water policy makers have begun to consider reforming the pricing of irrigation water as one of the main policy instruments for dealing with the water scarcity problem in the coming years (Feng, 1999; Wang, 1997; Shen, 2006; Zhen, 2007).

5.6   *Farmers' views on their possible response to increasing water scarcity*

Farmers were asked how they were coping with water scarcity. In particular, the village leaders were asked "do you think that groundwater resources will become scarcer in the future?" and 80% of the village leaders answered in the affirmative. As to the question "how will you cope with groundwater scarcity in the future?" 81% of farmers said that they would dig deeper tubewells, 69% said they would switch to higher capacity pumps and only 50% said they would adopt some kind of water saving technology. Only 13% of respondents said they would adjust their cropping pattern and 6% said they would reduce their water use (Table 9). These responses are in stark contrast to what researchers often recommend for coping with water scarcity, namely, reduce water use and change cropping pattern to less water intensive crops.

Table 9.   Farmers' preferred response to groundwater depletion.

Farmers response to groundwater depletion	% of respondents who would adopt the response
Digging deeper tubewells	81
Installing higher capacity pumps	69
Reducing water use	6
Adopting water saving technology	50
Changing cropping pattern	13

Data source: Primary survey conducted in 20 villages in Baoding district in December 2006.

6   INSTITUTIONAL ARRANGEMENT FOR GROUNDWATER MANAGEMENT

6.1   *Institutions for groundwater management*

Over the past 50 years, China has built up a vast and complex water bureaucracy. In fact the system was designed to construct and manage surface water to prevent floods, which have historically devastated the areas surrounding major rivers, and to effectively divert and exploit water resources for agricultural and industrial development. Until recently, neither groundwater use nor water conservation in agriculture have been of major concern to policymakers (Wang et al. 2005).

In Mancheng county the water bureau was set up in 1950s and was engaged in the construction of surface water canals and tubewells. In 1986 local government realized the problems caused by groundwater over-exploitation, and a special office was set up to manage the groundwater resources. However, according to an official, groundwater in rural areas is still not being dealt with. In Qingyuan county, a water resources office was set up in 1978. In this office only two people are assigned to the water resources management department and their jobs are limited to propagate government water policies and the collection of water resources fees (there are no fees for agricultural water use). In 1989, a separate water management office was set up to deal with groundwater over-exploitation, but it concentrated its efforts in urban areas only. Thus, none of the formal institutions, at least locally, deal directly with managing groundwater use in rural area.

6.2   *Water laws including groundwater laws*

Water policy is ultimately created and theoretically executed by the Ministry of Water Resources in China (Wang et al., 2007b). China's first comprehensive Water Law was enacted in 1988. According to this Water Law, all water, including groundwater is vested in the state. This means that the rights to use; sell and price groundwater lies with the state. The law does not allow extraction of groundwater if such pumping is harmful to the sustainability of the resource.

Beyond formal laws, there have also been many policy measures set up in part to rationally manage the use of the nation's water resources. In most provinces, prefectures and counties there are formal regulations controlling the right to drill tubewells, the spacing of wells, and

Table 10.   Permission for digging new wells.

Different questions about farmers perspective on digging tubewells	The share of answer (%)	
	Yes	No
Did the village leader give permission for the new well?	93	7
Did the town leader issue permit for the new well?	34	66
Did the county leader issue permit for the new well?	28	72
Have you obtained the digging license on your well from water resources office?	18	82

the price at which water can be sold. The national government has also set up the necessary regulatory apparatus to allow for the charging of a groundwater extraction fee.

Despite the plethora of laws and policy measures, there has been relatively little emphasis in implementing those laws (Wang et al., 2005). For example, despite the nearly universal regulation that makes it mandatory for a farmer or a village to seek a permit from the authorities before drilling a well, only 18% of the well owners in the sample area had done so.

From Table 10 it can be seen that 93% of the new wells got permission from village leaders or village leaders who then knew about the construction of these tubewells. Only 18% of the wells owners procured a digging license from the county department before constructing a well or tubewell. This means well digging license plays a relatively small role in groundwater management.

7   SUMMARY AND POLICY IMPLICATIONS

7.1   *Summary*

Using field data from 20 villages and 60 farmers, it was found that since the late 1980s, groundwater has been continually over-exploited and water tables declining steadily. In the study area, the groundwater resource is the only source of irrigation for the villages. Use of groundwater started in 1953 and then quickly increased. By the 1980s, groundwater was in a state of over-exploitation. This over-exploitation the groundwater resulted in decline of groundwater tables at the rate of 0.9 m every year.

Farmers responded to this threat of declining water tables in various ways. The first direct response was to drill deeper tubewells. From 1995 to 2005, additional 312 tubewells were constructed, most of which were deep tubewells with depth of more than 60 m.

The second indirect response was change in ownership pattern of tubewells away from collective tubewells to privately owned tubewells. In 2005, 51% of all tubewells were privately owned as against only 18% in 1995. This move towards privatization was in part propelled by the need to provide more secure access to the declining resource. In order to overcome the capital constraints (deeper tubewells needed larger investments), farmers formed themselves into shareholdings and invested in tubewells jointly. Majority (96% in 2005 and 98% in 1995) of the tubewells are owned on a shareholding basis. It was also found in the study, that those villages in which farmers had invested in private tubewells, the area under cash crops has increased.

The third response to water scarcity was the adoption of a variety of water savings technologies: traditional technologies, household based technologies and community based technologies. During the period of 1995–2005, the adoption of household based technologies has increased considerably showing that there is a demand for water saving among the farmers. On the other hand, adoption of community based technologies such as under ground pipelines has declined, mostly because government investment in these infrastructures has declined.

Our study also found, in conjunction with other studies that while China has built complex bureaucracy and issued various regulations governing water use, none apply directly to agricultural use of groundwater. The only policy that has some bearing on groundwater management in rural areas is the requirement of procuring a permit or license for drilling a new well. However, this requirement is hardly adhered to and only 18% of the farmers reported having got permits from the county level water resources office.

7.2   *Policy implications*

Several policy implications follow from this work. First, there is an urgent need to control over-exploitation of groundwater in north China. There is a need to invest in better understanding of groundwater resource availability and the volume of abstraction should be kept within the limit set by total availability. Construction of deep tubewells should be closely monitored and stopped if needed, because these contribute to over-exploitation of the deeper aquifers.

Second, increasing the water price to include a groundwater resource fee and expanding various water saving technologies might be an effective tool for managing water scarcity. In the study area, the water price is relatively low and, therefore, farmers do not have enough incentive to save water. However, with increasing water scarcity, there is a willingness among farmers to adopt water saving technologies. This willingness should be capitalized upon and the government should invest in water saving technologies, especially of community based technologies such as underground pipelines.

Third, there is scope for investment in artificial recharge of groundwater in northern China which the government should take up. In the study area, direct rainfall recharge is the main source of replenishment to the aquifers. Therefore, projects that capture rainwater and recharge the aquifers would be beneficial. The 'South-North' water transfer project will not only help reduce the stress on groundwater, but it will also act as a source of recharge to groundwater.

Finally, there is an urgent need for creating a change in the mind-set of the water officials who still tend to think that rural groundwater use does not pose much of a problem and as such are pre-occupied with urban groundwater issues only.

REFERENCES

Blanke, A., S. Rozelle., and B. Lohmar, (2007), 'Water Saving Technology and Saving Water in China', *Agricultural Water Management*, 87:139–150.

Bureau of Hebei Water Resources (2000), 'Water Resource Bulletin of Hebei Province', China Waterpower Press, Beijing.

Crook, F. (2000), 'Water pressure in China: Growth strains resources', *Agricultural Outlook, Economic Research Service. January-February, 25–29. http://www.ers.usda.gov/publications/agoutlook/jan2000/ao268g.pdf.*

Feng, Y. (1999), 'On Water Price Determination and Fee Collection in Irrigation Districts', *Journal of Economics of Water Resource 4*: 46–49.

Hebei Economic Yearbook (2006), China Statistic Press, Beijing.

Lohmar, B., J. Wang, S. Rozelle, J. Huang, and D. Dawe (2003), 'China's Agricultural Water Policy Reforms: Increasing Investment, Resolving Conflicts, and Revising Incentives', *Agriculture Information Bulletin No. 782*, Market and Trade Economics Division, Economic Research Service, U.S. Department of Agriculture, Washington, DC.

Meinzen-Dick, R. (1996), Groundwater Markets in Pakistan: Participation and Productivity, *Research Report 105*, International Food Policy Research Institute, Washington, DC.

Ministry of Water Resources (2002), 'China Water Resources Bulletin' China Waterpower Press, Beijing.

Nickum, J. (1988), All is Not Wells in North China: Irrigation in Yucheng County. In: O'Mara G.T. (ed.) *Efficiency in Irrigation*. World Bank: Washington, DC, 87–94.

Pant, N. (1991), Ground water issues in eastern India, in R. Meinzen-Dick and M. Svendsen (eds), *Future Directions for Indian Irrigation: Research and Policy Issues*, Washington, DC: International Food Policy Research Institute.

Shah, T. (1993), *Groundwater Markets and Irrigation Development: Political Economy and Practical Policy*. Bombay, India: Oxford University Press.

Shen D.J. (2006), 'Theoretical Base of Water Resources Fee and its Pricing Method', *Shuili Xuebao*, 1 (37):120–125.

Strosser, P., and Meinzen-Dick, R. (1994), Groundwater Markets in Pakistan: An Analysis of Selected Issues. In Moench, M. (ed.) *Selling Water: Conceptual and Policy Debates over Groundwater Markets in India*. Gujarat, India: VIKSAT, Pacific Institute, Natural Heritage Institute.

The Report of Water Exploitation in Mancheng (1949–1988), Mancheng County Water Conservancy Bureau, 1999.

The Report of Water Exploitation in Qingyuan (1949–1990), Qingyuan County Water Conservancy Bureau, 1999.

Wang, H. (1997), *Distorted Water Prices*. Beijing: China Water and Electricity Publishing House.

Wang. J., Z. Xu, J. Huang, and S. Rozelle. (2005), 'Incentives in Water Management Reform: Assessing the Effect on Water Use, Productivity and Power in the Yellow River Basin', *Environment and Development Economics* 10: 769–799.

Wang, J., J. Huang, Q. Huang, and S. Rozelle (2006), 'Privatization of Tubewells in North China: Determinants and Impacts on Irrigated Area, Productivity and the Water Table', *Hydrogeology Journal*. 14: 275–285.

Wang, J., J. Huang, A. Blanke, Q. Huang, and S. Rozelle (2007a), 'The Development, Challenges and Management of Groundwater in Rural China, in Giordano, M. and K.G. Villholth (ed), *The Agricultural Groundwater Revolution: Opportunities and Threats to Development*', Comprehensive Assessment of Water Management in Agriculture Series, CABI, pp: 37–62.

Wang, J., J. Huang, Z. Xu, S. Rozelle, I. Hussain, and E. Biltonen (2007b), 'Irrigation Management Reforms in the Yellow River Basin: Implications for Water Saving and Poverty', *Irrigation and Drainage*. 56: 247–259.

Xu Y. (2002), "Impact of Land Use Change on Groundwater Resources in the Southern Hebei Plain." PhD thesis, China Academy of Sciences.

Yang, H., and A. Zehnder (2001), 'China's Regional Water Scarcity and Implications for Grain Supply and Trade', *Environment and Planning A* 33: 79–95.

Yang, H., X. Zhang, and A.J.B. Zehnder (2003), 'Water scarcity, pricing mechanism and institutional reform in northern China irrigated agriculture', *Agricultural Water Management* **61**: 143–161.

Zhang, L. Wang, J., J. Huang, and S. Rozelle (2008), 'Development of Groundwater Markets in China: A Glimpse into Progress to Date', *World Development*, 36 (4): 706–726.

Zheng, T. (2007), 'Promoting Reform of Water Pricing System and Establishing a Better Operation Mechanism for On-Farm Water Structures', *China Water Resource* 23: 29–32.

Thematic issues on groundwater irrigation

CHAPTER 12

Anthropological perspectives on groundwater irrigation: Ethnographic evidence from a village in Bist Doab, Punjab

R. Tiwary
International Water Management Institute (IWMI), Sub Regional Office for South Asia, Hyderabad, India

J.L. Sabatier
Centre de recherché de Montpellier (CIRAD), Montpellier, France

ABSTRACT: An anthropological approach is used to analyse the relationship between social structure and irrigation in a traditional agricultural society. The ethnography for a village in Punjab (India) dependent on groundwater irrigation revealed a little known indigenous irrigation institution that has existed for a very long time in the Bist Doab area. The institution utilizes kinship relations as functional groups to execute instrumental economic functions such as irrigation. The social organisation of irrigation depends on a broad framework of principles and rules where linkages of land and water rights along with notions of equity and justice form the key for water allocation and use. The structural dimensions regulate ideal patterns of behaviour, numerous adaptations were visible in 'on farm' operations. The institution has been resilient enough to accommodate technological successions as well as changes in regional ecology over time.

1 INTRODUCTION

Anthropologists and sociologists have long been interested in studying the organisation of irrigation in traditional societies (Hunt and Hunt 1976; Leach 1961; Gray 1963). These studies reveal evidence of a close association between social and cultural life and the physical features of irrigation. A wide range of activities needs to be performed for irrigation and these require coordinated action by groups of people. Successful irrigation institution requires that rules and norms for sharing rights and responsibilities need to be developed. These in turn need to be socially and culturally acceptable to the community as a whole. In many traditional societies, elements of existing social institution help to shape the behaviour of the irrigators. These elements may be derived from kinship, family or marriage. In many social contexts the political and religious institutions also perform a key role. Consensus over rules and norms leads to a recurrent pattern of irrigation behaviour. This shapes the structural form of an irrigation system. However, a durable irrigation system cannot afford to be rigid as it needs to constantly interact with the changing environment and the irrigation systems have to successfully negotiate these variations. A durable irrigation system shows a considerable amount of adaptations and thus remains functional. This paper documents one such irrigation system in the Punjab state of India.

2   PURPOSE OF THE STUDY

The objective is to understand, analyze and explain details of structural as well as functional aspects of a traditional irrigation system in Bist Doab region of Indian Punjab. The institution is about ground water irrigation where rights and responsibilities of irrigation are shared among households. There is little information on actual behaviour patterns, rules and norms regarding sharing of irrigation and are largely missing from government records. However, a vibrant, intricate, socially embedded, and culturally dynamic world has always existed at farm level. This indigenous irrigation institution reflects linkages of kinship with physical aspects of irrigation. An ethnographic enquiry was carried out to analyse these linkages. The topic of enquiry is multidisciplinary, which in turn required multi-disciplinary tools of investigation. The irrigation system has both structural and functional aspects. Structural aspects are the ideal normative order of organisation of irrigation i.e. principles, rules and sub-rules. These were derived from the field observations and information collected from various sources (farmers, well drillers, *patwaris* (land record officials), *Sarpanch* (elected head of the village) and others). The functional aspects are the actual operations at field leveland these may or may not conform with the ideal order. Farmers adapt to changing external conditions and modify the structural aspects at operational level. The functional component has been revealed through profiling of tube well owners, focused group discussions, actual behaviour patterns learned from key informants and government records in changing source and technology of irrigation.

3   BHAJJAL VILLAGE

3.1   *Geographical features*

The village chosen to investigate ethnography around shared groundwater irrigation in Punjab is located in Garhsankar Block of Hoshiarpur district of Punjab (Fig. 1). The region has a sub-humid climate. The maximum temperature of 41°C is recorded during June and minimum temperature of 6°C is recorded during January. Annual long term average rainfall is about 800 mm, however, although erratic year by year. 80% of the rainfall occurs during the monsoon (June to September), and the remainder is recorded in the winter months of October to January. The soil is alluvial in origin and is a sandy loam, medium textured, and characterized by medium fertility. The Garhsankar Block of the Hoshiarpur District has been declared as 'grey' block as far its groundwater status is concerned, which according to Central Groundwater Board (CGWB) is defined as a block where the percentage of groundwater extraction to recharge is more than 65% but less than 85%. The villagers claim that there has been an average 3.3 m decline in the water table elevation per decade for the last thirty years.

3.2   *Sociological features*

Bhajjal village has a total area of 192 ha according to the 1991 Census. The population of the village is 1013 and there are 168 household (GOI, 1991). It is a multicaste village and

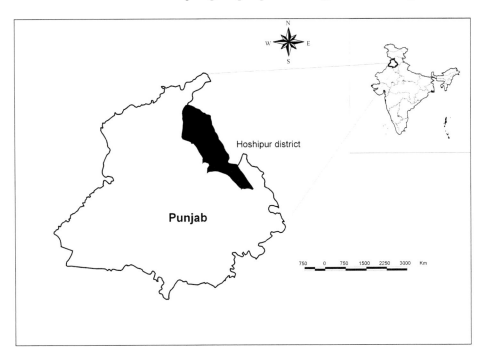

Figure 1.   Location of the study site in Punjab state of India.

Jat Sikhs are the dominant caste in the village. They are the cultivator caste. Bist Doab region in Punjab is well known for its small landholdings. 34% of household have 2 to 5 khet (1 khet = 0.404 ha) which makes them small farmers according to India's land holding classification. Households owning less than 2 khets and 5–10 khets respectively constitute only 10% of the total. Those with more than 10 khet constitute 12% (IWMI-DSSA, 2006) of all cultivator households. It is likely that such households are mainly joint families that are looking after the land of their sons or brothers settled abroad. There are numerous service castes like Chira (traditionally water carriers), Tarkhan (carpenter caste), Brahmans and schedule caste (SC), locally called *Adharmi*.

Brief knowledge about kinship system of Punjab will help us to unravel the social organ-isation of shared irrigation in this region. Among *Jat* Sikhs, we find patrilineal descent and patrilocal residence pattern. They follow caste endogamy and clan (*Got*) exogamy (Hershman, 1981). The caste *Jat* Sikh is divided into different *Gots*. In social life of Punjab, Got forms the structural and functional unit. The *Gots* are divided into *Khandan* which can be called maximum lineage. A *Khandan* will have many *Tabbar* which are minimal lineage. The minimal lineage is a group of living agnates, the eldest of whom can trace patrilineal descent from a common grandfather or great grandfather. The minimal lineage is composed of a group of households which, two or three generations previously, had formed a single household. In other words it is product of recently divided joint household. The *Tabbar* breaks into different *Ghar* which is synonymous to households. When *tabbar* faces parti-tion and house site is divided—members of the minimal lineage will occupy immediately adjacent households at old house site. Quite often, when land is divided among members of a minimal lineage; they frequently occupy farming strips of land with common boundaries.

There is a tradition that the members of the minimal lineage often share rights in a common well (*sanjha khuh*) which was built by a common ancestor. This tradition of kinship based shared irrigation has been practiced since days of Persian wheels (locally called *halts*). Passing reference of shared groundwater irrigation during Persian wheel days are found in fascinating village studies by Paul Hershman (1981) in Bist Doab regions of Punjab.

3.3   *Irrigation and shared tube wells*

The region has been dependant on groundwater since ancient times. The District Gazetteer of Hoshiarpur (1883–1884) reports that well irrigation was predominant in Garhsankar Tehsil during those times. In the 1960s, *kuccha* (unlined) and masonry wells fitted with Persian wheels as lifting device were the major method of irrigation. By 1991 the entire irrigated area of 128 ha in the village was dependent on tube wells and electric pumps (Census of India 1991). Currently there are 44 tube wells in the village. Earlier these electric tubewells were fitted with monoblock engine. However, after the year 2000 there has been shift to submersible pumps from monoblock engines, as a response to the decline in the water table.

The survey of tube wells in the village shows that 40 out of the current 44 tube wells are shared by more than one household. All of them are kinship based. Most of the tube wells were started as shared wells by immediate ancestors. Few started as individual owned wells. If somebody set up a tube well 30 years back, his family size eventually increased and the *tabbar* broke into separate households. The land was divided among inheritors, however, following the well established tradition and the tube well a remained shared property. In some cases households from different *gots* came together (kins or caste men) to install a tube well. In the next generation their tabbar grew but finally broke up into two or more households.

Agricultural lands are fragmented and individual plots are small, while investment costs for installing tubewells are high. Faced with such small land holding and the high cost of

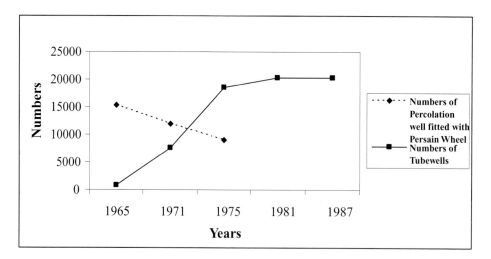

Figure 2.   Shift in sources of irrigation in Hoshiarpur district, Punjab, India.
Source: District Census Handbook, 1961 and 1991.

tubewell installation, the villagers had already a history of water sharing, and the basic principles and rules of sharing already existed.

3.4 *Cultural history of shared irrigation*

References of *Kuh* with *halt*, in the past, refer to wells with a Persian wheel. Most of the *Kuh* were shared. In 1961, the village had 21 *halts*, and most of them were shared. Out of the 44 current tube wells, 9 started as shared *halts*. In these shared Persian wheels the investment was made according to land ratio in the command area and this ratio also determined most of the rights and obligations. *Wari or Bari* (water turns) were allocated among partners according to their land ratio in the command area. There was gradual phasing out of Persian wheel during 1969–1975, as electricity connections spread rapidly in rural areas of Punjab. Electricity came to Bhajjal and the surrounding region in 1969.

3.5 *Changed technology and ecology*

Since 1970, and the arrival of the electric submersible pump, there has been a decline in the groundwater table elevation which has been more rapid in the last two decades. Villagers reported that on an average there has been decline of 3.3 meters per decade in last three decades. The villagers had no option but to chase the declining water table with deeper borings and since 2000 to submersible pumps of larger capacity (Table 1 and Table 2). However, the principles of water sharing remained the same.

Multidisciplinary tools and techniques were used to understand the various aspects of social organisation,. These included village schedules, focused group discussions (FGDs), documentary research (census, gazetteers, monographs), profiles of tube wells, constructing genealogies, interviews with historians and focused group discussion with 30 shared tube wells owners (shareholders). Out of these 30 shared tubewells, 20 shared tube wells were selected to create detailed irrigation profiles. The profiles were based on several research questions such as when and how the tubewell started, who were the original partners, inheritance (genealogy) and transfer of water rights, technology shifts, details of tube wells, water rights, scheduling patterns, oral/written contracts, how maintenance is carried out, payment of electricity bills, conflict and cooperation, role of *bada sanjha* (the largest shareholder, if any), asymmetry in water relations and cropping pattern for the year 2006. These profiles illustrate the principles, rules and sub-rules of shared irrigation. They also revealed that shared irrigation has its own conflicts and limitations. A typology has been attempted to show variations within the broad social organisation.

Table 1. Changing irrigation technology.

Technology	Depth to water table till these technologies were viable	Time period of use
Persian wheel	Up to 10 m	Until 1970
Electric monoblock pumps	Up to 25 m	1970–2000
Electric submersible pumps	Beyond 25 m	2000 onwards

Table 2.   Sources of irrigation and their percentage to net irrigated area,
Garhsankar Block, 1965 to 1991.

Source of irrigation/Year	1965	1991
Government canal	1.4	4.3
Tubewell	19	95
Wells	79	0.5
Total	100	100

Source: Census of India 1961 and 1991.

4   HUMAN ORGANIZATION OF SHARED TUBE WELLS

Profiles of shared tube well owners were created through focused group interviews. These
revealed a pattern of water sharing and the rules of sharing groundwater irrigation. These
recurrent behaviour patterns have been divided into principal, rules and sub rules. There are
minor modifications in operational details which are the sub-rules. The basic principle in
sharing the water is the land ratio. Water is also allocated by turns, *waris or baris*, which
reflect the *haq* (or right) of the shareholder. Water turns are usually of one day and are
calculated according proportion of land owned in the command area of the tube well by
the shareholder. This *haq* or water right is recognised by the partners and other villagers
and thus has customary legitimacy.

Whenever several landowners own adjacent plots of land and decide to install a shared
tubewell, the exact location of the tubewell is decided on the basis of a consensus. Usually
these landowners are kinsmen who currently form individual households or *ghar* and plots
are their result from the division of ancestral property. The shareholders in land also has
rights over common wells (if the well existed before partition). The shareholders can also
make collective decisions about upgrading the equipment. Almost all the old wells in
Bhajjal village were upgraded to submersible pumps.

This same principle is extended to water rights that are allocated to shareholders. The
ideal rule says that if there are four partners sharing one tube well and they have 0.4 ha each
then they are allocated a schedule of one day each successively. If the ownership ratio is
2:1:1:1, then the owner of 0.8 ha of land gets water for two consecutive days and the others
will get water for one day each. Water turns start automatically. Schedules are flexible
when the demand for irrigation is low. One shareholder may get an extra day of water or
few hours beyond the stipulated schedule if he so requests. But schedules are less flexible
during peak demand as the pressure on all the shareholders is high.

Irrigation schedules can be exchanged on a mutual basis between two parties, but for
various reasons, two households even belonging to the same kin may not always cooperate
with each other. The reasons may lie in past conflict or jealousy. Sociologists have also
reported disputes over shared irrigation (Hershman1981). These common properties are
source of conflict. Disputes occur over the boundary lines between the fields, and most
critically, in the case of farmers over the right to water and irrigate the fields at critical times.

The rule says that the electricity bill is also to be paid according to land ratio of land
owners. Change to the electric submersible tube needed anew rule to allow for sharing
of electricity bills according to the principle 'each according to his/her land'. There are,
however, variations over operational strategies in abiding with this rule. In some tube well

systems, the shareholders pool their respective shares, while in other cases they rotate the payment of the bill. Maintenance bills are also shared according to land ratio. The ideal rule says that each shareholder is equally responsible for maintenance works, but the interviews revealed that normally only one shareholder is accorded with this responsibility because of personal attributes like honesty or an ability to negotiate good rates with technicians. This person then tells other partners about incurred expenses which need to be shared.

The partners also bear the cost of maintaining the main distribution channels. Minor channels are constructed and maintained by beneficiaries (among shareholders) within a particular section of the command area only. Cleaning of the channel is done by partners whose land will be irrigated by a particular section of the channel.

Most of the arrangements are oral. The partners take decisions about the water schedule and rotation on mutual and oral basis. No register of schedules is maintained as each partner is aware of the principles and rules of the system. The sharing of expenditures on maintenance is also done through internal discussions records of maintenance expenditure are not usually kept.

4.1   *Inheritance of water rights*

The land ownership in the command area does not remain static over time. The land gets divided between the sons but the share in the tube wells remains a common property. Division of land in the original command leads to internal division of water rights. While describing the inheritance pattern in Punjabi society, Hershman states that the general mode of inheritance is patrilineal, but "There are some types of property which are indivisible such as Persian well, or an alleyway" (Hershman, 1981, page 67). In such cases the rights of access and usages are inherited patrilineally just like any other property and it so happens that members of a minimal lineage may have shares in a particular well, built by a common ancestor. The group discussions revealed that water rights are described in wills along with the land. By way of example, two brothers A and B having a common well fitted with Persian wheel were irrigating their respective plots each of eaqual area. In the next generation A has three sons A1, A2, A3 and B has two sons B1 and B2. Then A1, A2 and A3 will get water for three consecutive days and B1 and B2 will get three days of water as their fathers had equal share in water rights. In fact most of the *tabbars* continue to utilize the water right as the division of the water right will create very small schedules.

4.2   *Transfer of water rights*

The total land irrigated in the command area of the shared tube wells may also change ownership either through sale or purchase or gift of land. The attached water rights also get transferred to new land owners. However the reference point for well sharing remain the proportions of land owned by the shareholders in the original command area There are unwritten 'rules' of transfer of water which has social acceptability and sanctions. These rules also follow the rule of 'each according to his/her land'. The ideal rules practiced are:

1. If land from outside is brought into the existing system by purchase, inheritance, gift or otherwise and if this land was not in the original command area, then the shareholder who bought this land and included it in the existing shared tubewell command area would not get an additional right to water and his/her share of water from the shared tubewell would remain unchanged.

2. If, however, transfer of land occurs within the original command of a particular tubewell system, then the attached water right will increase or decrease in the same proportion, that is the new owner will receive the attached water right.
3. If one of shareholders sells land (partly) to a non shareholder then the new owner gets his share of water from the existing shareholder from whom he purchased the land, and others shareholders will not any have responsibility in sharing water with new entrant.

Just as there are well specified rules of entry, rules of exit from shared tube well system are also specified. One partner may leave the system to set up his own tube well, in order to have larger share of a water right for irrigation security. As soon he can arrange capital for a new tubewell, he may leave the shared system with the demand that the monetary value of the existing water right be refunded to him. There is a monetary value of water right in the shared tube well system and in lieu of exit; the exiting shareholder demands this value to be paid. Usually there are two reasons for an existing shareholding tubewell, either the water share was too small for adequate irrigation and thus the shareholder opts out for an individual tubewell, or there were recurrent disagreements among the shareholders.

The shareholder who wants to exit from the system begins by asking for compensation from the rest of the shareholders. After a few rounds of consultation among the shareholders themselves, the elders are consulted. Elders are usually respected members from the village. They talk to different shareholders and a price is fixed for the water right in consultation with all other shareholders. If partners agree to pay, then the shareholder makes a formal exit. Existing water rights and schedules are then readjusted among the remaining shareholders. The village land record also shows the readjustment in land records. There are also instances when a particular shareholder voluntary leaves a system without any claim or payment of compensation. Some well off farmers in the village are known to have foregone the water right in the old system when they installed new tubewells for individual use only. Whenever new issues emerge, the shareholders seek solution by consensus. This is not to say disputes do not occur but such disputes are amicably resolved as far as possible.

5   CONCLUSION

The village study provides details of a lesser known institution of shared irrigation which has been prevalent for long time. The shared irrigation system is based on linkages of land and water rights. Each according to its land is the organizing principle. The human organization around shared groundwater irrigation depends upon twin aspects of structure and function. Structural aspects denote well developed system of principles and rules which shape the irrigators behaviour. The durability of irrigation pattern can be explained by strong collective adherence to these structural aspects. The functional aspects on the other hand reveal the 'lived in' aspects of order and social adaptation to ecological and technological changes over a period of time.

REFERENCES

Census of India (1961). *District Census Handbook No 9 Hoshiarpur*, Census of India, Government of India, New Delhi, India. (Published in 1968).

Census of India (1991). *District Census Handbook, Series 20, Hoshiarpur: Village and town wise Primary Census Abstract*, Census of India, Government of India, New Delhi, India.

Gazetteer of the Hoshiarpur District (1883–84). Sang-e Meel Publications, Lahore, Pakistan, 2001.

Gray F. Robert (1963). *The Sonjo of Tanganyika: An Anthropological Study of an Irrigation-Based Society*, London Oxford University Press.

Hershman P. (1981). *Punjabi Kinship and Marriage*, Hindustan Publishing Corporation, Delhi.

Hunt Robert C., and Eva Hunt (1976). Canal Irrigation and Local Social Organisation, *Current Anthropology*, 17 (3).

IWMI-DSSA (2006). *Social Profile of the Bhajjal Village*, International Water Management Institute (IWMI), Colombo and Department of Sociology & Social Anthropology (DSSA), Punjabi University, Patiala, draft report submitted to IWMI Colombo, unpublished.

Leach E.R., (1961). *Pul Eliya, A Village in Ceylon*, Cambridge University Press.

CHAPTER 13

Social regulation of groundwater and its relevance to the existing regulatory framework in Andhra Pradesh, India

R.V. Rama Mohan

Centre for World Solidarity (CWS), Secunderabad, India

ABSTRACT: Exponential growth of tube wells and uncontrolled extraction of groundwater has resulted in a rapid decline of the groundwater levels in Andhra Pradesh, India. Repeated failure of bore wells has caused serious economic loss to the farmers. Andhra Pradesh Water Land and Trees Act (APWALTA), enacted in 2002, failed to check the growth of wells. The free power policy in the state contributed to further exploitation of groundwater. APWALTA proved itself to be a "disabling" law, approaching the problem through controls and penalties. A grass-roots project in three villages promoted sharing of groundwater among farmers and saving of groundwater using micro-irrigation. "Social Regulations" such as no new wells brought a change in the attitude of farmers to "cooperate" rather than to "compete" for groundwater. The paper recommends a blend of two approaches to make APWALTA an "enabling" and "constructive" policy.

1 INTRODUCTION

Groundwater has emerged as a vital resource for irrigation in Andhra Pradesh, especially in the semi-arid regions like Telangana and Rayalaseema (Figure 2). Low and erratic rainfall, coupled with recurring drought conditions prompted farmers to shift to groundwater irrigation from rainfed farming. Reliability of groundwater sources and the spread of tube well technologies triggered rapid growth of tube wells (also known as bore wells) in Andhra Pradesh over the last 15 years. But, indiscriminate drilling of wells and over-extraction of groundwater has adversely affected the groundwater resources in the state. Declining groundwater levels in many areas resulted in drying of existing open and shallow tube wells. Failure of wells at the time of drilling is another cause of loss of investments and a debt burden on farmers which in turn adversely affects their livelihoods.

Andhra Pradesh Water Land and Trees Act (APWALTA) was enacted in 2002 to check the growth of wells and groundwater extraction in the state. The Act was further improved in 2004 and 2005 by introducing features such as an insurance scheme for failed wells and a single window system for permissions to drill new wells. While APWALTA has a top-down structure that regulates the groundwater use in the state, there are few grass-root interventions in Andhra Pradesh that attempt to empower communities in water management and promote local governance of groundwater (van Steenbergen, 2006).

This paper analyses the performance of APWALTA over the last 4 years (2002 to 2006) based on the data available from secondary sources. The paper also describes the social regulation of groundwater implemented in three villages within the regions of Rayalaseema and Telangana by the Centre for World Solidarity (CWS)—a NGO located in Andhra Pradesh.

2   GROUNDWATER IRRIGATION IN INDIA AND ANDHRA PRADESH

Groundwater has been recognized as an important water source of irrigation since ancient times in India. Shallow open wells were dug and initially manual lifting devices were used to extract water. Large numbers of open wells were used as supplemental source of irrigation in semi-arid regions of the country.

With the advent of tube wells and pumping technologies during the 1980s, extraction of deeper groundwater began. More and more farmers preferred irrigation from groundwater due to its reliability and control on the resource. Availability of institutional and non-institutional credit also made these technologies easily available to farmers. The area irrigated under surface water sources (such as canals and tanks) has either remained stagnant or even declined, while the area under wells and tubewells has increased steadily since 1951 (Figure 1).

In line with the national trends, Andhra Pradesh (Figure 2) also witnessed growth of groundwater extraction structures and area under groundwater irrigation. Table 1 gives the number of different types of wells in Andhra Pradesh and India. In Andhra Pradesh, between 1993–94 and 2000–01, the number of wells increased by 58.4% with an average increase of 101000 wells per year (Table 2). During this period, the net area irrigated from groundwater increased by 9.0% with a corresponding reduction of 10% under surface water sources (Figure 3).

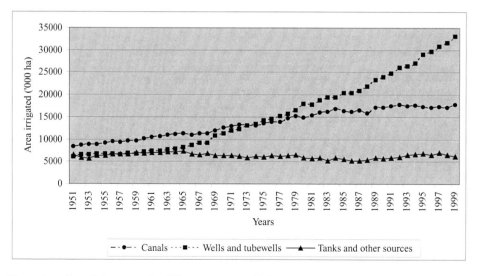

Figure 1.   Growth in area under different sources of irrigation, 1951–1999.

Source: Department of Agriculture and Cooperation, Ministry of Agriculture, India (several years) (http://agricoop.nic.in/statistics/sump2.htm) downloaded on 20th January 2007.

Shallow and deep tube wells played a major role in this growth of irrigation from groundwater (Table 2). This trend continued after 2001 and the number of agricultural electricity connections was reported to be 2.37 millions by 31 March 2005 and the same increased to 2.44 millions as of 31 March 2006, with an increase of 66 500 wells in one year (APTRANSCO, 2006). This rapid growth of bore wells and unregulated exploitation of groundwater resulted in decline of groundwater levels to alarming levels in several pockets of Andhra Pradesh and an increasing number of dug wells dried up. The Groundwater Department, Government of Andhra Pradesh, classified 187 micro-basins as over-exploited (groundwater development defined as groundwater extraction as a percentage of the net renewable recharge >100%); 82 micro basins in critical condition (groundwater development >90%); and another 203 micro basins as semi-critical

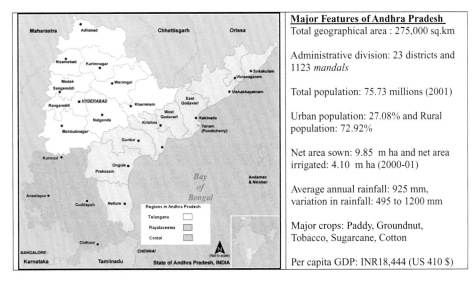

Figure 2. Location map and major features of Andhra Pradesh.
Sources: Directorate of Economics and Statistics, Ministry of Agriculture., Government of India (2000–01), http://dacnet.nic.in/eands/ downloaded on 23rd January 2007, Census of India (2001), http://censusindia.gov.in/ downloaded on 23rd January 2007.

Table 1.   Number of wells in Andhra Pradesh and India during 2000–2001.

Type of well	Number of wells (2000–01) (in thousands)		
	Andhra Pradesh (AP)	India	Wells in AP as % of wells in India
Dug Wells	1185	9617	12.3
Shallow Tube Wells (Less than 70 m deep)	656	8355	7.8
Deep Tube Wells (More than 70 m deep)	87	530	16.4
Total	1928	18502	10.4

Sources: Ministry of Water Resources, Government of India, 2001.

Table 2.   Number of and increase of total wells in Andhra Pradesh between 1993–94 and 2000–01.

Type of well	Up to 1993–94 (in thousands)	Up to 2000–01 (in thousands)	% increase in number of wells	Average increase of wells per year (in thousands)
Dug wells	939	1185	26.2	35
Shallow tube wells (Less than 70 m deep)	246	656	166.6	58
Deep tube wells (More than 70 m deep)	32	87	171.8	8
Total	1217	1928	58.4	101

Sources: Ministry of Water Resources, Government of India, 2001.

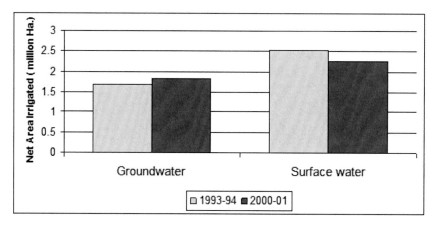

Figure 3.   Net area irrigated in Andhra Pradesh during 1993–94 and 2000–01.
Sources: Ministry of Water Resources, Government of India, 2001.

(groundwater development >70%) out of the total 1229 micro basins in Andhra Pradesh (Figure 3) (Groundwater Department, GoAP, 2006). This classification was based on the groundwater levels recorded from selected observation wells, and the pumping intensities and cropping patterns in each micro-basin. The average decline in the water table elevation over the period 1999–2003 was 2 m in Andhra Pradesh (Department of Disaster Management, GoAP, 2003).

Drilling of multiple bore wells by individual farmers due to repeated failures is a common phenomenon in the interior drought-prone Rayalaseema and Telangana regions of Andhra Pradesh. Indiscriminate sinking of bore wells and repeated failures have been reasons for the debt burden and resultant suicides of farmers in Andhra Pradesh in recent years. As an example, one farmer in Anantapur rural *Mandal*, a chronic drought-prone area with a relatively low average annual rainfall of 495 mm, drilled about 100 bore wells in the last 10 years (The Hindu, 2004). In Anantapur district, the number of agricultural bore wells increased from 93 000 in year 2003 to 136 030 in year 2005, an increase of 46% in just two years (Government of Andhra Pradesh, 2005).

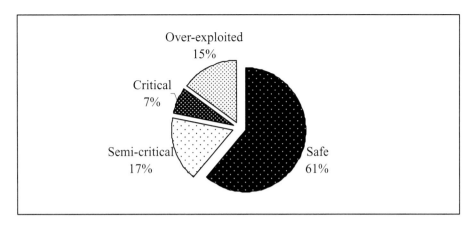

Figure 4.   Status of groundwater resources in various micro-basins in Andhra Pradesh during 2004–2005.
Sources: Groundwater Department, GoAP, 2006.

3   POLICY AND REGULATORY FRAMEWORK IN ANDHRA PRADESH

Repeated droughts and alarming depletion in groundwater resources during the late 1990s prompted the AP State Government focus on efficient water conservation and management. The last decade has witnessed the promulgation of various acts and guidelines related to water management (Table 3). One measure is the Andhra Pradesh Water, Land and Trees Act (APWALTA) (2002). This Act repealed earlier legislations such as Andhra Pradesh Ground Water Act (Regulation for drinking water purposes), 1996 and Andhra Pradesh Water, Land and Tree Ordinance, no.15 of 2000.

The APWALTA, enacted in April, 2002, is a comprehensive act covering surface and groundwater resources. The act aims to promote water conservation, enhance tree cover, and regulate the exploitation and use of surface and groundwater. The act empowered the State Government to appoint a state level authority, namely, Andhra Pradesh Water, Land and Trees Authority. The Authority has the mandate to promote water conservation, enhance tree cover, regulate exploitation of water, make regulations for the functioning of the authorities at District and *Mandal* level (administrative units within Districts), and advise the Government on various legislative, administrative and economic measures including the strengthening of public participation.

Provisions related to groundwater, the so-called 'Groundwater protection measures' are (Government of Andhra Pradesh, 2002 & 2004):

- All groundwater resources in the State shall be regulated by the Authority.
- Owners of all wells and water bodies shall register their wells/water bodies with the Authority.
- Prohibition of water pumping in any particular area in the State.
- Prohibition of and penalty for drilling new bore wells in any particular area in the State.
- Any person shall obtain permission for drilling a new bore well (other than drinking purposes) within 250 m of a public drinking water source.

Table 3.   Key events in water policy evolution in Andhra Pradesh.

1996	Enactment of Andhra Pradesh Ground Water Act (Regulation for drinking water purposes)
1997	Enactment of Andhra Pradesh Farmers' Management of Irrigation Systems Act to promote participatory management of irrigation systems in the state
1999	Spelt out Vision 2020, emphasizing the importance of water management and participatory approaches to irrigation management for sustainable growth in the agriculture and fisheries sectors
2000	Andhra Pradesh Water, Land and Tree Ordinance (no.15)
2002	Enactment of Andhra Pradesh Water, Land and Trees Act (APWALTA)
2002	Release of Guidelines for Watershed Development in Andhra Pradesh based on the national guidelines (1994) and recommendations of the reviews done time to time during later years
2003	Andhra Pradesh Water Vision defining a broad policy framework for water management in the state

- Declaration of over-exploited areas and ban on sinking new wells in these areas (other than drinking purpose). Periodic review of status and provision to revoke the declaration.
- Protection of drinking water sources by prohibiting extraction of groundwater from existing wells in the vicinity that are adversely affecting the drinking water source. Compensation to the owner for closure of existing well.
- Prohibition of water extraction for sale from an over-exploited water source or aquifer or residential areas or premises of multi-storied buildings in urban areas.
- Specification of well spacing and depth for sinking new wells to curb unhealthy competition to tap water from deeper layers of groundwater.
- Registration of drilling rigs by rig owners.
- Issue guidelines and impose conditions for rainwater harvesting measures in residential, commercial and other premises and open spaces to improve groundwater resources.
- Formulate guidelines for recycling and reuse of waste water by industrial, commercial users and local bodies.
- Prohibition of groundwater contamination from any source, including industrial, domestic and aquaculture/agriculture.
- Regulate sand mining to prevent depletion of groundwater and protect public drinking water supply sources.

Specific rules and procedures were periodically issued through Government Orders within the broad provisions of the Act. APWALTA rules and procedures were revised comprehensively in 2004. For example, a farmer can now submit an application to get a permission to drill a new well to a single office (the *Mandal* Revenue Office) and the decision would be announced within 15 days.

The State Government came up with a notification in February 2005 listing 4003 over-exploited villages (15 %) out of the total 26 586 villages and banned further exploitation of groundwater and sand mining in these villages. The regional classification of these villages in Andhra Pradesh. (Table 4) implies no new permissions to drill bore wells will be entertained. It was also stated in the notification that, the status of groundwater exploitation in these villages will be reviewed every six months and necessary modifications done.

The Government of Andhra Pradesh introduced well failure insurance scheme for the bore well owners in March 2005. Farmers who have a permit to drill a bore well are eligible to claim insurance subjected to a maximum limit of INR10 000 if the well turns out to

Table 4.　Number of villages notified as over-exploited in Andhra Pradesh.

Region in Andhra Pradesh	No. of *mandals* covered by notified villages	No. of notified villages
Coastal	64	395
Rayalaseema	169	1378
Telangana	232	2230
Total	465	4003

Sources: Andhra Pradesh Water Land and Trees Act 2004 (Amendment), Government of Andhra Pradesh.

be dry. Farmers were required to pay INR1200 towards the insurance premium, in addition to the geological survey charges of INR1000 (INR 500 for small and marginal farmers) towards the cost of site investigation by a qualified hydro-geologist. From 2006–2007, the Government decided to revise the insurance scheme for failed bore wells through the Commissioner, Rural Development, by maintaining a corpus fund independent from the insurance industry. All those bore wells which were drilled after obtaining necessary permissions under APWALTA are eligible for insurance compensation of INR10 000 or actual expenses, whichever is less, and the farmers need not pay any premium for obtaining the insurance cover (Government of Andhra Pradesh, 2006).

In June 2005, the State Government defined the minimum spacing to be maintained between drinking and irrigation wells for new wells. The minimum spacing for shallow tube wells is 260 m from an existing irrigation well and 250 m from an existing drinking water well.

Free power to agricultural pump-sets was introduced by the new State Government during late 2005, to replace the previous tariff system based on the the pump capacity. Although free power was supplied to farmers, the Government has been limiting the number of hours of power supply from 9 hours to 6–7 hours a day due to shortage of power, especially during *Rabi* crop season (the dry season, from November to February). The free power scheme was a major trigger for the steep increase in number of electric pumps (see Section 2) and deepened the power crisis during *Rabi* and summer months of 2006–2007. Hence, there is a direct conflict between the policy promoted by the APWLTA and the power supply policy in terms of incentives for the farmers to exploit groundwater.

4　IMPLEMENTATION EXPERIENCE OF APWALTA

APWALTA has been in force since 2002. Significant changes in the operations of the Act came about in 2004. During 2005, insurance of failed bore wells was introduced to protect farmers.

4.1　*Registration of existing wells and permissions for new wells*

Since inception in April 2002 to January 2006, a total of 2.2 million existing wells and 2178 drilling rigs were registered under the Act. Permission was granted to dig a total of 5389 new wells by January 2006 (Table 5).

Table 5.   Achievements of APWALTA.

	Apr. 2002–Jan. 2006	Jan. 2005–Jan. 2006
Existing wells registered	2.2 million	0.7 million
Drilling rigs registered	2178	284
New bore wells given permission	5389	2500

Sources: Progress Reports on Implementation of APWALTA, Ministry of Rural Development, GoAP, 2006, http://rd.ap.gov.in/WALTAReports.htm downloaded on 18th December 2006.

Table 6.   Fees and penalties under APWALTA.

Registration of existing wells	INR 10 (US 0.22 $)
Permission of drilling new well	INR 100 (US 2.2 $)
Registration of drilling rigs (August 2004 onwards)	Reduced from INR 10 000 to 1000 per year (US$220 to US$22
Failed well insurance premium	INR 1200 per well (US$26.67) later premium waved off in 2006
Illegal drilling of wells (Nov 2005 onwards)	Increased from INR 1,000 to 100,000 (US$22 to US$2,200)

Sources: Progress Reports on Implementation of APWALTA, Ministry of Rural Development, GoAP, 2006, http://rd.ap.gov.in/WALTAReports.htm downloaded on 18th December 2006.

The number of wells reported to be drilled legally during 2005–2006 (2,500) is much less than the average annual increase of 101 000 in Andhra Pradesh during 1993–2001 reported by the Ministry of Water Resources (Table 2). In contrast to this, the number of individual agricultural electrical service connections increased by 66 458 during 2005–2006 (APTRANSCO, 2006). Comparison of these statistics from different sources indicates that the majority of new wells in Andhra Pradesh were drilled without obtaining permission from APWALTA and APWALTA failed to check the proliferation of new wells in the state.

4.2   *Fees and penalties*

APWALTA specified the fees for registration of existing wells and applications for new wells. The Act also spelled in detail the penalties to be imposed for violating the law, such as, illegal drilling of a bore (Table 6).

The number of cases of illegal drilling of wells in the period Jan. 2005 to Jan. 2006 (corresponding to US$5680) is about 258, assuming a rate of US$22 per well. This seems to be very low compared to the 2500 well drilling permissions that were issued and the 66 500 wells completed. APWALTA transfers 90% of the revenue generated from penalties for illegal drilling of wells to the respective *gram panchayats* (local self-governments at village level) to take up awareness generation activities on the provisions of the Act. Table 7 shows recent data on progress of implementation of the Act, during the period May 2006 to October 2006.

Table 7.   Monthly progress of implementation of APWALTA, May to October 2006.

Month (Year 2006)	Number of wells registered	Number of applications sanctioned for new bore wells	Number of bore wells reported to be drilled	Number of successful bore wells	Number of wells to be failed	No. of insurance claims settled	Revenue generated in terms of insurance premium from farmers (INR millions)
May	50	0	0	0	0	0	NIL
June	64	53	42	22	0	0	0.005 (US$100)
July	99	161	32	22	0	0	0.0084 (US$ 168)
August	128	170	41	31	0	0	0.021 (US$ 420)
September	144	73	41	31	0	0	NA
October	178	79	101	37	8	1	NA
Total	663	536	257	143	8	1	

Sources: Ministry of Rural Development, Govt. of Andhra Pradesh Website http://rd.ap.gov.in 18th December 2006.

The sum of successful and failed bore wells (151) does not match with the number of total bore wells reported to be drilled during the period (257). The number of failed wells is under-reported. The number of insurance claims for failed wells settled may also be under-reported. Few of the farmers opted for insurance cover as the charges were felt to be high.

Following are the major observations on the APWALTA implementation experience:

- The Act was successful in registering most existing wells (2.2 million compared to 2.4 million electrical connections in January 2006 and March 2006, respectively), but did not check the growth in numbers of new wells.
- APWALTA primarily aims to regulate and control the drilling of new bore wells in water-stressed areas either by banning new bore wells or by laying out the procedures for obtaining permissions. Data from different sources on average increase of wells in Andhra Pradesh indicate that the illegal drilling of wells was hardly controlled
- The community has no recognized role in water governance, at local or regional level. People are mainly revenue generators for the state by paying registration charges for bore wells and drilling rigs, permission fees for new wells and penalties for violating the law. But farmers were not informed properly on the new regulations and alternative solutions to curb groundwater use.
- APWALTA does not promote constructive steps or incentives towards conserving water, or using water more efficiently nor offers any disincentive for unlimited extraction of groundwater. Thus, APWALTA is not an "enabling law" that provides incentives for not going for a new bore well and encourages farmers to save and use groundwater efficiently.
- While existing bore well owners continue to exploit groundwater, farmers who do not have bore wells are denied opportunity to access groundwater within a specified distance from the existing drinking and irrigation wells.
- APWALTA "empowers" the administration to control, regulate and manage groundwater with extensive and extraordinary provisions. One such provision is the power to regulate or limit the extraction from any well that is detrimental to the public interest. The Act does not propose any reforms in entitlements, such as altering the easement rights; defining entitlements of individual farmers or spell out electricity tariff strategies to control groundwater use.
- There is a perceptible gap in keeping track of happenings on the ground after giving sanctions to new bore wells and systematic reporting of progress achieved.

5   SOCIAL REGULATION IN GROUNDWATER MANAGEMENT

An action research project called "Social regulation of groundwater management at community level" was initiated in 2004 in three villages in Andhra Pradesh by the non-governmental organization Centre for World Solidarity (CWS). Then villages area Secunderabad, Andhra Pradesh. The project covers 665 families in 3 villages in 3 districts of Andhra Pradesh. The project aims to promote local regulation and management of groundwater resources with equitable access to all families in the communities. The project cost was around INR 1.2 million (US$26667) for each of 2 years.

The three project villages are in the semi-arid regions Rayalaseema and Telangana of Andhra Pradesh with repeated occurrence of drought (Figure 4). In all three villages, rain-fed agriculture is the norm, but groundwater is an important contributor to irrigation on

6–35% of the land (Table 8). Erratic rainfall and recurring drought conditions prompted farmers to use groundwater, which is more reliable and controllable. Groundwater provided the irrigation source during prolonged dry spells within the rainy season.

Initially, open wells were dug and electrical centrifugal pumps were used to extract groundwater. Farmers started drilling bore wells from the early 1990s and shallow open wells gradually dried up due to falling groundwater levels. Over the last 15 years, the number of bore wells grew rapidly in these villages. Due to indiscriminate drilling of bore wells and unscientific groundwater exploration, many bore wells failed either at the time of drilling or during later years. Furthermore, drilling bore wells as deep as 100 m resulted in drying of shallower open and bore wells. This phenomenon resulted in huge loss of investments to farmers and seriously affected the livelihoods of farmers dependent on irrigated farming.

The project interventions began with a participatory assessment of the water resources status in the three villages. Participatory Rural Appraisal (PRA) methods were used to map the resource status and existing water utilization pattern for different purposes, such as: drinking, domestic, irrigation. Growth of groundwater-based irrigation and trends in groundwater levels over a period of time were thoroughly discussed and analysed in community level meetings, wherein women and men from all households participated. A series of meetings and interactions helped to identify the issues: frequent failure of bore wells and increasing debts of farmers due to investment on new bore wells.

The competition between neighbouring farmers often leads them to drill bore wells as close as 2 m apart. For instance, in Madirepally village, three neighbouring farmers dug 13 bore wells in an area of 0.2 ha over a period of four years in competition to tap groundwater (Figure 6a). The project realized that there is need for changing the attitude of farmers from "competition" to "cooperation" and to increase the "water literacy" among the farmers for efficient use of water.

A number of training programmes, exposure visits and awareness raising meetings were organized by CWS in the project villages. Further public awareness and education was carried out through posters, pamphlets and advertisements. Participatory hydrological monitoring of rainfall and groundwater levels in selected bore wells was done regularly and shared and discussed at village meetings to increase the understanding of farmers on the behaviour of groundwater in relation to rainfall. A volunteer from the community measured rainfall from a simple manual rain gauge station installed in the village and recorded the static water levels in 10 sample bore wells using an electronic water level indicator. These data were displayed on a village notice board and updated periodically.

The last $2^1/_2$ years of facilitation has resulted in the community realizing the ill-effects of indiscriminate drilling of bore wells and use of groundwater. The community agreed on the following 'social regulations' and interventions in the village:

- No new bore wells to be drilled in the village.
- Equitable access to groundwater to all the families through well sharing.
- Increasing the groundwater resources by conservation and recharge.
- Efficient use of irrigation water through demand-side management.

Small groups of farmers were formed in all the project villages between a bore well owner and 2 or 3 neighbouring farmers who did not own bore wells. Bore well owners were motivated to share by explaining that drilling new wells in the vicinity of their wells may

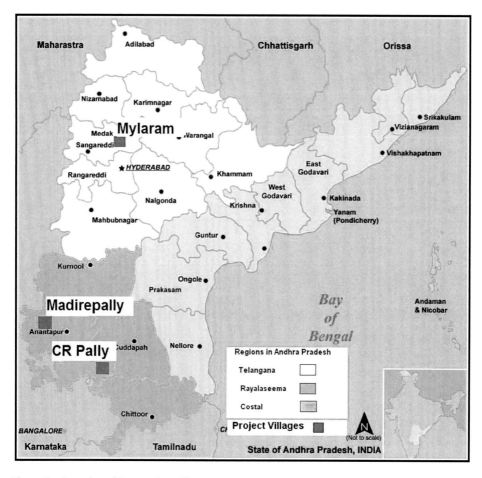

Figure 5.   Location of three project villages.

Figure 6a.   Bore wells drilled in competition (before the project intervention).

Figure 6b.   Neighbouring farmers sharing water (after the project intervention).

Table 8.   Details of project villages.

Administrative and demographic details			
Village	Madirepally	Mylaram	CR Pally
Gram Panchayat	Akuledu	Mylaram	CR Pally
Mandal and District	Singanamala, Anantapur	China Koduru, Medak	Tanakallu, Anantapur
Region	Rayalaseema	Telangana	Rayalaseema
No. of households	166	211	288
Population	721	1326	1390
Land resource details (in hectares)			
Total geographical area	298	1151	1069
Cultivable area	278	990	680
Irrigated land in *Kharif*[a] 2006	136	110	99
From surface water	90 (13.1%)	116 (4.7%)	120 (7.1%)
From bore wells	245 (35.7%)	156 (6.4%)	125 (7.4%)
Purely rain-fed	352 (51.2%)	2174 (88.9%)	1435 (85.4%)
Water resource status			
Average rainfall	495 mm	750 mm	545 mm
Surface water harvesting structures (tanks and ponds)	7	7	5
Small check dams	4	–	–
Groundwater extraction structures			
Open wells (functional during *Kharif* 2006)	57 (0)	135 (0)	60 (06)
Bore wells Open wells (functional during *Kharif* 2006)	137 (65)	115 (60)	60 (34)
Drinking water bore wells	2	3	5
Bore wells with hand pumps for domestic use	3	3	7
Crops			
Crops grown	Paddy, Ground Nut, Sweet lime, Maize, Vegetables	Ground Nut, Paddy, Cotton, Maize, Chilli	Ground Nut, Paddy, Sunflower, Vegetables, Millets, Mulberry

[a] Wet season, from June–Oct.
Sources: CWS, Primary Survey.

render them dry due to competitive extraction of groundwater. Instead, sharing a portion of water from his well helps his neighbours and at the same time secures his access to water. Sharing water with their neighbours will be a "win-win" situation benefiting both the bore well owners and water receivers.

Small farmers were given priority in formation of groups (Table 9). Group members were encouraged to save water by using micro-irrigation kits (sprinklers) and share water from the existing wells rather than drilling new bore wells in the vicinity of existing ones. Surface water harvesting and retaining structures were renovated, and existing dry open wells were converted into recharge wells. Regular monitoring and recording of hydrological information, such as rainfall, water levels in 10 sample wells and water storage in all surface

Table 9.   Water sharing groups in the three project villages by March 2007, (CWS, Primary Survey).

	Madirepally	Mylaram	CR Pally
No. of functional bore wells	65	60	34
No. of bore wells under sharing system	33	16	17
No. of sharing farmers[a]	68	48	36
Cultivated area under shared water system in *Kharif* 2006 (ha)	42	36	24
Percentage of area under shared water system to total cultivable area	15.0	3.7	3.6

[a] Includes the well owners.

Table 10.   Relative groundwater draft in the three project villages.

	Extraction as a % of recharge		
Village	2004–05	2005–2006	2006–07
Madirepally	121	95	87
Mylaram	326	295	261
CR Pally	147	108	97

Source: CWS, Primary survey.

water bodies was done since 2004. Groundwater abstraction for different purposes, such as irrigation, drinking and domestic was calculated based on the data collected from the villages.

Groundwater abstraction as a percentage of net available groundwater was calculated on a yearly basis from 2004–2005 (Table 10). While recharge to groundwater increased from 2004–05 to 2005–06 due to better rainfall conditions and water harvesting measures, farmers also reduced abstraction using micro irrigation kits and water saving crops. In Madirepally, in spite of relatively low annual rainfall (297 mm) during the year, abstraction was also reduced due to the increase in micro-irrigation kits and use of water saving crops.

While creating access to groundwater to 86 farmers who did not own bore wells, the project was successful in reducing the groundwater extraction in the project villages. The project aimed at bringing all functional bore wells in the three villages under the water sharing system by 2008.

CONCLUSION

APWALTA has no role for people in groundwater governance. This is primarily a top-down legislation that tries to "control and regulate" the behaviour of people through procedures and penalties. Contradicting policies, such as free electricity policy to agricultural pump-sets, caused rapid growth in the number of wells over the last $1^1/_2$ years and contributed to negating the results of APWALTA.

Social regulations and community control is a bottom-up approach that is based on collective and voluntary action. This approach offers incentives for farmers not to drill new wells, by providing secure and equitable access to water and water-saving technologies. However, social regulations and sharing also operate within the broader water policy framework and power subsidy regime. Subsidies on micro-irrigation, to the extent of 66%, was very positive and contributed to the success of the social regulations approach in the project. Similarly, the absence of a free power or subsidized cheap power policy, would have made it difficult to convince bore well owning farmers to share water. The free power policy should be re-assessed, and adapted sharing mechanisms tested so that water-deprived farmers could pay appropriate fees to the well owners for a share of the water.

Therefore, both these approaches to groundwater regulations have their own limitations. Social regulations at community level cannot be an isolated alternative to the macro level groundwater policies and approaches. Instead, a blend of both is required to address these limitations and to manage the groundwater resources in a sustainable way.

REFERENCES

Census of India (2001). http://censusindia.gov.in/ (downloaded on 23rd January 2007).

APTRANSCO (2006). Performance and Statistics, Andhra Pradesh Transmission Corporation, http://aptranscorp.com/pact01.pdf (downloaded on 23rd January 2007).

Directorate of Economics and Statistics, Ministry of Agriculture., Government of India (2000–01), http://dacnet.nic.in/eands/ (downloaded on 23rd January 2007).

Department of Disaster Management, (2003). A Document on Management of Drought. Government of Andhra Pradesh.

Government of Andhra Pradesh. (2002). Andhra Pradesh Water Land and Trees Act (APWALTA).

Government of Andhra Pradesh. (2004). Andhra Pradesh Water Land and Trees Act (Amendment).

Government of Andhra Pradesh. (2005). Hand Book of Statistics (2004–2005), Anantapur District, Andhra Pradesh.

Government of Andhra Pradesh. (2006). Failed Bore Well Compensation Scheme, G.O. Rt no.1099.

Ground Water Department, (2006). Groundwater Resources (2004–2005), Andhra Pradesh (Volume I & II), Government of Andhra Pradesh.

Ministry of Rural Development (2006). Progress Reports on Implementation of APWALTA, Government of Andhra Pradesh, http://rd.ap.gov.in/WALTAReports.htm.

Ministry of Water Resources (2001). *Report of the 3rd Census of Minor Irrigation Schemes*, Government of India.

The Hindu (2004). Lack of clarity on insurance worries ryots, news article dated 10th November.

van Steenbergen, F (2006). Promoting Local Management of Groundwater, *Hydrogeology Journal*, 14 (3): 380–391.

CHAPTER 14

Using the living wisdom of well drillers to construct digital groundwater data bases across Indo-Gangetic basin

S. Krishnan
CAREWATER, INREM Foundation, Gujarat, India

A. Islam
Indian Council of Agricultural Research Complex for Eastern Region, Patna, India

D. Machiwal
Soil and Water Engineering Department, CTAE, MPUAT, Rajasthan, India

D.R. Sena
Central Soil and Water Conservation Research and Training Institute (CSWCRTI), Gujarat, India

K.G. Villholth
Geological Survey of Denmark and Greenland (GEUS), Copenhagen, Denmark

ABSTRACT: The low density of current groundwater instrumentation networks in developing countries, which are both cost and management intensive, are an impediment to informed management of groundwater in these countries. By way of contrast, local knowledge of groundwater which is often perceptive has greater spatial coverage and can be obtained at a relatively lower cost. One efficient way to tap such local groundwater knowledge is through well drillers. In the Vaishali district of Bihar state in eastern India, a new methodological approach is used to identify and sensitize well drillers towards creating a local groundwater database. A localized database for a single village is created using the knowledge and current practice of the local drillers. Though subjective with various sources of uncertainty, the knowledge is verifiable and cost-effective. There is a potential for up scaling this approach to create accurate regional groundwater databases at low cost.

1 INTRODUCTION

Knowledge is an important key to better management of any resource. A critical problem associated with management of groundwater, especially but not exclusively in developing countries, has been the limited established knowledge of the resource at the local level. The existing monitoring networks established by scientific institutions and current management policies fail to highlight local scale information and inhibit appropriate water resource management. Such science-based information may even contradict common sense-perception of local groundwater conditions.

Different schools of thought debate strategies to address the current groundwater management problems. One school professes that it is the larger policies of energy, agriculture and trade that influence how groundwater is used and, therefore, one needs to focus on these policies. Another school professes that beyond these policies, local communities need to get together to assess, monitor and control the use and protection of groundwater in their own environment. In both, however, know-how of the current state of the resource and how it is affected by policy or local action is essential. Understanding the way a system responds to stimuli is essential for both pro-active and reactive measures. But, such knowledge of groundwater requires considerable investment in detailed studies and monitoring. Are there other ways in which local information on groundwater can be captured?

This paper explores and argues that it is possible to capture local information at a much lower cost by involving key stakeholders in the groundwater resource. Where there is extensive use of groundwater, there is also a reasonable density of wells, and where there are wells, local professionals are likely to be present. These include the well drillers and they are usually located in the village itself or in small towns close to villages. Often, they belong to local communities such as fishermen, carpenters and masons for whom the transition from their traditional occupations to well drilling is a natural process. Drillers are the main source of information for farmers on issues related to groundwater and between them they foster exchange of information across their area of operation. The drillers individually and collectively possess critical knowledge about local lithology, groundwater flow and groundwater quality conditions and could create local databases on groundwater, given some external help. The questions to be asked are:

i Do the well drillers really possess appropriate knowledge?
ii If so, how can it be extracted and collated into local information databases on groundwater?
iii Can the knowledge be verified, i.e. by comparing this information database with science-based information that already exists in some form?
iv How can this knowledge be disseminated?

von Hayek (1974) talks about the relevance of local and specific information in any complex science. The problem with many complex subjects is that global or conceptual pictures are easily made, but local detail is less easily available. However, it is this local knowledge that often adds substance and makes a concept useful. Unless this local knowledge is efficiently used, any global or theoretical knowledge remains an overview. Both traditional scientific knowledge and traditional local knowledge have their own benefits and disadvantages (Table 1). Ideally, a fusion of both is needed to achieve the best in terms of such as scale, tools used, spatial coverage, precision, repeatability, communication and purpose.

1.1   *Previous studies on local hydrologic knowledge*

Man acquires intuitive knowledge that is essential for his occupation, such that a farmer can sense the arrival of rainfall from the movement of clouds and wind. This intuition is built over time from experience and differs from one profession to another and from one professional to another depending on experience, interest and ability of deduction and abstraction. If such professions continue over generations, especially within families and communities as traditional occupations, knowledge gained gets passed by word of mouth and becomes part of common sense. Knowledge of local hydrology in communities closely associated with agriculture can be quite detailed.

Table 1.   Comparing science-based knowledge and local knowledge about groundwater.

Characteristic	Science	Local knowledge
Scale	Large scale, general, conceptual *Aquifers*	Smaller scale, specific, practical *Can describe nature of local flow*
Tool	Designed instruments, limited, focused, recorded *Rain gauge, Water level recorder, drill logs*	Many undefined instruments, unfocussed observation, mostly unrecorded *Different sensors, word of mouth, passing of information through generations*
Spatial coverage	Time and space sparse, interrupted time-series *Depends on monitoring network*	Dense in space and time, long term observations *Every individual is an observer*
Precision	More precise, errors more objective and amendable *Results from repeated measurements*	Perceptive, individual, errors difficult to evaluate *Every individual has different perception, possible bias*
Repeatability	Repeatable measurements *Can use same monitoring equipment at different places*	Possibly poor repetition *Cannot expect similar perception and experiences for same observation*
Communication	Easy to translate and communicate *Somewhat standardized terms, such as porosity*	In local language and need to be interpreted Terms such as *Kankar, Pathar, Khara Nadi*
Purpose	Observations useful for scientific interpretation and modeling *Measurements such as hydraulic conductivity*	Observations of importance to daily life and water use *How fast does water fill into a well?*

Note: *Kankar*: gravel; *Pathar*: stones; *Khara*: saline; *Nadi* river.

A study of a village in Rajasthan reported groundwater irrigation and water management practices in this arid region based upon a rich knowledge of local water resources (Rosin, 1993). This study, spanning 25 years of observation, looks at how local water harvesting structures are built with knowledge of siltation, runoff, recharge to groundwater, salinization processes and groundwater flow. Rosin documents the case of reasonably large diversion of a water resource and the discussion within the community leading to a variety of alternative actions. Rosin observed that the community was able to envisage the consequences of future actions of water diversion much better than engineers and their speculations were vindicated by later results. He proposes that hundreds of years of groundwater use had naturally brought the community to a stage where it was able to conceptualize groundwater through external signals. For example, farmers could hold their ear to the ground during shallow drilling and visualize the direction of flow. Their knowhow of local topography and flow was enough to indicate which were the areas of shallow groundwater and surface water stagnation and, therefore, possible locations of salinization of water.

CSE (2001) reported examples from across India of traditional practices of water management. These traditional water harvesting structures show a sound understanding of local hydrological processes and intuitive knowledge of geology. The *Surangams* of Kerala are tunnel-like structures constructed to collect groundwater and channel it out to wells. These ancient structures reveal a sound understanding of groundwater dynamics and an ability to use this knowledge for engineering activities. Examples of such structures have been found in various cultures in arid regions throughout the world under various local names, such as *karez*, *qanat*, *foggara*, and *falaj*.

Shah (1993) in a study of a coastal village of Junagadh district of Gujarat described how farmers built up their own picture of local groundwater hydrology through observation of water level dynamics during pumping. These observations lead them to conceptualize their own understanding of the aquifer structure in their local language. This was counter to the proposal of a hydrogeologist who based his conclusion upon a conventional hydrogeological survey of the area, proposing a leaky aquifer model. Later studies confirmed the local perception of the aquifer structure that there is lateral outflow rather than vertical leakage.

Sengupta (1993) documented different cases of proper planning for local water resource development and the aggregate effect of many small water harvesting and extraction structures on a regional level. He suggested that there must have been some sort of regional level planning at basin level in the past and ancient cultures may have survived thanks to such integrated planning of water resources.

Shaw and Sutcliffe (2003) in their documentation of ancient small dams in the Betwa basin of central India found that the size of these structures was linked to the runoff from their catchment. This link led to the belief that the builders of these structures followed some variant of the rainfall-runoff curve during their design of these structures and that there were sound observations of local hydrology.

NIH (1999) describes hydrology in ancient India and mentions verses from ancient poetry that exhibit knowledge of hydrology and groundwater. Bio-indicators for groundwater exploration were in use and various rules of thumb linked ground observations to the anticipated depth to the water table, for example by the presence of certain plant species. Verses such as 'The flow of water through earth is like blood flowing through veins', shows mature understanding beyond mere poetic imagination.

Recent development and research studies are showing an increasing appreciation of local knowledge in constructing hydrogeological databases. There has been work on combining remote sensing with ethnographic studies to study local hydrology (Jiang, 2003). NGO-facilitated programmes in Andhra Pradesh have organized efforts in participatory hydrologic monitoring by farmers to increase their awareness and local knowledge of groundwater and to better aid community-based groundwater management (Rama Mohan, Chapter 13). A worldwide programme, GLOBE, focusing on environmental data monitoring utilizes school children as a resource to record observations such a rainfall, water quality and humidity, thereby reinforcing existing local knowledge. In Bangladesh, well drillers have been asked to locate local layers of arsenic enrichment in subsurface sediments in locating sites for new wells and depth of water extraction for safe water supplies (Jonsson and Lundell, 2004). The Honey Bee network operating from Ahmedabad, India, records and disseminates innovations in environmentally friendly agricultural practices made by local farmers by organizing annual walks through different regions. Another innovative programme organized by the Arid Community Technologies (ACT) group in Kutch region, India, trains villagers to document local geological information in their own language. All these show that there is a developing appreciation of local knowledge and efforts are being made to harness this knowledge for better resource management.

2   METHODOLOGY AND TOOLS

This study was part of the capacity building programme called 'Groundwater Governance in Asia: Theory and Practice' in which water professionals from five different Asian countries participated in an intensive training and research programme between 2006 and 2007. Part of the work was performed at four different field sites spread across the Indo-Gangetic basin. In each area, three villages were chosen for in-depth study. The topics studied were related to groundwater-based irrigation, institutions linked with groundwater management and the present study exploring local knowledge on groundwater via well drillers. The task was to identify 3 to 5 well drillers working in each area and interview them in a structured manner using pre-developed and tested questionnaires. Several parameters of importance to local groundwater conditions and utilization were included in this interview. Furthermore, the drillers were requested to supply 5 to 10 borehole logs from recent drilling operations. A pre-determined format for entering litholog information was designed. In some cases, drillers incorporated their own procedure for storing and recording lithologs. It was found that drillers were keen to participate; the challenge, however, was to identify perceptive drillers who could relate with the study objectives and who had a long experience of working in the area.

The overall procedure followed was:

A. Obtain current piezometric maps in the study villages at a scale of 1:50,000.
B. Obtain temporal trends in groundwater level in the past two decades in the different villages.
C. Obtain the lithology and structure of the local aquifer (up to a depth of 100 m and 5 to 10 kms horizontally) using well logs provided by the drillers.
D. Look at the rates of well failure, present and historic.
E. Obtain a preliminary picture of the water quality.
F. Collect any papers or studies performed on the local groundwater hydrology by nearby research institutions for comparison with the data obtained from A to E.

The work focuses on parts C and F.

2.1   *Locating perceptive and experienced well drillers*

It is important to locate well drillers who are able to summarize and synthesize knowledge from their work. In the course of their profession, they often visit the same or nearby plots over a number of years. Their personal level of curiosity, involvement and memory (and means of recording, if any) decides how able they are to reconstruct the drilling sequences and to align and interconnect them spatially. Not all drillers are able to demonstrate the same level of perceptiveness with respect to geology and groundwater hydrology. Therefore, an initial screening is necessary.

2.2   *Extracting semi-statistical information and pattern-based sketch of the aquifer*

Often, the first response of the well drillers is to convey an overly complex picture of their experience. However, their understanding can be better extracted using a constructive dialogue. In the case of aquifer lithology, the data are in the form of frequencies of occurrence of different layer types, and the spatial links between them in the form of a rough image. The first step is to be able to extract this image from the mind of the driller in a semi-statistical, non-geo-referenced form.

2.3   *Constructing a conceptual digital image of the aquifer using well driller information*

Once the relevant semi-statistical data have been extracted and a conceptual sketch of the aquifer obtained, a digital conceptual image is produced and presented to the driller for verification. There may be a couple of iterations after which a final picture is derived. It is important to get the approximate scales correct, i.e. the vertical and horizontal dimensions. This conceptual image is localized and its extent depends on the work area of the driller being interviewed.

2.4   *Verifying the conceptual aquifer picture using scientific data*

To further test the accuracy of the image, a comparison with scientific-based information available at a broader scale is made using reports published by government agencies and scientific research institutes. These reports can be used to check any obvious errors in the local pictures and help confirm the scale in which it has been created.

2.5   *Obtaining local well logs from current drillings by the well drillers*

Another way to test the accuracy of the conceptual image is to carry out trial drilling using the well drillers in the area. It is important to be present in the field during the drilling season. In many parts of monsoon-dominated south Asia, the drilling season would be after the monsoon and generally before the second-crop season i.e. around December–January.

2.6   *Geo-statistical analysis of the data*

Once all these data are collected, they can be analyzed statistically and in a spatial form. Various techniques are present for such analysis and geostatistics offers tools to perform these analyses in a spatial form (Goovaerts, 1997).

2.7   *Using the conceptual image as a training image and anchoring it to the local well logs*

The conceptual image is known as a training image that contains the essential patterns of the local area. These essential patterns need to be extracted from this image and then fitted and adjusted along with the newly-derived well log data. The case study will describe an application of the (Sequential Normal Equations SIMulation) SNESIM algorithm. This algorithm summarizes the patterns and statistics of the training image and anchors them to actual well log data so as to construct a consistent picture of the aquifer (Strebelle, 2001). The algorithm maintains the geological conceptual picture and at the same time honours the collected well logs. The theory behind this technique is known as multiple-point geostatistics, which has been developed as an improvement over techniques that use only correlations considered incapable of capturing complex spatio-temporal patterns. Since the conceptual picture and the well logs are not enough to fully constrain the aquifer picture, the method provides not one but several of these aquifer pictures all of which are consistent with the training image as well as the well logs. These multiple pictures are all likely depictions of the actual aquifer and together they provide a measure of the uncertainty of the mapping process.

2.8   *Creating probability maps of lithology*

The case study will illustrate how multiple equiprobable aquifer depictions, when used together, span the uncertainty of the process of generation of a localized aquifer model. A set of such equally likely depictions allows maps of the probability of occurrence of specific layers within the subsurface to be derived.

3   DESCRIPTION OF SITES

This study was conducted in four sites located across the Indo-Gangetic alluvial plains in India and Nepal. The areas were Hoshiarpur district (Punjab, India), Vaishali district (Bihar, India), Murshidabad (West Bengal, India), Jhapa district (Nepal). In each of these districts, three contiguous villages were selected for the study. In this paper, we will describe in detail the study conducted in Vaishali (Figure 1).

The Vaishali district (area: 2036 km^2) lies north of the Gandak River, which flows into the Ganges River from the north. It is an area with extensive and deep alluvial aquifers composed of interspersed layers of clay, sand and gravel. There is a long history of settled civilization in this area. One of the study villages is said to be the ancient cultural centre of Buddhist society more than two millennia back. The topography is affected by the meandering rivers of the region with small depressions (called *chaur*) formed by abandoned ox-bow lakes. These depressions can be from 20–30 m to several kilometers in diameter. The topographic variation and these depressions influence the soil type, vegetation and the groundwater table. In some depressions, water collected during monsoon remains throughout the year, whereas in others water is pumped out, mostly for agricultural use, within 4 to 5 months. The natural groundwater table can be only 2–3 m below ground surface close to the centre of the depressions in the dry season but can have depths of 10–15 m in more upland areas. The annual rainfall in the area is 1121 mm (CGWB 1993, NIH 2000).

Figure 1.   Location of Vaishali district, Bihar, eastern India.

4   APPLICATION TO THE VAISHALI EXAMPLE

4.1   *Locating perceptive and experienced well drillers*

Five well drillers were interviewed for the Vaishali case study. Table 2 describes these drillers, all of whom except one are currently practicing. All drillers are involved in drilling for domestic wells (known locally as *Chapakal*) and irrigation wells. The mode of drilling is based on manual techniques using bamboo pipes as casings. Well drilling is a significant livelihood option for some castes or communities such as the *Malla* in the study villages. Groups of 10–15 drillers tend to stay together but operate over a wide area. Such communities have made a natural transition from their traditional occupation, in this case fishery, to drilling, which is considered a water-based occupation.

From our experience, the drillers were keen to participate. The important task was to identify perceptive drillers who could relate with the project objectives and had a good experience of working in the area. An initial and structured scoping of the drillers, it was decided that Keshav Ram was suitable for further questioning about the aquifer structure. He and his father Gopal Singh had a wide and localized understanding of the hydrogeological setting.

4.2   *Extracting semi-statistical type information and pattern-based sketch of the aquifer*

The biggest difficulty in extracting lithological information from a driller is that he is often not able to summarize his experience in a tangible form. The response is often that, "There is too much variation from farm to farm. At one farm, we get water at 10ft depth and at the next farm, there is no water till 50 ft." Beyond this initial expression of uncertainty, additional questioning can reveal a general understanding. The process was to first identify the general aquifer trends and then specify these trends with specific depths and sediment characteristics. The progression of information is:

1. The main types of layers are sand (*Balu*), concrete (clay, sand, and gravel mixture), clay (*Mitti*), and gravel (*Pathar*). Here note that 'concrete' is a term used locally by drillers to describe a geological layer that is similar in appearance to man made concrete used in construction. Since this paper is about understanding local semantics of groundwater and linking it with conventional scientific understand, we chose to use the word 'concrete' in the same sense as used by the local well drillers and hence it should not confused with the literal meaning of concrete as we know it.
2. Sand and clay can be both black and yellow.

Table 2.   Well drillers interviewed in the Vaishali case study.

Name	Age	Village	Drilling since	No of wells drilled
Gopal Singh	70	Bedauli	1956	3,500
Keshav Ram	45	Bedauli	1977	4,500
Balini Sahini	45	Chakwezo	1985	1,500
Rajesh Ram	42	Purkhali	1980	500
JayKishore Paswan	45	Vaishali	1985	2,000

3. Sand and clay are the pre-dominant materials, with roughly 70% clay and 10–20% sand in the overall sequence of layers.
4. Within the first 100 ft:

    a. Concrete layer is found 1 to 6 times. It is 6 inches to 2 ft thick and has a maximum of 5 ft thickness.

    b. Sand layer is first encountered at around 30 ft–50 ft , then at 80 ft–100 ft. Each concrete layer is followed by sand, but not always.

    c. With depth, propensity of occurrence of concrete and sand layers decreases and clay layer increases.

From this initial information, an initial image can be generated, which upon iteration turns into a pattern-based sketch of the aquifer, at a scale corresponding to the area of operation of the drillers (approx. 3–4 km). Figure 2 shows the non-directional 2-D vertical cross-section of the aquifer generated through this process.

4.3   *Constructing a conceptual digital image of the aquifer using well driller information*

The conceptual sketch is digitized as a 300 × 400 pixels image with three facies representing clay, calcrete and sand (Figure 3). Each pixel corresponds to 10 m horizontally and 0.2 m vertical resolution. This image was presented to the well driller who verified its overall structure. Here, it is important to note the subjectivity involved in this process of training image generation which could lead to uncertainty in model generation. Here, only the essential characteristics of the structure are incorporated in the sketch.

(a)

(b)

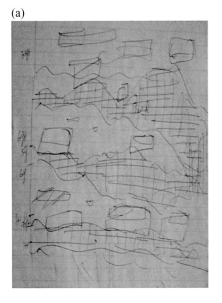
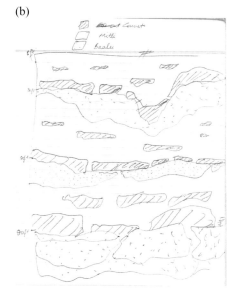

Figure 2.   (a) Initial and (b) Final, iterated conceptual pattern-based sketch of the aquifer from well drillers' knowledge.

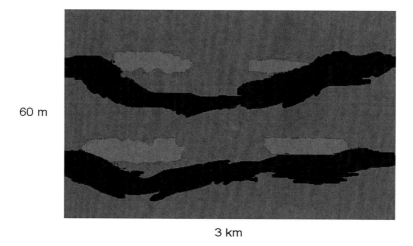

60 m

3 km

Figure 3.   Digital image of the aquifer, created from well driller information (black: sand; light grey: concrete and dark grey: clay).

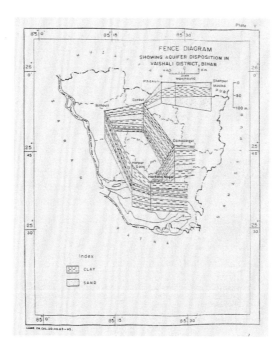

Figure 4.   Fence diagram of lithology for Vaishali district (CGWB, 1993).

4.4   *Verifying the conceptual aquifer picture using scientific data*

The conceptual image is validated from other information obtained from surveys by government agencies in the study area. A survey of the Vaishali district itself produced a fence diagram shown in Figure 4, which is at a much larger scale (CGWB, 1993). Verification

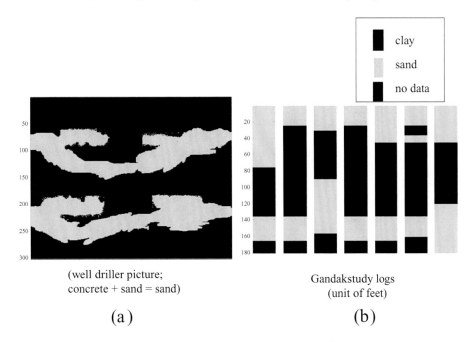

(well driver picture;
concrete + sand = sand)

(a)

Gandakstudy logs
(unit of feet)

(b)

Figure 5.   Comparison of conceptual image (a) with logs provided in the Gandak report (b) (NIH, 2000) (See colour plate section).

against such large scale information is not possible. However, this government survey depicts a similar pattern of multiple aquifers interspersed by clay.

Another science-based survey conducted in the Gandak canal irrigation command area (NIH, 2000) reported seven logs, the locations of which are not precisely recorded. The lithology is divided only between sand and clay. Nevertheless, these data are valuable references for the conceptual picture (Figure 5).

The comparison shows that a similar conceptual idea is given by these logs and some justification of the scale of the image is now possible. Note that with our image produced using the well driller's experience, we add more detail in the form of the 'concrete' layer to that given by NIH (2000).

4.5   *Obtaining local well logs from current drillings by the well drillers*

The next step is obtaining local logs of wells being drilled as part of the field activity in December 2006 and January 2007. The surveyed drillers were requested to record the different lithologies encountered, their depths and the location of drilling. After initial training, the drillers were able to carry out this activity by themselves and, nine logs were recorded in and around the village of Chakramdas in Vaishali district, covering an area of approx. 8 km^2 (Figure 6).

These logs are used to construct a fence diagram (Figure 7). The key assumption is that the horizontal continuity is isotropic, i.e. does not show any preferred direction of continuity. This assumption however, might not be true, but this assumption is made in order to demonstrate the simpler 2D analysis. There is no difference in methodology when making a 3D analysis, except that the training image (see below) has to be made in 3D.

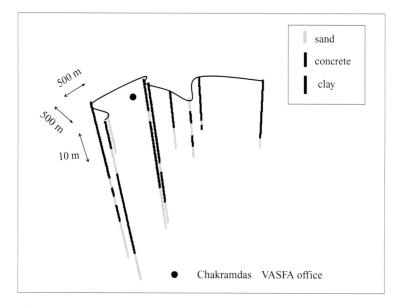

Figure 6.   Collected well logs in and around Chakramdas village (See colour plate section).

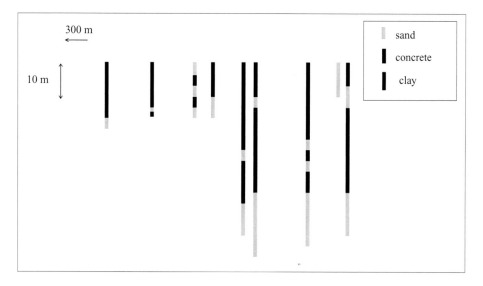

Figure 7.   2-D planar representation of the well logs in Chakramdas village (See colour plate section).

4.6   Geo-statistical analysis of the data

There is a proportion of 45% for clay, 35% for sand and 20% for concrete. The spatial correlation between clay, sand and concrete layers can be shown by using variograms in different directions.

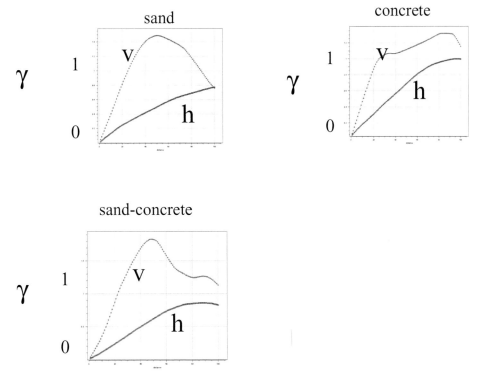

Figure 8.   Variograms and cross-variograms extracted from drilling logs of Chakramdas village (h: horizontal, v: vertical).

Figure shows that the sand layers are connected much better horizontally than vertically. The concrete layers show a similar lesser degree of horizontal anisotropy. The cross-correlation between sand and concrete is also greater in the horizontal direction than the vertical direction.

4.7   *Using the conceptual image as a training image and anchoring it to the local well logs*

Geological structures are often too complex to be described by statistics such as histograms and variograms or covariances. A digital representation such as Figure 3 provides a richer and more useful representation of the geology since it incorporates a higher level of statistics beyond the histogram and correlations. Figure 3, however, is one of many possible outcomes or representations of the aquifer structure. Applying the statistical anchoring procedure involving the SNESIM algorithm and Figure 3 as the training image, several equally likely images are produced. Figure 10 shows two such aquifer depictions. Note the similar locations of the second sand layer in both these depictions since the well log data strongly affirms that, but note the different placement of the other sand and concrete formations within the clay background.

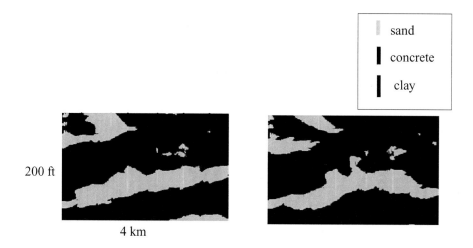

Figure 9.   Two equally likely depictions of the aquifer, anchoring the collected well logs to the training image (See colour plate section).

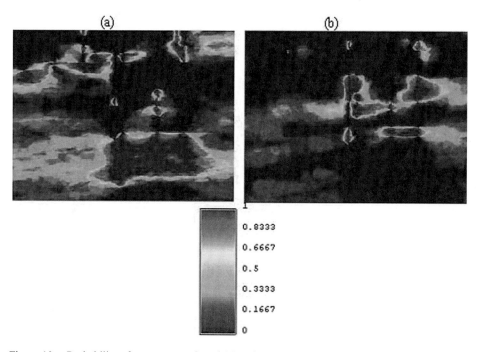

Figure 10.   Probability of occurrence of sand (a) and concrete (b) based on averaging of 15 equally likely depictions (See colour plate section).

4.8   *Creating probability maps of lithology*

In order to use these simulations to obtain a measure of the uncertainty, it is necessary to generate many different simulations and look at the frequency of the occurrence of the various layers. Fifteen equally likely aquifer depictions have been generated and at each location the probability of occurrence of sand and concrete is made (Figure). Probabilities of other facies (clay and gravel) were likewise computed.

Two distinct layers of sand occur in this aquifer (from Figure (a)) of which the second layer is much more probable. Some layers of concrete (from (b)) are more probable than others and they are located right above the second sand layer of the aquifer depicted by the large red path in the lower part of Figure (a).

There are multiple levels of uncertainty:

- Subjective conceptual picture: From the point of constructing the rough picture from the well driller, to verifying it from scientific information, there is subjectivity involved. Another training image constructed in this manner would give a different picture of the aquifer. Recording and interpretation of logs: The logs have been recorded by the drillers themselves. The length of the drilling rod decides the vertical resolution of measurement. The accuracy of the horizontal positioning is also subject to errors.
- Algorithm used for constructing aquifer model: There are assumptions involved in the methodology and algorithm used. Specific assumptions on pattern recognition from the training image, if changed can give rise to a different aquifer picture.
- Data density: The greater the number of well log data collected, the more accurate is the picture of the aquifer. Given the subjective conceptual picture and given an algorithm such as SNESIM for constructing the aquifer, the uncertainty arising from data scarcity can be measured as the probability of occurrence of say, sand or concrete at any particular location. Note that this measure of uncertainty is subject to uncertainty of the two previous steps above and, therefore, cannot be considered absolute. It is at best indicative.

5   DISCUSSION AND CONCLUSION

It has been possible to answer all four questions that were initially posed. First, well drillers do possess a good amount of information on local groundwater hydrogeology. It is important to select the right kind of driller(s) who are experienced, collaborative, knowledgeable and systematic in their approach. Not only do they possess knowledge of lithology and water bearing layers, but their perception of groundwater quality in localized settings is also valuable.

Second, the question was answered of how to extract lithological and other information that drillers possess and translate it into a local digital database of groundwater lithology, using the case of Chakramdas village of Vaishali district in Bihar state of eastern India as an example.

Third, it was possible to compare and combine this information with science-based knowledge and thereby verify the correctness of the local knowledge and improve the overall picture of the local aquifer setting. Furthermore, using advanced geostatistical software, it was possible to create multiple equally likely aquifer maps and thereby give a measure of uncertainty to the process and the probability of occurrence of specific lithological layers.

The cost of using existing information from local well drillers and supplementing it with new drilling logs performed by these drillers is more cost-effective than using traditional exploratory drilling (Figure 12). The cost of processing the logs in the laboratory is about 90% of the total cost. The actual cost per log can be estimated at around Indian rupees

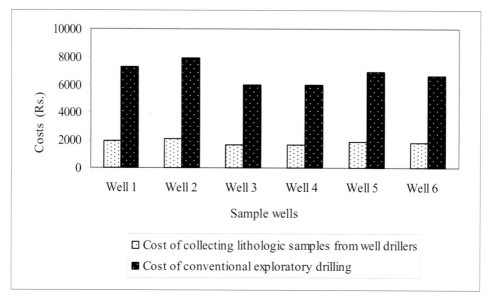

Figure 11. Comparative costs of obtaining lithology information using well driller experience, current drilling by drillers, and traditional exploratory drilling.

(INR) 100–150[1]. If one imagines expanding this methodology to a district for a period of 1 year, then with 25 well drillers, each supplying 20 logs, there can be a database of 500 logs which will cost INR 75,000 (around US$1900) for data collection. This is approximately equivalent to the cost of only 10 traditional drill logs.

In order to use this approach more widely, and thereby answer the last question posed, this technique should be more broadly disseminated and adapted. In this process, diverse skills and methodologies may have to be developed and combined, from interviewing and interacting with local well drillers in remote villages to applying advanced software-based technologies.

Looking in general at the subject of local hydrological knowledge generation, there is a potential for interrogating many more local persons and institutions than just well drillers. Local schools have been used previously as environmental monitoring stations (see Global Learning and Observations to Benefit the Environment or GLOBE website) and they also provide another low-cost means of creating databases. The concept of village informa-tion centres in Tamil Nadu is also a means of realising these ideas (M.S. Swaminathan Research Foundation). The use of information technology in mass knowledge transfer has also been attempted by different initiatives (Almost All Questions Answered (AAQUA), Virtual Academy for the Semi-Arid Tropics, VASAT).

It is important with all these means of alternative knowledge generation to keep in mind that there are maybe some tradeoffs. As long as this perspective is maintained and appropriate corrections as suggested in this paper are made, locally based knowledge

---

[1] 1 USD ~ INR 40 as in May 2008.

generation promises to be a favourable direction for a host of environmental data generation activities in many developing countries.

REFERENCES

CGWB (1993), *Hydrogeology of Vaishali district*, Central Groundwater Board, Patna, India.

CSE (2001), *Dying wisdom: Rise, fall and potential of India's traditional water harvesting systems*, (eds) Agarwal, A., Narain S. and I. Khurana, Centre for Science and Environment, New Delhi.

Goovaerts, P., (1997), *Geostatistics for natural resources evaluation*, Oxford University Press.

Jiang, H., (2003), 'Stories remote sensing images can tell: Integrating remote sensing analysis with ethnographic research in the study of cultural landscapes', *Human Ecology*, 31 (2): 215–232.

Jonsson, L. and Lundell, L. (2004) 'Targeting safe aquifers in regions with arsenic-rich groundwater in Bangladesh: Case study in Matlab Upazila'. *Minor Field Studies No. 277*, Swedish University of Agricultural Sciences, Uppsala, Sweden. 60 p.

NIH (1999), *Hydrology in Ancient India*, National Institute of Hydrology, Roorkee, India.

NIH (2000), *Development of conjunctive use model for lower Gandak basin (Part 1)*, National Institute of Hydrology, Roorkee, India.

Rosin, T., (1993), The tradition of groundwater irrigation in northwestern India. *Human Ecology* 21 (1): 51–86.

Sengupta N., (1993), 'Lessons of indigenous management methods', paper presented at Workshop on Water Management—India's Groundwater Challenge. VIKSAT/Pacific Institute Collaborative Groundwater Project, Ahmedabad, India.

Shah T., (1993), *Groundwater markets and irrigation development: political economy and practical policy*. Oxford University Press, India.

Shaw, J. and Sutcliffe, J.V., (2003), 'Ancient dams, settlement archeology and Buddhist propagation in central India: the hydrological background', *Hydrological Sciences Journal*, 48 (2): 277–291.

Strebelle, S., (2001), 'Conditional simulation of complex geological structures using multiple-point statistics', *Mathematical Geology*, 34 (1): 1–22.

von Hayek, F.A., (1974), The Pretence of Knowledge, Nobel Prize Lecture for Economics. http://nobelprize.org/nobel_prizes/economics/laureates/1974/hayek-lecture.html (accessed on 11th February 2008).

WEBSITES AND OTHER SOURCES

Almost All Questions Answered (AAQUA), http://aaqua.persistent.co.in/ (accessed on 11th February 2008).

Global Learning and Observations to Benefit the Environment (GLOBE), http://www.globe.gov, The GLOBE program (accessed on 11th February 2008).

Honey Bee network, http://www.sristi.org/honeybee.html (accessed on 11th February 2008).

M.S. Swaminathan Research Foundation (MSSRF), http://mssrf.org/iec/index.htm (accessed on 11th February 2008).

Virtual Academy for the Semi-Arid Tropics (VASAT), http://www.vasat.org (accessed on 11th February 2008).

CHAPTER 15

Crop per volume of diesel? The energy-squeeze on India's small-holder irrigation

T. Shah
*International Water Management Institute (IWMI), Anand field office,
Gujarat, India*

A. Dasgupta
DHRIITI, Assam, India

R. Chaubey
Independent consultant, Bihar, India

M. Satpathy
PRADAN, Bhubaneshwar, India

Y. Singh
Independent consultant, Uttar Pradesh, India

ABSTRACT: India's small-holder irrigation is in the grip of an energy-squeeze on top of the general cost-price squeeze that is a source of so much stress in Indian agriculture. Particularly hard hit are marginal farmers and share croppers in eastern India who depend upon pump owners for renting pumps. As pump prices have remained static, the rental rates for pumps have risen in parallel with every rise in the diesel price because of the monopoly power of pump owners at village-level. Pump rentals rise when the diesel price rises but stay static when the diesel prices fall. This paper synthesizes the results of 15 village surveys from West Bengal, Assam, Orissa, Uttar Pradesh, Haryana, Punjab, Madhya Pradesh, Gujarat, Maharashtra and Kerala aimed at investigating the impact of the energy-squeeze and the coping strategies adopted by the small holders.

1   INTRODUCTION

The years of 1975–2000 were the golden age of small-holder irrigation in South Asia. Until then, much irrigation in the region was gravity flow, and confined to the command areas of canal systems and traditional irrigation structures such as tanks, ponds and *ahar-pyne* systems. Since 1975, a spontaneous boom in private investments in small boreholes and mechanized diesel and electric pumps revolutionized irrigated agriculture taking it beyond the traditional command areas. This happened at a time when growing population pressure made it imperative for marginal farmers to intensify their farming to ensure family food and livelihoods security. Growth of local, informal, but fragmented pump irrigation service markets, through which the poor could access irrigation from pump owners, vastly expanded the productivity and equity impacts of this irrigation boom. Government policies supported the pump irrigation revolution through expansion of institutional credit, through a variety of subsidy schemes on borings and pumps, through support to farm electrification and electricity subsidies. While pumps and boreholes emerged as the mainstay of small-holder irrigation, new concerns emerged about the threat of groundwater depletion, and about the adverse impacts of electricity subsidies on the viability of the electricity industry. Management of the pump irrigation economy has become one of the trickiest water policy issues in the region.

Since 2000, however, evidence suggests that the region's groundwater economy has begun shrinking in response to a growing energy squeeze. This energy squeeze is a combined outcome of three factors: [a] progressive reduction in the quantity and quality of power supplied by power utilities to agriculture as a means to contain farm power subsidies; [b] growing difficulty and rising capital cost of acquiring new electricity connections for tubewells; and [c] an eightfold increase in the nominal price of diesel during 1990–2007, a period during which the nominal rice price rose by less than 50%. In a 2002 survey of over 2,600 tubewell owners carried out in India, Pakistan, Nepal terai and Bangladesh, respondents unanimously ranked 'energy cost and availability' as the top challenge to their farming, far above 'groundwater depletion', 'high rate of well failure', and ' rising groundwater salinity' (Shah et al., 2006). Since then, diesel prices have risen a further 70%; no surprise then that the diesel price squeeze on small-scale irrigation is heading towards a crisis in all the countries of South Asia but is particularly visible in eastern India and Nepal terai where the ratio of rice to diesel price is critical (Table 1).

Of even greater significance for the poor is the response of pump rental prices to the rise in diesel prices because the poorest of India's peasantry depend on water markets for securing their irrigation. Because water markets are natural oligopolies (Shah 1993), pump owners use diesel price increases to raise their pump rental rates in parallel with every major rise in diesel price despite the fact that pumps themselves have become cheaper during 1990–2007. Figure 1 shows the changes in the nominal price of diesel versus the price of pump irrigation in Mirzapur, Uttar Pradesh. Between 1990–2007, diesel prices here have risen from INR 4.6/l to INR 34.8 (1 USD ∼ INR 40 as in May 2008); but the rate buyers incur per hour of pump irrigation has increased from INR 23–25/hour to INR 90–95/hour, far larger than needed to cover the increase in fuel cost. Another characteristic of this relationship has been the downward stickiness of pump irrigation prices; every time there is a big increase in diesel price, pump irrigation price tends to jump; however, the reverse is never the case.

As a result, pump rental fee relative to farm produce price—which is what matters to the marginal farmers and share croppers—have risen even faster than the diesel price relative

Table 1.   Farm gate rice price relative to diesel price in countries of South Asia.

	Diesel price/litre (USD) February 2007	Farm gate rice price per kilogram (USD) February 2007	Kilograms of rice needed to buy a litre of diesel
India	0.86	0.16	5.7
Pakistan	0.60	0.19	3.2
Bangladesh	0.50	0.13	3.9
Nepal terai	0.86	0.15	5.7

Source: Field research results by IWMI researchers.

to rice and wheat prices. In Deoria, eastern Uttar Pradesh, a marginal farmer could buy an hour of pump irrigation for the farm gate price of a little over 3 kg of rice and wheat in 1990; today, this ratio is 10 kg of wheat and 12 kg of rice (Figure 2).

Electric tubewells, subject to flat horse-power linked tariff, are cheaper to operate than diesel pumps; their owners also sell pump irrigation at much lower rates compared to diesel pump owners. Therefore, new electricity connections are avidly sought after. However, most

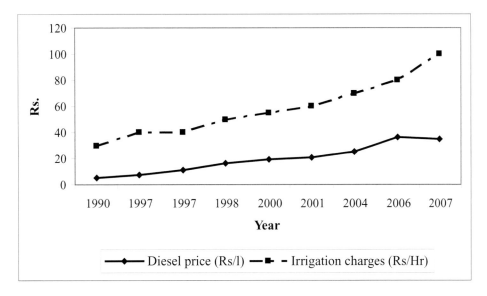

Figure 1.   Diesel price rise and pump irrigation price: Mirzapur, UP.

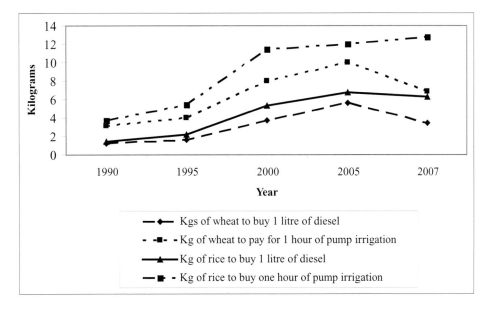

Figure 2.   Deoria: Relative price of diesel pump irrigation with respect to farm gate prices.

Table 2.   Cost of irrigating a hectare of sugarcane in village Akataha, district Deoria, eastern UP (INR/ha).

	Diesel pump	Electric pump
Own irrigation source	4050	92.5
Purchased pump irrigation	9450	2700

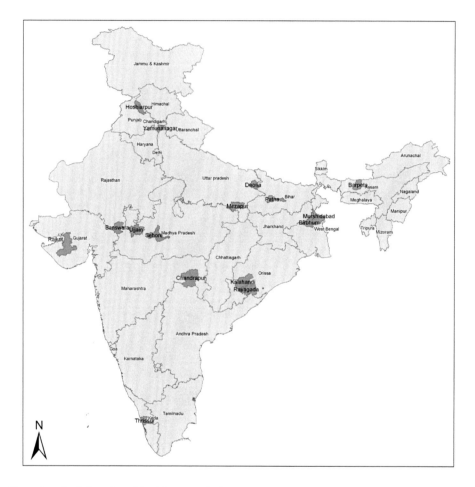

Figure 3.   Study location of the 15 case study sites.

states—which in the early 1960's gave district collectors monthly targets for the number of tubewells to be electrified—now operate an embargo on new electricity connections to tubewells; and where they are issued, the entire cost of taking the power line to the tubewell—of poles, cables and transformers—is charged to the farmer. This has made new electricity connections scarce as well as costly. Even so, existing electric tubewell owners, and marginal farmers who are close enough to their tubewells to buy pump irrigation from them are luckier compared to diesel pump owners and their buyers (Table 2). Since farmers who can buy pump irrigation from electric tubewell owners incur lower cost than using their own diesel pumps, diesel pump owners in Uttar Pradesh today prefer purchased irrigation from electric tubewells than irrigating with own diesel pump.

This paper summarizes the results of 15 village studies carried out in different parts of India, using field-based researchers to investigate impacts of the energy squeeze on small-holder irrigation (Table 3 and Figure 3). The aim of the studies was to explore, identify and document rather than to measure and quantify. The work suggests that the groundwater economy in many parts of India, especially in the east, is shrinking and that marginal farmers have been badly hit by the energy squeeze.

2 DECLINING WATER MARKETS?

Most social impacts of the energy squeeze on small holder irrigation and the agrarian poor work through the groundwater markets. Around 1990 and before, when diesel was an eighth the price it is today, and farm power supply better than today, electric tubewell owners were natural oligopolists in a highly competitive market (Shah, 1993). Flat electricity tariffs, which reduced their marginal cost of pumping almost to zero, created a powerful incentive to maximize pump irrigation sales, and in the process pare down the prices. Diesel pump operators were able to offer some competition because the price of diesel was low and their portability allowed diesel pumps to irrigate where electric tubewells could not. Numerous field-based studies showed that such local groundwater markets emerged as the mainstay of ultra-marginal farmers and share croppers, especially in eastern India and Bangladesh. In Bangladesh, Fujita and Hussain (1995) noted that 'the economic value of land . . . has decreased in a relative sense' in farm income generation and 'opportunities for the landless and near-landless to climb the social ladder [have] expanded greatly'. In Uttar Pradesh, Niranjan Pant wrote: ". . . the smallest farmers with land-holdings up to 0.4 ha are the largest beneficiaries of the groundwater markets as 60% of the farmers of this category irrigated their wheat crop by water purchased from the owners of private Water Extraction Devices. . . " (Pant, 2005) Shah and Ballabh (1997) based on a study of water markets in six villages of North Bihar concluded that these had opened up new production possibilities for the poor which left them better off than before, and thereby imparted a new dynamism to the region's peasant economy. Even, Wilson (2002:1232), otherwise critical of profiteering by water sellers in Bihar, wrote: "extension of irrigation through hiring out [mobile diesel pump sets] to small and marginal holdings is in fact the major factor accounting for the further increase since 1981–82 in cultivated area irrigated at least once to approximately 73% in 1995–96. Those hiring in pump sets are overwhelmingly small and marginal cultivators; they cultivate an average of 0.5 ha (compared with an average of 1.6 hectare) cultivated by pump set owners). . . ". Most recently, Mukherji (2007a) in an extensive study of water markets in West Bengal reaffirmed their myriad benefits to the agrarian poor. Water markets, and indeed groundwater irrigation itself, have been a source of much succor to the agrarian poor. Studying rural poverty ratios across the Indian states between 1973/4 and 1993/94, Narayanmoorthy (2007:349) concluded that, "there is a significant inverse relationship between the availability of groundwater irrigation and the percentage of rural poverty. . . "

With soaring diesel prices and shrinking power supply to tubewells, the situation has rapidly changed for the worse. Pump irrigation markets, which boomed during the 1980's and 1990's and probably served a larger area than all the public irrigation systems in India

Table 3. Three most important responses of farmers to energy squeeze in study villages.

	Village study location	Most important response	2nd most important response	3rd most important response
1	Kendradangal, Birbhum, West Bengal	Decline in pump irrigated boro rice area	Marginal farmers & share croppers exit farming	Kerosene/crude as a diesel substitute
2	Kaya, Murshidabad, West Bengal	Shift to low-water using crops	Chinese pumpsets	Kerosene as a diesel-substitute
3	Ferozpur Ranyan, Haryana	Give fewer irrigations; same crop pattern	Water conveyance through pipes	Exodus of marginal farmers from farming
4	Purana Pradhan, Khurda, Coastal Orissa	Install electric pump or buy from electrified borewells	Switch to high-value crops	Move out of pump irrigated agriculture
5	Badhkummed, Ujjain, Madhya Pradesh	Turned to electric pumps	Decline in diesel pump irrigated area	Irrigate fewer times
6	Berkhedakurmi, Sehore, Madhya Pradesh	Increase in irrigation with electric pumps	Decline in area under diesel pump irrigation	Switch from sugarcane to wheat and gram
7	Lilapur, Rajkot, Gujarat	20–25% decline in rabi (winter crop) irrigation	Increased irrigation interval	Small bed and alternate furrow irrigation
8	Jawrabodi, Vidarbha, Maharashtra	Increased irrigation interval	Optimizing on rainfall/ life-saving irrigation	Reduced irrigated area

9	Keotkuchi, Barpeta, Assam	Diesel pumps run on kerosene	Decline in pump irrigation	Farmers quitting farming
10	Dharamgarh, Kalahandi, Orissa	Increased use of manual canal irrigation and lifting	High-value crops	Longer irrigation interval
11	Shergarh, Hoshiarpur, Punjab	Farmers lease out lands to Bihar laborers	Distress shift to off-farm livelihoods	Optimizing water application
12	Veerpur, Banswara, Rajasthan	Kerosene used to run diesel pumps	Longer irrigation interval	Pump irrigation concentrated on vegetables for market
13	Simra, Phulwari, Bihar	Return to rainfed paddy in kharif (monsoon crop) and pulses in rabi	Pump irrigation concentrated on summer onion for market	Share-cropping irrigation with purchased declining
14	Akataha, Deoria, Eastern UP	Increased dependence on flow irrigation	Pump irrigation concentrated on high value crops	Longer irrigation interval
15	Abakpur Mobana, Mirzapur, Uttar Pradesh	Pump irrigation concentrated on cash crops	Irrigation interval longer	Water saving crops.

(Mukherji, 2008), are shrinking rapidly. The groundwater irrigation economy is also declining. During the 1980's and 1990's farmers in northern and eastern India purchased diesel pumps often as stand-by's for their increasingly unreliable electric pumps. But diesel has now become unaffordable, especially for water buyers, and the preference for electric tubewells has increased. However, electric tubewells are unable to fulfill expectations because electricity supplies as well as connections are dwindling.

In eastern India, Nepal terai and Bangladesh, electric tubewells are rare. Where they do exist their owners find their monopoly power enhanced, which they use to increase their share in the groundwater markets and irrigation surplus, and they are able to moderate the energy squeeze on marginal farmers especially when the power supply situation is good and the tubewell owners pay a flat electricity tariff. This was the case in Uttar Pradesh, West Bengal and Orissa. So unequal is the competition that owners of diesel pumps prefer to purchase irrigation from electric tubewell owners rather than use their own diesel pumps. In Uttar Pradesh, a 5 hp electric tubewell connection costs INR 410 per month but can generate up to INR 9,000/month as gross income from water sale. In Birbhum, West Bengal, the high price for electric pump irrigation enables the electric submersible pump owners to get their own irrigation free of cost and make some profit as well. Here, the flat tariff paid by electric submersibles increased from INR 5460/year to INR 8950/year between 1990–2007; in response, irrigation rates charged for *boro* (rice grown in summer season) rice doubled from INR 3375/ha to INR 6750/ha. This rise was much smaller than the rise in the cost of purchased diesel pump irrigation. While electric submersible owners make money, the marginal farmers of Bengal can only grow *boro* rice if they can buy irrigation water from an electric shallow/mini-deep tubewell owner.

The West Bengal government policy seems designed to minimize new connections for electric tubewells. To promote *boro* irrigation, the government had a scheme to issue temporary seasonal connections. In 2003, temporary connections were offered to Birbhum farmers for *boro* rice at INR 7000 for 3 months. In one village, seven diesel pump owners took advantage of this, but the next year, the tariff was increased to INR 18000, and the *boro* season electrification ended. Permanent connections are preferred by all but take 3–4 years to get approved and cost INR 125,000–140,000 for poles, 11 KV cables, a 10 KW transformer and a meter. The only farmer in this study village who has so far afforded such a mini-deep connection had 2.8 hectare of his own land and 2 hectare of neighboring lands to command.

The ability of flat-tariff paying electric tubewells to moderate the impact of the diesel price squeeze is undermined by an inadequate supply of new electricity connections for irrigation, the prohibitively high cost of installing new connections and the low amount and quality of the power supply to agriculture. New electricity connections are available in Uttar Pradesh; but the demand was subdued because the farmer has to pay for the cost of laying the cable, poles and transformer, which may add up to INR 120,000 or more. In Kalahandi villages in Orissa, the electricity supply is plentiful, and electric tubewell costs a seventh of the cost to operate a diesel pump of comparable output. However, an electric pump 500 m away from the village may cost INR 40,000 in cables and poles besides the cost of well, pumpset pump house, starter, and other equipment. As a result, in this study village, we found only six large farmers owned electric pumps while small farmers manage with own or rented diesel pumps. The large farmers are able to earn INR 30,000–35,000 net/year from their tubewell in crop-sharing contracts, which implies a satisfactory rate of return on their capital investment. In West Bengal, even if the farmers were willing to incur such high cost, connections are discouraged by the concern of the State Water Investigation Department about over-exploitation of the groundwater resource.

The only place where the energy squeeze left farming unperturbed was water-abundant Kerala. Diesel pump irrigation disappeared from Kerala in the 1970's as the government laid electricity infrastructure throughout the state. The state has invested large sums in creating paddy irrigation infrastructure, but labour and land shortages and soaring farm wage rates, land use in Kerala is rapidly shifting away from paddy toward plantation crops, mainly rubber, banana, areca nut and coconut. Much plantation economy is built around homesteads where dug-wells, augmented by bores drilled through the bottom of the wells, provide the domestic and garden supply. Farmers lift the small quantity of water needed to water their trees manually or with electric motor-pumps. The energy squeeze is not a serious issue here, yet the government supplies 7.5 l/month of subsidized kerosene per hectare to small holders to soften the energy shock. A 1.5 hp kerosene pump can lift 25 m^3 of water and irrigate 0.4 hectare in 4 hours. The energy cost of irrigation here must be less than 5% of the value of the output it supports compared to 25–35% in northern and eastern India.

3   RETURN TO RAINFED FARMING

In the 1970s irrigation was the leading input in agricultural growth (Ishikawa, 1967). Since 1975 small-holder agriculture boomed with supplemental irrigation made possible by diesel and electric pumps. However, the energy squeeze is now forcing farmers, especially the marginal farmers and share croppers, to economize. In groundwater-rich eastern Uttar Pradesh and Bihar, marginal farmers are withdrawing from wheat and sugarcane cultivation because they cannot afford the cost of supplemental irrigation with rented diesel pumps. In Gujarat as well as Vidarbha, case studies show that farmers dependent on rented diesel pumps are giving up *rabi* wheat and replacing it with rainfed gram and other pulses. In West Bengal (and Bangladesh) small farmers are compelled to give up *boro* rice cultivation.

In the village of Kendradangal in Birbhum district of West Bengal, electric tubewells, generally owned by influential, upper caste farmers, covered most of the village lands barring a small pocket of 70 ha owned by Schedule Caste families. Since 1985, when the *boro* rice revolution overran Bengal, the electrified parts of the village experienced a productivity boom, and the Schedule Caste families irrigated *boro* rice with the help of 25 diesel pumps. By 2005 only nine diesel pumps remained in use by the Schedule caste families, and only three in 2006.

In the canal villages in Kalahandi in Orissa, with diesel pump irrigation rates soaring from INR 25/hour in 1995 to INR 60 in 2007, pump rental markets has shrunk. Many *mali* farmers in this high-water table area took to manual irrigation of vegetables by pots or by lifting water using *dhenkuli* from a depth of 3 m in their 1 m diameter open wells. Moreover, farmers renting diesel pumps shifted to diesel or turned to rainfed cultivation of groundnuts and black gram while expanding vegetable cultivation with pump irrigation for the nearby town. Similar transition from pump irrigated crops to rainfed crops was noted in drier areas.

4   SHARE CROPPERS

The groundwater boom had powerful labour absorption impacts on agriculture. In the Murshidabad village of Kaya, the decline in *boro* paddy and jute cultivation depressed the demand for labour. Leasing small parcels of land for a fixed annual rent has been an important way for the landless families to employ family labour in order to ensure food security. However, with rising diesel prices, this practice has all but disappeared.

Instead of cash tenancy, crop-sharing for water is on rise in some areas. In the Rajkot village of Saurashtra, Gujarat, water buyers depend on renting diesel pumps only for supplemental irrigation in kharif. Renting diesel pumps for rabi crops, once widespread, has completely disappeared. Electric tubewell owners under Gujarat's new Jyotirgram Scheme get 8 hours of uninterrupted, full voltage power under a fairly high flat charge of INR 850/hp/year (Shah and Verma, 2008). These owners are aggressive sellers of pump irrigation during rabi. The common arrangement is crop sharing rather than cash sales: the land owner provides land and labour.

The rise in diesel prices has increased the rental value of surface irrigated land wherever surface irrigation is reliable. In Kalahandi villages in Orissa, electric pump owners generally provide irrigation service on share-cropping basis and earn INR 30,000–35,000 thousand annually from water selling. In a standard contract, the pump owning large farmer contributes land and irrigation usually for groundnuts while the tenant contributes labour; both parties share other costs and output on a 50:50 basis. If a small farmer contributes land and labour and the pump owner contributes just irrigation, then the latter absorbs all costs of other inputs—mainly seeds and fertilizer—and both share the output equally.

5   AGRARIAN EVOLUTION

There is a tendency for small and medium farmers to migrate out of unviable irrigated farming and for even poorer households to reverse-migrate into irrigated farming. This was evident in Keotkuchi, in Assam. In this flood prone village, *kharif* paddy, always at the risk of being washed out, and farmers grow mustard, potato or vegetables soon after *kharif* paddy and then grow their main crop of summer paddy. This procedure was helped during the 1990s when government supplied a large number of diesel pumps at subsidized rates. But now, summer paddy is on a decline due to the diesel prices and most farmers in Keotkuchi who could find off-farm work have taken it, and left farming either to the large farmers or the share croppers. The village is surrounded by villages full of hard-working but landless Bangladeshi Muslims whose priorities are food security by growing their own rice and putting their free family labour to productive use. They bought the spare diesel pumps from Keotkuchi's previous farmers, and lease their paddy land in summer. The other classes of farmers who have survived the energy-squeeze are large farmers who can invest in electric pumps, diesel pumps, tractors and generators.

A departure from farming was noted in the more mechanized agriculture of Punjab, Haryana and Madhya Pradesh. Here, diesel prices have affected small-holder farming through its impact not only on pump irrigation but also on hire rates of other machine services, mainly ploughing and threshing. With water tables down to 20 to 25 m, 50 to 100 m deep tubewells with submersible pumps are needed to access groundwater, and the investment required may exceed INR 120,000. Since tractors are often used to run generator sets, farmers who have tractors and deep tubewells with submersible pumps enjoy economies of scale in the agrarian economy. This works against small farmers who depend on rentals of all machines. Since electric tubewell owners get hardly enough electricity to irrigate their own fields, their customers have to contend with 'generator-set irrigation' which may cost up to INR 1100/day to water 4–5 hectares.

In West Bengal, help has come to the 'energy squeezed' farmer from an unlikely quarter in the form of the Chinese kerosene-cum-diesel pump. *Boro* rice is far more intensive in working capital, labor and irrigation but it is land-saving and therefore appealing to marginal farmers and share croppers. It offers 7 mt/ha of rice yield against barely 1–1.5 mt/ha rainfed *aman* (kharif) rice. Growing a small parcel of *boro* rice may liberate a farming family from subsistence worries for the whole year, and is therefore prized by the poor. The Chinese pumps are cheaper to buy, costing INR 7000 and INR 8500 for 3.5 and 5 hp against INR 16000 for a 5 hp Kirloskar. The Chinese 5 hp pump runs for 2 hours from a litre of diesel which a Kirloskar 5 hp burns up in an hour or less. Finally, while a Kirloskar needs a bullock cart to move around, the Chinese pump can be easily carried by a farmer on his shoulders.

There is also a new trend, throughout India, of using subsidized PDS[1] kerosene meant for cooking for running irrigation pumps. Against the fact that it reduces the life of the engine, poor farmers see two advantages in using kerosene: first, PDS kerosene, subsidized as a cooking fuel, is cheaper than diesel; second, used with Chinese pumps, it yields more water per litre. In many parts of eastern India, collecting PDS quota subsidized kerosene meant for cooking and storing it for irrigating a Rabi or summer crop has increasingly become a standard operating procedure for many poor households.

The rise in pump irrigation costs has forced farmers to search for diesel-efficient irrigation options, including crop choices, irrigation techniques and fuel options. In Rajkot villages in Gujarat, for example, farmers are adopting small-bed irrigation in winter crops such as cumin, gram and wheat, and alternate furrow irrigation for cotton. These can save 20–25% diesel but reduce crop yield by 100 kg in cotton as well as wheat.

A system of rice intensification was introduced in some areas as a water-saving technology; but after trying it for a few seasons, farmers found it labour intensive and it was abandoned. However, many small farmers did switch to the practice of dividing their farms in to small basins, roughly of 200 m[2], at different elevations for more efficient water, and diesel, use.

Curiously, in several of our study areas, small farmers have responded to the diesel price squeeze by adopting even more diesel-intensive crops, mostly vegetables and sugarcane. In one village, highly profitable sugarcane cultivation has replaced some of the wheat and paddy. This reflects farmers moving from a low-input-low-output mode to high-input-high-income one in order to survive the rising costs of farming.

The primary driver of the high-risk, capital intensive cropping strategy is the need to maximize the crop (and cash) per volume of diesel used. In Purana Pradhan village in Khurda district of coastal Orissa, it was found that vegetables cost a lot more to cultivate in cash inputs than *kharif* or summer paddy; but these also offer greater cash returns (see table 4). Some years ago, sugarcane was widely irrigated by diesel pumps; but now vegetables are the most important irrigated crop by diesel pumps in this village because they yield the highest income per volume of diesel. Poorer farmers, whose main concern was

---

[1] Public Distribution System which issues kerosene as cooking fuel to ration card holders.

Table 4.   Costs and returns from paddy and vegetables, Khurda, Coastal Orissa.

	Cost of cultivation (INR/hectare)	Net return (INR/hectare)
Kharif paddy	8650	6175
Summer paddy	14820	7410
Vegetables	123500	74100

food-grain security for the family have had to learn the new skills of vegetable cultivation and of marketing it to maximize their household income.

6   CONCLUSION

Small-holder irrigation in India is under pressure from an energy squeeze from deteriorating farm power supplies, embargo on new electricity connections and an eight-fold increase in diesel prices since 1991. The Government of India's Accelerated Irrigation Benefits Programme is millions of rupees annually in surface irrigation the total area of which is declining. But the real challenge Indian agriculture faces today is of helping the small-holder irrigators cope with the energy squeeze. The fifteen village studies from different parts of India illlustrate the various impacts of the energy squeeze particularly on the poor.

What can be done to ameliorate the energy squeeze? Several ideas emerge from the farmers themselves. Promoting fuel-efficient diesel/kerosene pumps of the Chinese variety can ease the costs. Making the PDS kerosene allocation to a wider distribution of poor farmers, as in Kerala, would also help. The idea of providing subsidized diesel to farmers, is also a possibility. Improving manual irrigation technologies and better management of surface water bodies for gravity flow irrigation can relieve the stress from the energy squeeze. Helping marginal farmers acquire their own pumps can help save them from the monopoly rents currently imposed within pump irrigation prices.

However, all these are short term strategies. The longer term solution lies in improving electricity supply to agriculture. Increasing diesel pump density helps the poor water buyers little, but increasing electric pumps under flat tariffs can improve the net returns from farming for poor water buyers by 20–25 percent (Kishore and Mishra 2005). This is true for all of eastern India (Mukherji 2007b for West Bengal). The political economy of power subsidies that has emerged in India over the past three decades has encouraged state governments and power utilities to view agriculture as the pariah. Today, irrigation contributes as much to farm value creation as land and by giving the agrarian poor preferential control over electricity connections and groundwater, they will have the opportunity that land reforms did not provide.

REFERENCES

Fujita K. and F. Hossain 1995. Role of Groundwater Market in Agricultural Development and Income Distribution: A Case Study in a Northwest Bangladesh Village, *The Developing Economies* 33 (4): 442–463.

Ishikawa, S. 1967. *Economic Development in Asian Perspective*, Economic *Research Series 8.* Tokyo: Kinokuniya Bookstore Co.

Kishore, A. and Mishra, K.N. 2005. Cost of energy for irrigation and agrarian dynamism in Eastern Uttar Pradesh. Anand: IWMI-Tata Water Policy Program.

Mukherji, A. 2007a. *Political economy of groundwater markets in West Bengal, India: Evolution, extent and impacts.* PhD thesis, University of Cambridge, United Kingdom.

Mukherji, A. 2007b. The energy irrigation nexus and its impact on groundwater markets in eastern Indo-Gangetic basin: Evidence from West Bengal, India, *Energy Policy*, 35 (12): 6413–6430.

Mukherji, A. (2008). Spatio-temporal analysis of markets for groundwater irrigation services in India, 1976–77 to 1997–98, *Hydrogeology Journal*. http://dx.doi.org/10.1007/s10040-008-0287-0.

Narayanamoorthy, A. 2007. Does groundwater irrigation reduce rural poverty? Evidence from Indian states", *Irrigation and Drainage* 56: 349–361.

Pant, N. 2005. 'Control and Access to Groundwater in UP' *Economic and Political Weekly*, 40 (26): 2672–2680.

Shah, T. 1993. *Groundwater Markets and Irrigation Development: Political Economy and Practical Policy*, Bombay: Oxford University Press.

Shah, T. and Ballabh, V. 1997. *Water Markets in North Bihar: Six Village Studies in Muzaffarpur District.* Economic and Political Weekly, 32 (52): A183–A190.

Shah, T. and Verma, S. 2008. Co-management of Electricity and Groundwater: An Assessment of Gujarat's Jyotirgram Scheme. *Economic and Political Weekly*, 43 (7): 59–66.

Shah, T., Singh, O.P. and A. Mukherji. 2006. Some aspects of South Asia's groundwater irrigation economy: analyses from a survey in India, Pakistan, Nepal Terai and Bangladesh", *Hydrogeology Journal*, Vol. 14 (3): 286–309.

Wilson, K. 2002. Small cultivators in Bihar and 'New' Technology. *Economic and Political Weekly*, Vol. 37 (13): 1229–1238.

CHAPTER 16

Managing the energy-irrigation nexus in West Bengal, India

A. Mukherji
International Water Management Institute (IWMI), Colombo, Sri Lanka

P.S. Banerjee
Independent Consultant, Kolkata, India

S. Daschowdhury
Independent Consultant, Santiniketan, West Bengal, India

ABSTRACT: South Asia in general and India in particular is heavily dependent on groundwater for supporting its largely agrarian population. Quite predictably, in such pump lift based economy, the fortunes of the energy and irrigation sectors are closely entwined. This has often been called the 'energy-irrigation' nexus. There are two major sources of energy for pumping groundwater, namely, electricity and diesel. Current discourse in the field has, however, looked exclusively at the 'electricity-irrigation' nexus to the exclusion of 'diesel-irrigation nexus'. This chapter looks at both these aspects. In doing so, it makes two propositions. First, a high flat rate electricity tariff encourages pro-active and competitive water markets whereby the water buyers—mostly small and marginal farmers—benefit. Second, the low rate of rural electrification has forced the majority of farmers to depend on diesel for groundwater pumping and steep increase in diesel prices over the last few years has resulted in further decline in farm incomes.

1   INTRODUCTION

Most of South Asia today is dependant on groundwater irrigation for supporting its predominantly agrarian economies. South Asia is the world's single largest user of groundwater accounting for almost 210 km^3 of withdrawals every year (Mukherji and Shah 2005, Shah et al. 2003). It is no surprise that in such pump lift irrigation based economies, the fortunes of the groundwater and energy sectors are closely entwined. This relationship between the two sectors has been often called the 'energy-irrigation nexus' (Shah et al. 2003) or the 'energy irrigation conundrum' (Dubash 2007).

The scope of the term energy is broad, yet discussions have been limited to the 'electricity-irrigation' nexus. Much of the debate so far has been about the pros and cons of two different modes of electricity pricing, namely, rational flat tariff system and pro-rata metering. While one school of thought has propounded the superiority of rational flat tariff on grounds of equity and administrative ease (Shah 1993, Shah et al. 2003), another group has alleged that such a tariff system leads to loss of efficiency and endangers sustainability of both groundwater and electricity (Briscoe 2005). Pre-occupation with the electricity tariff is understandable because the electricity sector provides the only formal point of contact with an otherwise informal groundwater economy in south Asia.

Table 1. Total replenishable groundwater resources per unit of net cropped area and geographical area, May 2004.

State	Gross replenishable GW per unit of NCA (MCM/'000 ha)	Rank
Assam	9.03	1
West Bengal	5.55	2
Tamil Nadu	4.83	3
Uttar Pradesh[#]	4.78	4
Bihar[*]	4.53	5
Punjab	4.43	6
Kerala	3.51	7
Orissa	3.31	8
Andhra Pradesh	3.31	9
Madhya Pradesh[@]	2.68	10
Haryana	2.38	11
Maharashtra	2.13	12
Gujarat	2.13	13
Karnataka	1.56	14
Rajasthan	0.78	15

Source: Central Water Commission (data downloaded from website www.indiastat.com on 28th February 2006). [#] includes Uttar Pradesh and Uttaranchal, [*] includes Bihar and Jharkhand, [@] includes Madhya Pradesh and Chattisgarh. NCA = Net cultivated area.

The other major energy source for groundwater pumping is diesel but it remained outside the discussions on the 'energy-irrigation nexus'. This is because diesel prices were heavily subsidized until recently and areas which are now highly dependant on diesel operated water extraction mechanisms (WEMs) (the eastern and central parts of India), which in the 1980s and early 1990s were less dependant on groundwater irrigation than their counterparts in northern, western and southern India. However, over the last 10 years and so, this situation has changed. There has been a gradual removal of subsidies on diesel and there has been intensive groundwater use in the eastern states. There is an urgent need to broaden the scope of work on the 'energy-irrigation' nexus and move beyond the narrow 'electricity-irrigation' focus. This chapter attempts to broaden the scope of debate by looking at both the faces of 'energy-irrigation' nexus. In doing so, it draws on empirical evidence based on extensive fieldwork carried out in 2004–05 in the state of West Bengal, India.

2   WEST BENGAL: A LAND OF ABUNDANT YET SCARCE GROUNDWATER

West Bengal (Figure 1), an eastern state of India located within the Ganga-Meghna-Brahmaputra basin is a land of plentiful rainfall (1200 to 2500 mm annual rainfall) and alluvial aquifers which hold some 31 billion cubic meters of groundwater (WIDD 2004) with a water table some 5 to 10 m below ground level in 95% of villages (3rd MI Census, GOI 2001). In terms of per unit availability of groundwater, West Bengal ranks second (million cubic meters (MCM) of groundwater/1000 ha of net cultivated area).

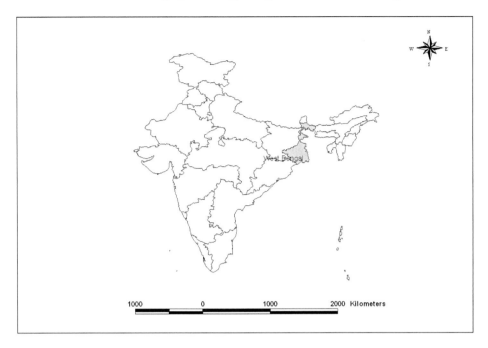

Figure 1.   Location of West Bengal in India.

The level of groundwater development in the state varies from as high as 84.6% in Nadia district to as low as 5% in Jalpaiguri district, the average for the state being 41.3%. Thus, none of the 17 districts fall in the 'over-exploited' category of the Central Groundwater Board. Groundwater estimation carried out jointly by the State Water Investigation Directorate (SWID) and the Central Groundwater Board following the GEC 97 methodology, found as many as 231 blocks (or 86%) of the blocks to be 'safe', while 37 blocks were declared 'semi-critical' and only 1 block was put in the 'critical' groundwater category (Ray Chowdhury 2006). Contrast this to recent statistics from the state of Punjab: of the 137 blocks in Punjab, 103 blocks (or 75%) are over-exploited; five blocks are in a 'critical' stage, four blocks are in the 'semi-critical' stage and only 25 blocks are in the 'safe' category (Takshi 2006). In the state of Gujarat: 45% of 184 blocks were in the category of 'over-exploited', aquifer in the year 1997, a number that has almost certainly gone up by now (Hirway 2000, p. 7).

Table 2 juxtaposes the net groundwater available for irrigation with that of tubewell density (number of tubewells/1000 hectares of cultivable land). The results show that tubewells density is very low even in the high groundwater potential blocks, showing that there is an under-exploitation of the groundwater resources in the state and that there is further scope for groundwater utilisation without jeopardising groundwater sustainability.

While groundwater resources are relatively abundant in the state, the state policies are geared towards discouraging farmers from groundwater irrigation. Farmers in West Bengal pay one of the highest flat electricity tariffs anywhere in India. The state also has the distinction of having the lowest number of electrified WEMs in the country. Only 12.2% of all WEMs in West Bengal are electrified (3rd MI Census GOI 2001) as against a national average of 50%. The process of electrification of pumps is governed by strict SWID regulations

Table 2.   Net groundwater available for irrigation (MCM/1000 ha of cultivable land) versus density of
tubewells (No. of tubewells/100 ha of net cultivable land) in 2000–01: A block level cross-tabulation.

Sr. No.	Category	Number of blocks	Percentage to total
1.	High groundwater potential- High tubewell density	68	26.0
2.	High groundwater potential- Low tubewell density	113	43.1
3.	Low groundwater potential- High tubewell density	3	1.1
4.	Low groundwater potential- Low tubewell density	78	29.8
5.	Total	262	100.0

Source: Based on 3rd MI census (GOI 2001).
High groundwater potential: >5 MCM of net groundwater/1000 ha of cultivable land;
Low groundwater potential: <5 MCM of net groundwater/1000 ha of cultivable land;
High tubewell density: >20 tubewells/100 ha of net cultivable land;
Low tubewell density: <20 tubewells/100 ha of net cultivable land.

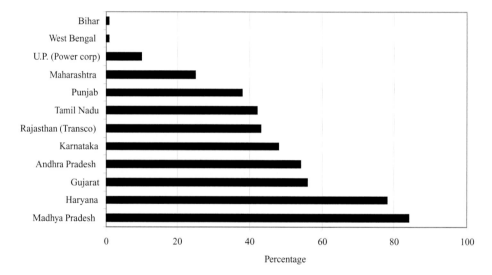

Figure 2.   Electricity subsidy to agriculture as percentage of gross fiscal deficit 2000–01.
Source: Briscoe 2005:24.

making the entire process lengthy and cumbersome. All these facts, viz. abundance of
groundwater, high electricity tariff, difficulty in getting new electricity connections and
low rates of rural electrification (which makes West Bengal's groundwater economy largely
diesel dependent) put the state in contrast with other Indian states such as Punjab, Haryana,
Gujarat and Tamil Nadu. In these states, in spite of precarious groundwater conditions,
farmers get concessions from the state in terms of electricity subsidies.

In West Bengal the 'energy-irrigation' nexus is different than in other Indian states. The
issue of electricity subsidy to agriculture and its impact on state fiscal deficits while being
very important in most states of India, is not an issue in West Bengal, given that only 12.5%
of all pumps are electrified and electricity tariffs are quite resulting in lower state deficits
(see Figure 2). However, the impact of rising diesel costs on irrigation is a major issue in
West Bengal given that 90% of all WEMs in the state are diesel operated.

3 HIGH FLAT RATE ELECTRICITY TARIFF AND GROUNDWATER MARKETS

In West Bengal, access to groundwater irrigation is either through ownership or through purchase of pump irrigation services. Of the 6.1 million farming households in West Bengal, only 1.1 million reported owning WEMs, while 4.6 million farming households reported using irrigation (NSSO 1999). Some of them certainly fell within the canal command areas, but an overwhelming number of 3.1 million households (or 50.4% of all farming households) reported hiring of irrigation services from other farmers. The current survey, conducted in 40 villages spread across 17 districts, shows that 70.5% of all households reported buying water from private WEMs, 75% of pump owners sold water to other, 33% of all WEM owners also bought water from others and 92% of all respondents were part of the groundwater market either as a seller or buyer or both. Groundwater markets in West Bengal have been described by Mukherji (2007).

Farmers in West Bengal pay a flat tariff of electricity. This rate at present is around INR 5500/year for centrifugal electric pumps and INR 6800/year for submersible electric pumps. There has been a rapid increase in the electricity tariff over the last 10 years or so (Table 3).

Proponents of a rational flat tariff argue that under a relatively high flat tariff regime, owners of WEM have a positive incentive to sell water to others. This is because under a flat tariff regime, the marginal cost of pumping approaches zero and an additional hour of pumping does not entail an additional cost. However, by selling water, the pump owner is able to recoup his or her electricity bill and also earn a profit. In the case of the diesel pump (which entails incremental costs for every hour of pumping) under a pro-rata or metered electricity tariff, this incentive will be absent and the spread of groundwater markets limited. This will be more so given the escalating diesel prices in recent years. The following hypotheses can be derived:

An electric WEM owner will sell water for a longer number of hours than a diesel WEM owner. Therefore, groundwater markets will be highly developed (where development of groundwater market is defined in terms of breadth and depth of water market transactions) in villages dependent predominantly on electric WEMs than in diesel WEM dependent villages.

3.1 *Role of motive power of pumps in determining actual irrigation hours sold*

Determination of the hours of water sold by WEM owners is based on a sample of 243 WEMs belonging to 220 water sellers from 40 villages. Simple tabulation of data (Table 4) showed that the number of hours of water sold by a WEM owner is a function of at least two variables, the motive power of the pump and the type of pump. This is because as already shown, the motive power of the WEM dictates the cost structure of water extraction and hence the nature of water transactions and an electric WEM owner will sell more water than a diesel WEM owner. The second factor that affects the number of hours of water sold is the type of WEM, i.e. whether the WEM is a surface mounted centrifugal or a submersible pump. The type of WEM thus reflects the hydrogeological condition in a region. Given the same pump capacity, the discharge of a submersible pump is higher than the discharge from a surface mounted pump. Owners of submersible pumps could have a higher amount of 'surplus' water available for sale than those of centrifugal pumps.

Table 3.   Change in flat rate electricity tariff in West Bengal, 1995 to 2003.

Year	Electricity tariff for shallow tubewells (INR/year/tubewell)		Electricity tariff for submersible tubewells (INR/year/tubewell)	
	North Bengal	Other districts	North Bengal	Other districts
1991	1100	1100	1100	1100
1995	1380	1700	1380	1700
1996	1660	2040	2500	3060
1999	2676	3284	4028	4932
2001	4064	5008	5080	6252
2003	4434	5460	5540	6810

Source: West Bengal State Electricity Board (WBSEB) records, various years.

Table 4.   Hours of pumping and hours of water sold to others by type of WEM, 2003–04.

Sr. No.	Type of WEM	Sample size	Average hours of pumping	Average hours of water sold to others	% of hours of water sold to total pumping hours
1.	Diesel shallow tubewell (DST)	189	250.8	91.3	36.4
2.	Diesel submersible tubewell (DSB)	7	411.6	201.6	49.0
3.	Electric shallow tubewell (EST)	73	1649.3	863.3	52.3
4.	Electric submersible tubewell (ESB)	65	2151.7	1715.3	79.7
5.	All	334	929.1	579.9	62.4

Source: Primary questionnaire survey in 40 villages, August to December 2004.

The demographic and land owning characteristics of the WEM owners might have significant impact on the volume of water sold. The area cultivated by the WEM owners could be negatively related to hours of water sold, as owner will direct most of the pumped water for self-use. The presence or absence of alternative sources of cheap irrigation (such as a government deep tubewell, river lift irrigation, canal or cooperative tubewells) might adversely impact the quantity of water sold by a WEM. The level of groundwater development and the long-term trend in groundwater level (denoted by categories of safe, semi-critical and critical blocks) might also affect the total volume of water sold, although the exact way this variable would affect the water selling decision of a water seller is an empirical question. A linear Ordinary Least Square (OLS) model where:

HOURS = fn{MOTIV, TWEM, GCA, ALTIRR, GWDEV}
Where, HOURS = Hours of water sold by an WEM owner in the year 2003–04
MOTIV = Dummy variable for the motive power of the pump, 0 if diesel,
    1 if electric
TWEM = Dummy variable for the type of WEM, 0 if centrifugal pump, 1 if submersible
    pump

Table 5.   Determinants of hours of water sold in 40 villages in West Bengal, 2003–04.

Sr. No.	Variables	Unstandardised coefficient B	Standardised coefficient β	t-value
1.	Constant	294.867*	–	4.420
2.	MOTIV (dummy)	879.972*	0.514	11.532
3.	TWEM (dummy)	702.358*	0.370	8.021
4.	GCA	−68.290*	−0.194	−4.908
5.	ALTIRR (dummy)	−105.445	−0.062	−1.595
6.	GWDEV (dummy)	332.600*	0.173	4.395
7.	Adjusted R^2	0.646		
8.	Sample size	43		

* Denotes significance at 1% level.
Source: Author's calculations based on questionnaire survey conducted between August to December 2004.

GCA = Gross cultivated area of the WEM owner (ha) in 2003–04
ALTIRR = Dummy variable for presence of alternate sources of irrigation in the village, 0 = No, 1 = Yes
GWDEV = Dummy variable for level of groundwater development and trend in water level, 0 = safe, 1 = critical and semi-critical.

The result of the regression equation is presented in Table 5. The motive power of the pump is the most important determinant of actual hours of water sold by a pump owner, followed by the type of pump owned. While land owning characteristics, sources of alternate irrigation and the groundwater resource condition in a village affect the amount of water sold by a WEM owner, the most important determinant of hours of water sold is the type of WEM and its motive power.

3.2   *Level of development of groundwater markets and the motive power of pumps*

The level of development of groundwater markets have been measured in terms of breadth and depth of transactions (Shah 1993). The breadth of groundwater markets refers to the horizontal spread of the market while depth refers to the vertical spread. While there are several measures of depth and breadth of groundwater markets, in this section using one indicator of each, the 40 study villages have been classified into three levels of development: highly developed, moderately developed and under developed. The indicator of breadth used here is percentage of gross irrigated area (GIA) in the village irrigated through privately purchased groundwater. The indicator of breath that has been used is percentage of gross income of households derived from water selling and buying. The level of development of groundwater markets has been related to the predominant type of WEM (defined by its motive power and technology of pumping equipment) in that village. Table 6 shows that all the villages that show high development of groundwater markets invariably have an electric WEM dominated irrigation economy.

This analysis shows that electric WEM owners with a flat rate tariff are likely to have a larger volume of water pumped than diesel pump owners and that villages with predominantly electric WEMs have highly developed groundwater markets in comparison with diesel WEM villages.

Table 6.   Village measure of breadth and depth of groundwater market.

Village code	District	% of GIA irrigated by purchased water	% of gross incomes derived from water sale and purchase	Breadth vs. depth comparison	Level of development	Predominant type of WEM by motive power and type of pump
BR02	Bardhaman	61.9	37.8	HB-HD	High	ESB
HW01	Howrah	50.0	38.1	HB-HD	High	EST
HG01	Hugli	69.5	52.9	HB-HD	High	ESB
HG02	Hugli	73.2	33.9	HB-HD	High	ESB
HG03	Hugli	64.9	32.4	HB-HD	High	ESB
HG04	Hugli	63.9	26.9	HB-HD	High	ESB
MS03	Murshidabad	63.3	63.8	HB-HD	High	EST
MS07	Murshidabad	58.4	50.4	HB-HD	High	ESB
MS05	Murshidabad	67.7	48.7	HB-HD	High	ESB
MS01	Murshidabad	68.3	47.8	HB-HD	High	ESB
MS04	Murshidabad	52.4	36.2	HB-HD	High	ESB
MS06	Murshidabad	54.5	29.7	HB-HD	High	ESB
MS02	Murshidabad	51.2	29.2	HB-HD	High	EST
NP01	N. 24 Prgs	56.6	45.9	HB-HD	High	EST
NP02	N. 24 Prgs	60.2	43.0	HB-HD	High	EST
NP04	N. 24 Prgs	61.6	37.6	HB-HD	High	EST
ND02	N. 24 Prgs	54.8	26.0	HB-HD	High	EST
ND06	N. 24 Prgs	51.5	25.8	HB-HD	High	EST
ND03	N. 24 Prgs	56.8	25.5	HB-HD	High	EST
BN01	Bankura	54.4	13.6	HB-LD	Moderate	DST
BR03	Bardhaman	70.0	19.8	HB-LD	Moderate	DST
BR01	Bardhaman	59.4	8.8	HB-LD	Moderate	DST
HG05	Hugli	58.3	24.0	HB-LD	Moderate	DST
JL01	Jalpaiguri	50.0	22.7	HB-LD	Moderate	DSB
KB01	Koch Bihar	50.0	22.7	HB-LD	Moderate	DST
ML02	Maldah	46.2	40.7	LB-HD	Moderate	DST
ML01	Maldah	24.7	25.2	LB-HD	Moderate	DST
MD01	Medinipur	56.7	18.0	HB-LD	Moderate	DST
MD02	Medinipur	25.0	31.5	LB-HD	Moderate	DST
NP05	N. 24 Prgs	50.4	21.2	HB-LD	Moderate	DST
ND01	Nadia	60.6	19.6	HB-LD	Moderate	DST
ND05	Nadia	40.7	25.9	LB-HD	Moderate	DST
BI01	Birbhum	21.1	7.1	LB-LD	Low	DST
DJ01	D. Dinajpur	26.9	21.2	LB-LD	Low	DST
NP03	N. 24 Prgs	42.2	12.2	LB-LD	Low	DST
ND04	Nadia	47.0	13.2	LB-LD	Low	DST
PR04	Purulia	28.7	13.6	LB-LD	Low	DST
PR03	Purulia	33.2	6.2	LB-LD	Low	DST
PR01	Purulia	48.8	3.7	LB-LD	Low	DST
PR02	Purulia	37.6	3.0	LB-LD	Low	DST

Source: Calculations by the author based on primary data collected during questionnaire survey, August to December 2004. HB = High breadth (>50%), LB = Low breadth (<50%), HD = High depth (>25%), LD = Low depth (<25%).
ESB = Electric submersible pumps, EST = Electric centrifugal pump, DST = Diesel centrifugal pump, DSB = Diesel submersible pump.

Table 7.   Size class classification of WEM owners and water buyers in West Bengal.

Sr. No.	Size class category	No. of WEM owners	No. of pure water buyers
1.	Sub-marginal (<0.5 ha)	56 (19.0)	142 (49.7)
2.	Marginal (0.51–1.0 ha)	80 (27.2)	85 (29.7)
3.	Small (1.01–2.0 ha)	90 (30.6)	46 (16.1)
4.	Medium (2.01–4.0 ha)	56 (19.0)	10 (3.5)
5.	Large (>4.01 ha)	12 (4.1)	3 (1.0)
6.	All	294 (100.0)	286 (100.0)

Source: Primary data collected during questionnaire survey, August to December 2004. Figures in parentheses are percentage to total.

3.3   Are groundwater markets beneficial?

Groundwater markets are described as the 'vehicle of poverty alleviation' (Palmer-Jones 2001) at one extreme to 'creating water lords' (Janakarajan 1990) at the other. Theoretically, there are several advantages of groundwater markets. Firstly, they lead to increased use of tubewell capacity, thereby encouraging efficient use of tubewells. Secondly, they increase access to irrigation for those farmers who cannot afford their own irrigation equipment. Thirdly, water markets encourage farmers—even small and marginal ones—to invest in tubewells with the potential of profiting from water sales. Fourthly, as a direct result of increased access to groundwater, cropping intensity, as well as the demand for labour, increase. The net irrigation surplus is higher with a groundwater market than without it. Finally, groundwater markets in water surplus regions such as West Bengal lower the water table and control water logging and flooding (Roy 1989).

In the absence of state provisioning of irrigation, groundwater markets have done a commendable job in distributing the benefits of irrigation to those who do not have their own means of irrigation. Most of them also happen to be small and marginal farmers (Table 7).

Second, due to existence of groundwater markets and their efficient functioning, those without their own mean of irrigation (i.e. the water buyers) are still able to achieve similar cropping pattern cropping intensities, crop productivity and even similar gross incomes as owners of means of irrigation (Table 8).

It was precisely the rise in groundwater irrigation through the operation of groundwater markets that propelled West Bengal to its high rates of agricultural growth during the 1980s and 1990s.

4   ESCALATING DIESEL PRICES, IMPACT ON GROUNDWATER MARKETS AND COPING STRATEGIES ADOPTED BY THE FARMERS

The electricity-irrigation nexus is only part of the energy irrigation nexus. The other and more important aspect of the energy irrigation nexus in West Bengal is the impact of rising diesel prices on the groundwater economy. This is specially so because almost 80–85% of all WEMs in the state are diesel driven. Escalating diesel prices have had at least three impacts on agricultural sector in general and functioning of groundwater markets in particular. First, the most immediate impact has been contraction in water market operations. Second, there

Table 8.   Impact of groundwater market: Evidence from West Bengal.

Sr. No.	Indicator	Pump owners	Water buyers
1.	Cropping intensity (%)	184.0	180.0
2.	Percentage area under water intensive *boro* paddy to GCA	24.1	22.8
3.	Percentage of area under profitable potato crop to GCA	8.0	8.1
4.	Productivity (kg/ha) of *boro* paddy	5005	5005
5.	Productivity (kg/ha) of potato	16135	17928
6.	Hired labour use (mandays/ha) for *boro* paddy	128	111
7.	Fertilizer use (kg/ha) for *boro* paddy	499	464
8.	Gross income from crop cultivation (INR/year/ha)	31075	28468
9.	Sample size (Numbers)	294	286

Source: Fieldwork in 40 villages in West Bengal, August 2004 to December 2004.

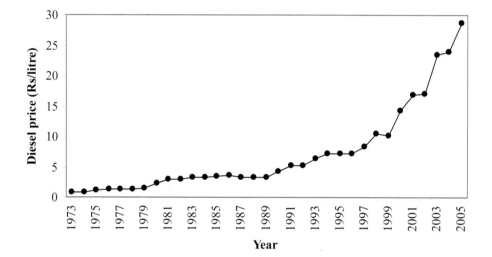

Figure 3.   Retail price of diesel in Kolkata, 1973 to 2005.
Source: Ministry of Petroleum and Natural Gas, Government of India, downloaded from site www.indiastat.com on 12 July 2006.

has been a shift away from water intensive (but profitable) *boro* paddy cultivation to rainfed crops, or vegetable and orchard crops. Third, in order to break even, farmers have had to become innovative.

4.1   *Impact of rising diesel prices on spread of groundwater markets*

The cost of water extraction is central to the economics of groundwater pumping and water selling. The major part of this cost is the energy costs. Since 1998 both diesel and electricity have gone up in price and this has influenced the functioning of water market. Figure 3 shows the changes in diesel prices since 1973 to 2005.

Increase in diesel prices have led to shrinkage of water market transactions in diesel dominated WEM villages. The contraction of the groundwater market operations and absence

Table 9.   Area under *boro* paddy and the pre-dominant WEM type in 40 villages of West Bengal, 2003–04.

Sr. No.	Area under *boro* paddy cultivation	Number of villages	Pre-dominant type of WEM (number of villages)*
1.	Less than 10%	12	DST (7), DNO (4), DSB (1)
2.	10–20%	8	DST (4), ESB (2), HNO (1), DNO (1)
3.	20–30%	10	ESB (7), EST (1), DST (2)
4.	More than 30%	10	EST (6), ESB (2), DST (2)

Source: Author's fieldwork in 40 villages in West Bengal, August to December 2004.
* DST = Diesel shallow tubewell, DNO = Diesel pumps not permanently attached to any tubewell, HNO = Honda and Chinese pumps not permanently attached to any tubewell, DSB = Diesel submersible tubewell, EST = Electric shallow tubewell, ESB = Electric submersible tubewell.

of any other affordable sources of irrigation negatively affects those who are dependent on diesel pumps to access irrigation, and this hits the poorest farmers the hardest. This has happened in West Bengal during the past five years or so.

4.2   *Changes in cropping pattern in response to rising diesel prices*

How are the farmers coping with rising diesel costs? One of the obvious ways in which they are doing this is through changes in cropping patterns away from water intensive *boro* paddy cultivation to less water intensive crops. While some of these less water intensive crops at times are more profitable (but risky) than *boro* paddy cultivation, it is not always so. For instance, in various parts of West Bengal, farmers have reverted to rainfed crops in the winter season.

Boro paddy is one of the most profitable crops in West Bengal. In 20 of the sample of 40 villages, *boro* paddy is cultivated in less than 20% of the gross cropped area of the village. Of these, 18 villages have diesel shallow tubewell (DST) type of water market. In all of these villages, area under *boro* paddy has declined sharply since 2001 in response to rising diesel prices (see Table 9).

Given that almost 80–85% of all water extraction devices in West Bengal are diesel operated (GOI 2001), and diesel prices have been going up, it serious repercussions on *boro* paddy cultivation in the state. It is widely acknowledged that increases in area under *boro* paddy coupled with productivity increases in *aman*, *aus* and *boro* paddy had propelled spectacular growth in agriculture in West Bengal (Rogaly et al., 1999). This scenario is likely to change given the very high diesel prices. In an unfavorable input output price regime, those farmers who depend exclusively on diesel pumps are at disadvantage to those who have access to electric WEMs. Table 10 shows net profitability from *boro* cultivation for electric WEM owners, diesel WEM owners and their respective buyers. It shows the very low profit for the diesel WEM owners and even lesser profits for those who buy water from diesel WEM owners. Ghosh and Hariss-White (2002) voiced this concern when they found a "deep crisis in rice economy"—a crisis that has since then deepened in intensity and severity.

Vegetable cultivation and orchards crops have emerged as viable alternatives to the water intensive *boro* crop. However, in view of inadequate marketing channels and lack of insurance and credit markets, cultivation of capital intensive vegetables and orchard crops

Table 10.   Cost of cultivation and net returns from boro paddy for diesel and electric pump owners and water buyers in West Bengal, 2003–04.

Crop	WEM type	Water transaction status	N	Yield (kg per ha)	Gross revenue (INR/ha)	Cost of cultivation (INR/ ha)		Net returns (INR/ha)	
						Cost of cultivation without including family labour	Cost of cultivation including family labour	Net returns without imputing family labour	Net returns after imputing family labour
Boro	Diesel	PO	55	5005	27527	17779	20109	9748	7418
Boro	Diesel	WB	28	5229	28760	23202	26967	5558	1793
Boro	Electric	PO	64	5453	29992	12333	14910	17659	15082
Boro	Electric	WB	61	5154	28349	15515	18152	12833	10197

Source: Author's calculations based on fieldwork in 40 villages in West Bengal, August to December 2004.

tends to be a risky venture. Vegetable cultivation is more profitable than paddy cultivation provided that the villages are well connected to the market and have a suitable soil type. These two conditions rarely coincide and the direct impact of increased diesel prices is a lowering of cropping intensity and changing cropping patterns away from remunerative irrigated crops to rainfed cropping systems.

4.3   *Technical innovations in face of rising diesel costs*

Farmers have been resorting to various innovations for cutting their total irrigation costs. For instance, traditional large diesel pumps are now seldom used. Instead, farmers use lightweight pumps (Honda pumps) or Chinese pumps smuggled from across the Bangladeshi border. Both these types of pumps consume only 300 ml to 500 ml of diesel per hour and alternatively, may be operated with kerosene. Traditional diesel pumps consume 800 to 1000 ml diesel an hour. All the diesel pumps purchased in or after 2001 in the sample villages were either Honda or Chinese pumps. Innovations were also aimed at reducing energy costs, including the use of cooking gas cylinders to operate pumps. This reduces the cost of water extraction by 50%–60%. Similarly, instead of using solely diesel, farmers increasingly mix kerosene (which continues to be subsidized) with diesel to operate their pumps, although continued use of kerosene reduces the effectiveness of pumps. The fieldwork showed that rural pump mechanics are constantly making changes in pump design (for example, altering the size of the fuel intake pipe etc.) in order to increase fuel efficiency. The rise in diesel prices has forced farmers to economize on the use of diesel by making pumps more fuel efficient.

5   CONCLUSIONS AND POLICY OPTIONS

The 'energy-irrigation' issues facing West Bengal are different from the issues in other states of India. In most other states of India the 'groundwater economy has boomed by bleeding the energy economy' (Shah et al. 2003: v). However, this is not the case in West

Bengal, where electricity subsidies form only a small part of the state fiscal deficits. Given the favorable hydrogeological conditions, groundwater markets have a positive impact on the state. This may not necessarily be true in western and southern parts of the country where groundwater resources are scarce and over-used. The rate of rural electrification has been very slow in West Bengal with the result that only 12.5% of all pumps in the state are electrified. Therefore, the benefits from a rational flat tariff system have only been partially realized. The negative implication of the 'diesel-irrigation' nexus has been much more serious than the positive implication of the 'electricity-energy' nexus. Escalation in diesel prices has had a negative impact on the spread of water market transactions thereby putting the livelihoods of millions of poor farmers in jeopardy.

The groundwater sector in West Bengal does not warrant extensive regulation, it does merit electrification. West Bengal is a state with the lowest percentage of electric pumps. Reintroduction of a capital subsidy for extending the rural electricity programme is needed. Special attention needs to be given to the poorer districts of North Bengal. West Bengal has one of the highest flat rate tariffs in the country—this has encouraged rapid development of groundwater markets in villages with electric WEMs and promoted competitive water selling. The water buyers earn higher net returns from cultivation in electric WEM villages (especially in ESB villages), than in diesel WEM villages. There is a recent proposal in the state for switchover to metering of irrigation tubewells. This might lead to shrinkage in water markets and the hardest hit would be the water buyers. There are peaks (during March–May) and dips (during monsoon months) in demand, but the supply of electricity remains uniform (12 to 20 hours on an average depending on the district) across the year. The state electricity board could restrict supply in the slacker periods so saving money.

There are some other ways in which state policies could help farmers with declining farm profits. One would be allowing import of cheap and fuel efficient Chinese pumps—currently they are being smuggled into India through the Bangladeshi border. Second, much on the same lines as in Bangladesh, the GoWB could supply diesel at a subsidized price to small and marginal farmers during the *boro* paddy season. More importantly, the government should encourage farmers to shift to lucrative vegetable cultivation by investing in proper infrastructure such as cold storages, processing plants and improved roads.

REFERENCES

Briscoe, J. (2005), India's water economy: Bracing for a turbulent Future, India Water Review, Washington D.C., The World Bank.

Dubash, N.K. (2007), 'The electricity-groundwater conundrum: Case for a political solution to a political problem' Economic and Political Weekly, 42 (52): 45–55.

Ghosh P.K. and Harriss-White B (2002), A crisis in the rice economy, *Frontline*, 19 (19), A Hindu group publication.

GOI (2001), *All India report on agricultural census, 2001–02, New* Delhi, Ministry of Agriculture, Government of India.

Hirway, I. (2000), Dynamics of development in Gujarat: some issues, Economic and Political Weekly, 33 (25): 1533–1544.

Janakarajan, S. (1990), 'Interlinked transactions and the market for water in the agrarian economy of a Tamilnadu village' in S. Subramanian (ed.) *Themes in Development Economics: Essays in Honour of Malcolm Adiseshiah*. New Delhi, Oxford University Press.

Mukherji, A. (2007), Political economy of groundwater markets in West Bengal: Evolution, extent and impacts, PhD thesis, University of Cambridge, U.K.

Mukherji, A. and T. Shah (2005) 'Groundwater socio-ecology and governance: a review of institutions and policies in selected countries', *Hydrogeology Journal*, 13 (1): 328–345.

NSSO (1999), *54th round: Cultivation practices in India, January 1998-June 1998*, Department of Statistics and Programme Implementation, Government of India, August 1999, New Delhi.

Palmer-Jones, R.W. (2001), 'Irrigation service markets in Bangladesh: Private provision of local public goods and community regulation', chapter presented at *Symposium on Managing Common Resources: What is the solution?* Held at Lund University, Sweden, 10–11 September 2001.

Ray Chowdhury, P.K. (2006), Framework and implementation of the West Bengal Groundwater Resources (Management, Control and Regulation) Act, 2005, in Romani S., K.D. Sharma, N.C. Ghosh and Y.B. Kaushik (eds.) *Groundwater Governance: Ownership of Groundwater and its Pricing*, Capital Publishing Company, New Delhi, Kolkata, Bangalore.

Rogaly, B., B. Harriss-White and S. Bose (1999), Introduction: Agricultural growth and agrarian change in West Bengal and Bangladesh in Rogaly et al. (eds) *Sonar Bangla? Agricultural Growth and Agrarian Change in West Bengal and Bangladesh*, New Delhi, Sage Publications.

Roy, K.C. (1989). Optimization of unconfined shallow aquifer water storage for irrigation, PhD thesis, Utah State University.

Shah, T. (1993), Water markets an irrigation development: Political economy and practical policy, Bombay, Oxford University Press.

Shah, T., C. Scott, A. Kishore and A. Sharma (2003), Energy-*irrigation nexus in South Asia: Improving groundwater conservation and power sector viability.* Research Report 70. Colombo, Sri Lanka, International Water Management Institute.

Takshi, K.S. (2006), Groundwater governance: Issues and perspectives regarding model bill application in Punjab State, in Romani S., K.D. Sharma, N.C. Ghosh and Y.B. Kaushik (eds) *Groundwater Governance: Ownership of Groundwater and its Pricing*, Capital Publishing Company, New Delhi, Kolkata, Bangalore.

WBSEB (several years), Annual Report of West Bengal State Electricity Board, Kolkata.

WIDD (2004), *Groundwater resources of West Bengal: An estimation by GEC-1997 methodology*, Water Investigation and Development Department (WIDD), Kolkata, Government of West Bengal.

OTHER SOURCES OF INFORMATION

www.indiastat.com on 12th July 2006 (Information on retail diesel prices in Kolkata, 1975–2005, Source: Ministry of Petroleum and Natural Gas, Government of India).

CHAPTER 17

Groundwater markets in the North China Plain: Impact on irrigation water use, crop yields and farmer income

L. Zhang, J. Wang and J. Huang
Center for Chinese Agricultural Policy, Chinese Academy of Sciences, Beijing, China

S. Rozelle
Shorenstein Asia Pacific Research Center, Freeman Spogli Institute for International Studies Stanford, CA, USA

Q. Huang
Department of Applied Economics, University of Minnesota, MN, USA

ABSTRACT: Although increasing attention is being paid in the literature to document the rise of groundwater markets and understand where and in what conditions they emerge, little empirical work has been done to measure the impact that groundwater markets have on irrigation water use, crop yields and farmer income. This is surprising given the potential effects—both positive and negative—that groundwater markets might be expected to have on the production and welfare of farm households. Based on a survey of 46 randomly sampled villages and 173 households in two provinces (Hebei and Henan Province) in 2001 and 2004, farmers buying irrigation water through groundwater markets significantly reduce water use compared with farmers who have their own tubewells. Little effect, however, is found on agricultural productivity or yields. Findings also demonstrate that groundwater markets in the North China Plain do not have a negative effect on income.

1   INTRODUCTION

The rise of groundwater markets in the agrarian economies of the developing world has begun to attract the attention of researchers, especially those that work on South Asia and China. Over the past several years, a number of authors have documented the expansion of groundwater markets (Shah, 1993; Strosser and Meinzen-Dick, 1994; Meinzen-Dick, 1996; Mukherji, 2004; Sharma and Sharma, 2004; Shah et al., 2006; Zhang et al., 2008). In general, papers in both South Asia and China have found that there has been an increase in the prevalence of groundwater markets, as measured in the percent of tubewell owners selling groundwater and in the percent of total water pumped being sold for irrigation. For example, Shah et al. (2006) show that the share of sample villages reporting water markets varies from 9.1 to 100% and the regional aggregate is 52.2%. The share of tubewell owners selling water in villages reporting water markets varies from 1.9 to 87% and the region aggregate is 35%. The share of hours sold of total pumping hours varies from 0.2 to 69% and the region aggregate is 12.7% in 12 sample regions in India, Pakistan, Nepal Terai and Bangladesh. Zhang et al. (2008) indicates that 44% of the villages in four provinces of the northern China study, covering 68 000 villages, have groundwater markets; water is sold from 18% of tubewells in these villages and these tubewell owners sell 77% of the water pumped from their tubewells.

There has been an attempt to understand why some villages have seen groundwater markets rise and others have not. For example, in India, Shah (1993) shows that the availability of water resources, the scale of irrigation technology and land fragmentation are correlated with the rise of groundwater markets. Strosser and Meinzen-Dick (1994) set up a theoretical framework that demonstrates (among other factors) the depth of the groundwater table and the population density of a community as important factors favoring groundwater markets. Shah (1993) and many others, such as Mukherji (2004), also indicate the importance of policy interventions in promoting or constraining the development of groundwater markets. In Zhang et al. (2008), groundwater markets are found to be more prevalent in areas that are increasingly dominated by private well ownership, facing increasing water and land scarcities, and policy interventions (such as subsidy, bank loan and well-drilling permit regulation policies). Clearly, groundwater markets are becoming increasingly important institutional bodies in the economies of many developing countries and are helping to provide water and ensure access in certain regions.

Despite the interest in groundwater markets there has been a noticeable absence of work seeking to measure the impacts of groundwater markets on the rural communities that they are serving. This is a puzzling since there are many reasons, *ex ante*, to be concerned that groundwater markets may not always have positive effects on different aspects of the rural economy. For example, although farmers having access to groundwater markets get better access to irrigation services, it may lead to lower water use per hectare since they may be paying a higher price for water than those that have their own tubewells and hence cannot afford to irrigate sufficiently. If that is the case, the yields of those that access irrigation services through groundwater markets may be lower than those that have their own pump. In this way, it is possible that productivity is compromised. Moreover, as argued in Meinzen-Dick (1996), since groundwater markets might supply water to buyers only after the seller has satisfied their own needs, and since purchasing water requires a cash expenditure (though in some areas alternative institutions have arisen to eliminate the need for a cash payment), it is possible that groundwater markets may adversely affect the income of some users and negatively affect the income distribution between rich and poor. In other words, groundwater markets could easily have effects on the use of water, yields and rural incomes in different settings. Because these effects are so fundamental, it is surprising that more work has not been done.

There are exceptions, however. For example, based on descriptive statistics, Shah and Ballabh (1997) found that water buyers in all the six villages in Muzzafarpur of North Bihar in India invariably achieve higher yield than water sellers. For some crops, such as potato, water buyers even achieve nearly twice the output of water sellers (including tubewell owners). In contrast, Fujita (2004) found that tubewell owners achieve higher yield in the production of *boro* paddy based on a survey in Bangladesh. He indicated that water sellers are not conscious enough in delivering water to buyers in time and in proper volume, while water buyers have a strong tendency to refuse payment of water charges, which are the major reasons for the difference in agricultural productivities. Unlike Shah and Ballabh (1997) and Fujita (2004), Meinzen-Dick (1996) used a multivariate analysis and concluded that water from farmer's own tubewells in Pakistan gives a higher return than that purchased from others in terms of both crop yields (such as wheat) and gross margins.

In other words, groundwater markets have increased access of the poor to groundwater, presumably decreasing poverty and increasing equity.

Although of interest, there are several shortcomings in the earlier work. First, it is focussed on South Asia. Second, whether using descriptive or econometric approaches, the analyses were not able to control the possible endogeneity that is inherent in this type of analysis. Because unobserved factors might affect both outcomes (such as water use and productivity) and the rise of groundwater markets, there needs to be an effort to isolate the net effect of groundwater markets on the outcome of farmers. Perhaps for this reason, the previous literature has found that results disagree in different study areas.

The overall goal of this paper is to measure the impact of groundwater markets on water use, yields and income. It is assumed that people who irrigate get higher yields and earn higher income and it is also important to note that this is even more true for the poor (Huang et al., 2006). This paper asks: does it matter "how" people get access to water? Specifically, if farmers buy water from groundwater markets instead of drilling and pumping water from their own tubewells or from collective tubewells does this affect their water use, yields and income?

2 METHODOLOGY AND DATA COLLECTION

The data come from the China Water Institutions and Management survey (CWIM), which was conducted by Center for Chinese Agricultural Policy (CCAP), Chinese Academy of Sciences (CAS) and University of California, Davis in December 2001 and 2004. In this survey, four separate questionnaires were administered—one for farmers, one for tubewell managers, one for canal managers and one for village leaders. Enumerators collected data and information from 338 households, 110 tubewell owners, 68 canal managers, and 80 village leaders in 80 villages in three provinces (Hebei, Henan and Ningxia Provinces). Since there is almost no groundwater irrigation in Ningxia Province, only part of data from Hebei and Henan provinces was used (100 households in 35 villages). Located in the North China Plain, the two provinces face serious water shortages and have the highest extent of groundwater irrigation of any province in China (about 78% of irrigated area is from groundwater).

The scope of the surveys was quite broad. Each of the questionnaires included more than 10 sections. Among the sections, there were those that focused on the village's resource base (both the scarcity of water and the amount of cultivated land), the evolution of the ownership of tubewells, the village's basic socio-economic conditions and government policies and regulations.

In addition, there was a section that focused specifically on groundwater markets. Groundwater markets are here defined as localized, community-level arrangements through which owners of tubewells sell pump irrigation services to other farmers of the village and neighbouring villages (i.e., they sell water to other farmers from their wells for use on crops). This paper only examines "private" water markets, or those groundwater transactions that are driven by individuals and groups of individuals that sink tubewells. In adopting such a definition, we are assuming that when village leaders (the collective) provide water to villagers, it is being done under *non-market* conditions and such a transaction is not considered a groundwater market transaction.

In the section of groundwater markets, a number of questions were designed to focus on the approach by which the farmers get access to groundwater on each of their plots. During the survey, three main types of approach to access to groundwater were identified: buying water to irrigate from groundwater markets (that is one farmer buys water from another); pumping water from one's own tubewell; and getting access to groundwater from collective tubewells. Enumerators asked each respondent how they sourced groundwater to irrigate their plots for each of the sample years.

The survey also collected information that was used to develop several measures to infer the effects of groundwater markets on water use, yields and income. In order to get relatively accurate measures of water use, the strategy was to interview all of those that were involved in the irrigation scheme: farmers, water managers and village leaders. Moreover, questions asked about crop water use were asked in a number of different ways: water use per irrigation, water use per hour or the number of hours per irrigation. This allowed several measures to be used to get accurate estimates of crop water use.

Other questions were designed to collect systematically information on crop production and income by plot and by crop for each cropping season. Cropping income is an estimate of each household's full net income on each plot, including revenues and expenses. As is standard in the household economics literature, home production is used for a household's own consumption at its market price. If the household did not buy or sell a product that it consumed itself, the average price from the village is used to value the good. In addition, farmers were asked about other sources of income for the household, including that from livestock, off-farm wage labour, earnings from the family's business enterprise, and other miscellaneous sources.

3   IMPACT OF GROUNDWATER MARKETS ON WATER USE, YIELDS AND INCOME: DESCRIPTIVE STATISTICS

3.1   *Impact on irrigation water use and yields*

The CWIM survey was conducted in two rounds. The first round was in 2001 and the second round of survey was in 2004. Because wheat is the most important crop in Hebei and Henan provinces, it was selected as an appropriate crop with which to explore the impact of groundwater markets on water use and production. Descriptive statistics of the main variables are shown in Appendix Tables 1 and 2.

The field surveys show that farmers in the North China Plain have three ways to access groundwater. First, some farmers irrigate land from collective tubewells. Second, some farmers buy water from groundwater markets, i.e. from other tubewell owners. Third, some farmers dig tubewells by themselves and source irrigation from their own tubewells. The results of descriptive analysis show that nearly half of the farmers sourced irrigation from collective tubewells in both 2001 and 2004 (Table 1).

The survey results show that groundwater markets in the North China Plain possibly reduce crop water use. Compared to other types of irrigation, farmers getting irrigation through groundwater markets use less water for wheat production (Table 2, column 1). For example, if farmers buy water through groundwater markets to irrigate wheat, water use per hectare is 3408 m^3, which is lower by about 5% than the use of water from own tubewells (3571 m^3) and 16% lower than that from collective tubewells (3943 m^3). In about 40% of all villages, there is more than one way in which farmers get access to irrigation. In one sample village, some farmers irrigate wheat from their own tubewells

Table 1. Ways of getting access to groundwater for farmers.

	Share of households (%)	Share of plots (%)
Buying water from groundwater markets		
2001	26	19
2004	23	22
Pumping water from own wells		
2001	29	27
2004	30	28
Getting water from collective tubewells		
2001	45	54
2004	47	50

Data source: Authors' survey in 100 randomly selected households and 74 wheat plots in 2 provinces (Hebei and Henan) of CWIM.

Table 2. Relationship between groundwater markets and crop water use, crop yield and farmer income.

	Water use for wheat (m³/ha)	Wheat yield (kg/ha)	Cropping income per capita (yuan)	Total income per capita (yuan)
Buying water from groundwater markets	3408	4843	902	2059
Pumping water from own wells	3571	4890	1482	2491
Getting water from Collective wells	3943	5237	1168	2529

Data source: Authors' survey in 100 randomly selected households and 74 wheat plots in 2 provinces (Hebei and Henan) of CWIM.

and other farmers irrigate wheat with water from groundwater markets. When comparing the two types of farmers, those farmers getting irrigation from their own tubewells use 12% more water than farmers buying water from markets (Table 3, column 2, rows 1 and 2). In addition, in four sample villages some farmers in each village get irrigation from collective tubewells, and other farmers get access to irrigation from groundwater markets. Those farmers in these villages that use water from collective tubewell use 24% more water than farmers that buy water from groundwater markets (rows 3 and 4).

So why is it that farmers that buy water use less water? One reason maybe that farmers that purchase water pay more for their water. If so, they would have an incentive to reduce water use. They may also use their water more efficiently. The survey results show that compared with those farmers depending on own tubewells or collective tubewells, farmers irrigating crops through groundwater markets pay more for their water (Table 4). The water price paid by water buyers for wheat is 0.38 yuan per m³. This is more than two times of the cash expenditure paid by tubewell owners or the fee paid by those farmers getting irrigation from collective wells (Table 4, column 1). The survey results confirm that water buyers

Table 3.   Crop water use, crop yield and water price in different groups within villages.

		Number of villages	Water use for wheat (m³/ha)	Wheat yield (kg/ha)	Water price (yuan/m³)
Villages in which farmers either buy water or use own well	Buy water	1	4746	7500	0.09
	Own well		5332	11000	0.04
Villages in which farmers either buy water or use collective well	Buy water	4	2992	5231	0.23
	Collective well		3711	5249	0.21

Data source: Authors' survey in 74 randomly selected wheat plots in 2 provinces (Hebei and Henan) of CWIM.

Table 4.   Water price and water fee among different group.

	Water price (yuan/m³)	Water fee (yuan/hectare)
Buying water from groundwater markets	0.38	1295
Pumping water from own wells	0.15	536
Getting water from collective wells	0.15	591

Data source: Authors' survey in 74 randomly selected wheat plots in 2 provinces (Hebei and Henan) of CWIM.

pay more than other farmers who do not depend on groundwater markets for irrigation (Table 3, column 4). This reflects that with increasing water price, farmers will respond by reducing water use, which is consistent with other results (Wang, et al., 2007).

Perhaps it is not surprising that crop yields possibly fall with the decrease in water use for farmers buying water from groundwater markets. The survey indicates that if farmers buy water from markets, wheat yields are lower than for those farmers depending on either their own or collective tubewells (Table 2, column 2). For example, if farmers irrigate wheat through groundwater markets, per hectare wheat yield is 4843 kg, which is a little lower than those farmers getting irrigation from their own tubewells (only about 1%). Statistical tests (t-tests) also show that the difference is not significant (t is 0.15). Compared with those farmers depending on collective tubewells, the wheat yield of water buyers is lower by 8%, but again the difference is not significant (t is 1.67). Within the same village, comparison of the wheat yield between water buyer and other farmers who do not depend on groundwater markets for irrigation, also shows that the wheat yield of water buyers is lower than the other two kinds of farmers (Table 3, column 3).

3.2   *Impact on farmer income*

Groundwater markets may have a negative effect on the income of farmers that buy water from them. Evidence from the survey reveals that farmers buying water have lower cropping income and total income than tubewell owners and farmers using collective tubewells (Table 2, columns 3 and 4). For example, per capita cropping income for water buyers is 902 yuan, 61% of tubewell owners (1482 yuan) and 77% of those farmers getting irrigation

Table 5. Cropping income and total income in different groups within villages.

		Number of villages	Cropping income per capita (yuan)	Total income per capita (yuan)
Villages in which farmers either buy water or use own well	Buy water	2	884	969
	Own well		985	1705
Villages in which farmers either buy water or use collective well	Buy water	5	896	1510
	Collective well		1695	2809
Villages in which some farmers buy water, some use own wells, and others use collective wells	Buy water	7	1056	2117
	Own well		1411	2798
	Collective well		1166	2458

Data source: Authors' survey in 100 randomly selected households in 2 provinces (Hebei and Henan) of CWIM.

from collective tubewells (1168 yuan) (column 3). However, the descriptive statistics have not controlled the effects of other variables on farmer income, so it is premature to conclude that groundwater markets will affect the income negatively.

4 IMPACT OF GROUNDWATER MARKETS ON WATER USE, YIELDS AND INCOME: MULTIVARIATE ANALYSIS

4.1 *Multivariate empirical models*

In order to identify the impact of groundwater markets on crop water use, crop yields and farmer income, a set of econometric models have been specified. The equations use a number of control variables (Meinzen-Dick, 1996; Fujita et al., 2001; Meinzen-Dick et al., 2002; Wang et al., 2006). The first econometric model to measure the effect of groundwater markets on water use can be written as:

$$w_{ijk} = \alpha + \beta_{ijk} + \gamma C_{ijk} + \delta Z_{ijk} + \varepsilon_{ijk} \tag{1}$$

where w_{ijk} represents wheat water use per hectare for the ith plot of household j in village k.

The variables on the right hand side of equation (1) are those that explain crop water use. B_{ijk} and C_{ijk}, our variables of interest, measure the way in which farmers gain access to groundwater for irrigation. If farmers irrigate their plots by buying water from groundwater markets, B_{ijk} equals 1; otherwise, it equals 0. Similarly, C_{ijk} equals 1 if farmers irrigate their plots by pumping water from collective tubewells and equals 0 otherwise. If farmers pump groundwater from their own tubewells, both B_{ijk} and C_{ijk} equal 0.

There are also a set of control variables Z_{ijk}, to represent other factors that affect water use. Specifically, the first category of control variables includes two variables which are included to assess the effects of the village's production environment on crop water use. Variables measuring the share of irrigated area serviced by groundwater and the degree of water scarcity in the village measured as a dummy variable are also incorporated. The second category of control variables represents household characteristics including age and education of the household head. Finally, the model includes variables related to the plot area, the plot soil type and the distance from the home to the plot as a way to control for the plot's characteristics. The symbols α, β, γ and δ are parameters to be estimated and ε_{ijk} is the error term.

In equation (1), there could be an endogeneity problem in the multivariate analysis that makes the true relationship between water use per hectare on the left side and access to groundwater on the right hard to determine simultaneously, or be affected by unobserved factors. If there is endogeneity for any reason, the estimated parameters would be biased. In order to estimate consistently the parameters in equation (1) when the explanatory variables B_{ijk} and C_{ijk} are endogenous, an instrumental variable (IV) approach is used as a way of solving the problem of endogeneity. To do so, prior to estimating equation (1), a set of variables are regressed on the irrigation access to water, B_{ijk} and C_{ijk}.

$$B_{ijk} = \lambda_1 + \rho_1 IV + \phi_1 Z_{ijk} + \mu_1 \tag{2}$$

$$C_{ijk} = \lambda_2 + \rho_2 IV + \phi_2 Z_{ijk} + \mu \tag{3}$$

where the predicted value of B_{ijk} and C_{ijk} from equation (2) and (3), $\hat{B}_{ijk}$ and $\hat{C}_{ijk}$, would replace B_{ijk} and C_{ijk} in equation (1). Equation (2) and (3) include Z, which are measures of the other exogenous variables (which are the same as those in equation (1)—e.g., measures of the village's production environment and household characteristics).

This IV approach, however, is only valid if the variables in the IV matrix in equation (2) and (3) have two properties: (a) IV must be exogenous, that is, it must be uncorrelated with the error term of equation (1); (b) it must be partly correlated with the endogenous explanatory variables. In many econometric problems, finding a variable with these two properties is challenging. The key instrumental variables in equations (2) and (3) that we use to solve the endogeneity problem are two variables that measure the policy intervention in drilling tubewells in village k. The first variable, fiscal subsidies for tubewells, is a dummy variable equal to one if there was a programme of fiscal investment in the village that targeted tubewell construction (and zero otherwise). This government programme, run by the local Bureau of Water Resources, is primarily targeted at individuals. The second instrumental variable, bank loans for tubewells, is also a dummy variable to control whether or not there was a programme through banks that gives preferential access to low interest rate loans for investing in tubewells. Unlike the fiscal subsidy programme, most bank loan programmes target local villages and leaders, and loans are supposed to be used for investment into collective wells. There is no reason to believe, however, that these policy interventions will have any independent effect on the crop water use.

Descriptive statistics and regression results suggest that the choices of instrumental variables (IVs) are satisfactory. First, looking for a variable that is correlated with the endogenous variable (B_{ijk} and C_{ijk}) but is not correlated with the outcome variable (W_{ijk}) except through its impact on B_{ijk} and C_{ijk}, requires correlations between the IVs and the unobservables that are causing the endogeneity. By definition, of course, this is impossible. But if any of the unobservables are correlated with the variables that are observed (and included as control variables in the analysis), one way to examine the validity of the IVs is to see if there is any correlation between the IVs and the control variables. In Appendix Table 4, dividing the sample into those villages that do not have fiscal subsidies for tubewell investment (column 1) and those that have (column 2), shows little difference in the level of the control variables (rows 1 to 9). The same is true when by dividing the sample into those villages in which there is not a programme through banks that gives preferential access to low interest rate loans for investing in tubewells (column 3) and those in which there is (column 4). The field work and interactions with local Bureau of Water Resources, showed that officials believed that these government programmes were randomly basewd; village leaders and farmers were almost never aware that they can influence these programmes. The two instrumental variables, fiscal subsidies and bank loans for tubewells, are logically exogenous and should have no partial effect on water use, except through the influence on the way in which farmers gain access to groundwater. Second, the IVs are partially correlated with the endogenous variable (B_{ijk} and C_{ijk}). The regression coefficients of the IVs are statistically significant in the regression results of equations (2) and (3) (Appendix Table 3, columns 1 and 2, rows 1 and 2). In other words, our IVs are correlated with the decision of farmers to select how to get access to groundwater to irrigate.

In order to answer the question of whether groundwater markets affect crop yields, the following econometric model has been specified:

$$Q_{ijk} = a + bW_{ijk} + cX_{ijk} + dZ_{ijk} + e_{ijk} \tag{4}$$

where Q_{ijk} represents the yield of wheat from the ith plot of household j in village k (which comes from the household survey). In equation (4), yield is explained by the variable of interest, W_{ijk}, which measures water use per hectare, X_{ijk}, which measures other inputs to the production process, Z_{ijk} which holds other factors constant, including characteristics of the production environment of the village, household and plot. Agricultural production inputs include measures of per hectare use of labour (measured in man days), fertilizer (measured in cost) and expenditures on other inputs, such as fees paid for custom services. The control variables for village, household and plot characteristics are the same as for equation (1). A variable is added that represents production shocks (measured as farmer-estimated yield reduction in percentage terms on a plot due to a flood, drought or other "disaster"). The symbols a, b, c and d are parameters to be estimated and e is the error term.

The impact of groundwater markets on crop yields is measured through the water use variable. If the regression results of equation (1) show that buying water from groundwater markets will make farmers reduce water use, and production responds positively to water use from the regression results of equation (4), then it can be deduce that buying water from groundwater markets will have an effect in the opposite direction on production.

In order to measure the effect of groundwater markets on income, it is necessary to establish the following econometric model to examine the relationship between income and other factors:

$$y_{jk} = \pi + \sigma B_{jk} + \omega C_{jk} + \psi Z_{jk} + \xi_{jk} \tag{5}$$

where y_{jk} represents either cropping or total income per capita for household j. The variables, B_{jk} and C_{jk}, are the same as in equation (1). Z_{jk}, is a set of control variables representing other factors that affect farmer income. Specifically, the first category of control variables measuring the village's production environment are the same as in equations (1) and (4). The second category of control variables represents household characteristics including age and education of the household head, and arable land area per capita of the household. π, σ, ω and ψ are parameters to be estimated and ξ is the error term.

In order to estimate consistently the parameters in equation (5) when the explanatory variables B_{jk} and C_{jk} are endogenous, the same method of instrumental variables as in equation (1) is used to solve the problem of endogeneity.

5   RESULTS

5.1   *Impact of groundwater markets on water use and yields*

In estimating equation (1) with the survey data, the econometric estimation performs well (Table 6, column 1). Most of the coefficients of the control variables have the expected signs and a number of the coefficients are statistically significant. For example, we find that after holding constant other factors, households that are in villages with less share of irrigated area serviced by groundwater resources use more water per hectare.

The econometric estimation also performs well when estimating the impact of groundwater markets on crop yields in equation (4) (Table 6, column 2). The Chi-square is 158.21. A number of the coefficients of the control variables are of expected sign and statistically significant. For example, the coefficient on the variable of production shock is negative and significant (Table 6, column 2, row 17). This means that droughts and floods negatively influence agricultural production.

More importantly, the results show that water use is significantly reduced for farmers that buy water from groundwater markets compared with those that have their own tubewells or that use collective wells. The coefficient of the variable of buying water from a private tubewell is negative and significant (Table 6, column 1, row 1). All other things held constant, farmers who buy water from groundwater markets use less water for wheat than tubewell owners. This result is consistent with the descriptive statistics. One explanation of this is that farmers that buy water from groundwater markets have more incentive to reduce crop water use since they pay more for their water. For tubewell owners, since their cash outlay for their water is relatively low, they have less incentive to save water. Such results also indicate the importance of increasing the water price to reduce crop water use.

The results show that although water use per hectare falls for farmers that buy water from groundwater markets, yields do not fall significantly. The coefficient on the water use variable is positive, but not significant (Table 6, column 2, row 3). However, from the regression results in the wheat water use model, when farmers buy water from groundwater markets, they reduce water use per hectare (column 1, row 1). This means that after holding other factors constant, even when groundwater markets reduce crop water use significantly, wheat yields will not be negatively influenced. The field work suggests that this is because those that buy water do not waste it because that they are paying more for their water.

5.2   *Impact of groundwater markets on income*

The econometric estimation also performs well when estimating the impact of groundwater markets on cropping income and total income in equation (5) (Table 6, columns 3 and 4). The Chi-square is from 24.75 to 42.21. A number of the coefficients of control variables are of the expected sign and statistically significant. For example, the coefficient on the

Table 6. Regression analysis of the impact of groundwater markets on crop water use, crop yield and farmer income.

	Log of water use per hectare for wheat	Log of wheat yield in kg per hectare	Cropping income per capita	Total income per capita
Groundwater markets				
Buying water from private tubewell	−3.501 (1.73)*		84.249 (0.05)	−718.512 (0.34)
Using water from collective tubewell	−2.552 (1.68)*		2,305.948 (1.51)	861.595 (0.44)
Production inputs				
Log of water use per hectare		0.036 (0.84)		
Log of labor use per hectare		−0.121 (2.85)***		
Log of fertilizer use per hectare		0.143 (2.97)***		
Log of value of other inputs per hectare		0.086 (2.17)**		
Production environment				
Share of village irrigated area serviced by groundwater	−0.662** (2.00)	0.274*** (2.63)	437.095 (0.74)	169.110 (0.23)
Village water scarcity indicator variable	0.109 (1.23)	−0.020 (0.49)	−102.536 (0.34)	−215.973 (0.56)
Household characteristics				
Age of household head	0.107 (2.00)**	−0.008 (0.46)	22.576 (0.31)	54.391 (0.60)
Age of household head, squared	−0.002 (2.30)**	0.000 (0.49)	−0.053 (0.07)	−0.384 (0.37)
Education of household head	−0.068 (1.16)	0.006 (0.65)	−59.787 (1.19)	42.633 (0.67)
Area of plot	−4.386 (1.02)	−0.360 (1.75)*		
Arable area per capita of household			9,412.560 (3.69)***	6,123.917 (1.89)*
Plot characteristics				
Loam soil	−0.070 (0.42)	0.042 (0.78)		
Clay soil	0.230 (1.41)	0.084 (1.71)*		
Distance to home	0.435 (1.70)*	−0.075 (1.80)*		
Production shocks				
Yield reduction due to production shocks		−0.015 (11.27)***		
Constant	9.956 (5.67)***	7.146 (10.89)***	−2,017.856 (1.19)	−644.721 (0.30)
Observations	148	148	200	200
Chi²	21.27	173.07	42.21	24.75
R²	0.01	0.57	0.10	0.09

Absolute value of z or t statistics in parentheses; *significant at 10%; **significant at 5%; ***significant at 1%.

variable of arable land per capita is positive and significant. This means that the larger the arable land that the households have, the higher the farmer income (both cropping income and total income).

Importantly, research results demonstrate that groundwater markets in the North China Plain do not have a negative effect on income. In the cropping income and total income equations, the coefficients on the variable of buying water from private tubewell are all not statistically significant (Table 6, columns 3 and 4, row 1). The results imply that when holding other factors constant, compared with tubewell owners, income of water buyers will not be lower. Zhang, et al. (2008) report that groundwater markets have provided water access to poorer farmers, and that households that buy water from groundwater markets are poorer than water-selling households. Combining the two pieces of research, which are from the same region of China, it can be concluded that groundwater markets in the North China Plain have made a great contribution to improving the welfare of the poor in rural areas.

In summary, groundwater markets help farmers get access to water (i.e. they do not need to invest in their own well), and when they do so, water use is reduced, but crop yields and income will not be negatively influenced. With increasing water scarcity, increasing water use efficiency is an important issue and has been addressed by policy makers. Therefore, the emergence of water markets is an effective way to provide irrigation services.

6   CONCLUSIONS

The impact of groundwater markets on water use, yields and income is significant in providing groundwater for irrigation in the North China Plain. The survey data show that when farmers buy water from groundwater markets, they use less water than those that have their own tubewells or that use collective wells. However, crop yields do not fall at the same time. In addition, the results show that groundwater markets in the North China Plain do not have negative effect on income.

The results demonstrate that groundwater markets are an effective way to provide irrigation services. However, many other environmental, social and economical effects still need to be investigated. For example, it is possible that groundwater markets could accelerate the fall of the groundwater table. If this is so, the question remains whether groundwater markets should be encouraged. In order to avoid damaging the environment, instead of directly trying to suppress groundwater markets, alternative policies that would control the drawdown of the water table (e.g., water pricing policies) should be promoted. In addition, efforts to allow groundwater markets to proliferate could help spread the benefits that come with greater access to irrigation as long as gaining access to irrigation positively affects yields and income.

ACKNOWLEDGEMENT

The authors would like to thank Steve Beare, Amelia Blanke, Anna Heaney, Mark Giordano, Tushaar Shah, Aditi Mukherji, Yang Liu, Karen G. Villholth and Jorge García for their insights and helpful suggestions. We also would like to thank the useful comments of four anonymous reviewers. We acknowledge financial support from the Knowledge Innovation

Program of the Chinese Academy of Sciences (KSCX2-YW-N-039), the National Natural Sciences Foundation (70733004) in China, the International Water Management Institute, the Food and Agriculture Organization of the United Nations, the Comprehensive Assessment of Water Management in Agriculture, and the Alcoa Foundation's Conservation and Sustainability Fellowship Program. We also acknowledge the support from the Program of China Women Economists. Scott Rozelle is a member of the Giannini Foundation.

REFERENCES

Fujita, Y., Hayami, Y., and Kikuchi, M. (2001). The conditions of collective action for local commons management: the case of irrigation in the Philippines. Paper presented at World Bank seminar, Washington DC, September 1 in India (Andhra Pradesh). IMT case study presented at FAO/INPIM International E-mail Conference on Irrigation Management Transfer, June–October 2001, available at ftp://ftp.fao.org/agl/aglw/.

Fujita, K. (2004). Transformation of Groundwater Market in Bengal: Implications to Efficiency and Income Distribution. The 18th European Conference on Modern South Asian Studies, Lund, Sweden, 6–9 July 2004.

Huang, Q., Rozelle, S., Lohmar, B., Huang, J., and Wang, J. (2006). Irrigation, Agriculture performance and poverty reduction in China. *Food Policy*, 31, (1), 30–52.

Meinzen-Dick, R. (1996). *Groundwater Markets in Pakistan: Participation and Productivity*, Research Reports 105, International Food Policy Research Institute, Washington, D.C.

Meinzen-Dick, R., Raju, K.V., and Gulati, A. (2002). What affects organization and collective action for managing resources? Evidence from canal irrigation systems in India. *World Development*, 30 (4): 649–666.

Mukherji, A. (2004). Groundwater Markets in Ganga-Meghna-Brahmaputra Basin: Theory and Evidence. *Economic and Political Weekly*, 39 (31): 3514–3520.

Shah, T. (1993). *Groundwater Markets and Irrigation Development: Political Economy and Practical Policy*. Bombay: Oxford University Press.

Shah, T., and Ballabh, V. (1997). Water Markets in North Bihar: Six Village Studies in Muzaffarpur District. *Economic and Political Weekly*, 32 (52): A183–A190.

Shah, T., Singh, O.P., and Mukherji, A. (2006). Some aspects of South Asia's groundwater irrigation economy: analyses from a survey in India, Pakistan, Nepal Terai and Bangladesh. *Hydrogeology Journal*, 14 (3): 286–309.

Sharma, P., and Sharma, R. (2004). Groundwater Markets Across Climatic Zones: A Comparative Study of Arid and Semi-Arid Zones of Rajasthan. *India Journal of Agricultural Economics*, 59 (1), 138–150.

Strosser, P., and Meinzen-Dick, R. (1994). Groundwater Markets in Pakistan: An Analysis of Selected Issues. In M. Moench (Ed.), *Selling Water: Conceptual and Policy Debates over Groundwater Markets in India*. Ahmedabad: VIKSAT-Pacific Institute-Natural Heritage Institute.

Wang, J., Xu, Z., Huang, J., and Rozzelle, S. (2006). Incentives to Managers or Participation of Farmers in China's Irrigation System: Which Matters Most for Water Savings, Farmer Income, and Poverty? *Agricultural Economics*, 34: 315–330.

Wang, J., Huang, J., Rozelle, S., Huang, Q., and Blanke, A. (2007). Agriculture and Groundwater Development in Northern China: Trends, Institutional Responses, and Policy Options, *Water Policy*, 9, No S1, pp: 61–74.

Zhang, L., Wang, J., Huang, J., and Rozelle, S. (2008). Groundwater Markets in China: A Glimpse into Progress, *World Development*, forthcoming. doi:10.1016/j.worlddev.2007.04.012.

APPENDIX

Appendix Table 1.   Descriptive statistics for major variables used for analyzing the impact of groundwater markets on water use and crop yield.

	Mean	Standard deviation	Minimum	Maximum
Dummy of buying water from private tubewells	0.20	0.40	0	1
Dummy of using water from collective tubewell	0.52	0.50	0	1
Dummy of fiscal subsidies for tubewell investment	0.22	0.42	0	1
Dummy of bank loans for tubewell investment	0.07	0.26	0	1
Share of village irrigated area serviced by groundwater	0.93	0.19	0	1
Village water scarcity indicator variable	0.35	0.48	0	1
Age of household head (year)	45	9	12	65
Education of household head (in years)	7	3	0	15
Area of plot (hectare)	0.17	0.13	0.02	0.60
Water use per hectare (m^3/hectare)	3731	1665	741	10120
Labour use per hectare (days/hectare)	121	63	20	356
Fertilizer use per hectare (yuan/hectare)	1304	584	381	3699
Value of other inputs per hectare (yuan/hectare)	1715	1551	366	14211
Dummy of loam soil	0.27	0.45	0	1
Dummy of clay soil	0.46	0.50	0	1
Distance to home (km)	0.76	0.52	0.01	2.00
Yield reduction due to production shocks (%)	6	14	0	90
Wheat yield per hectare (kg)	5051	1202	469	7500

Data source: Authors' survey.

Appendix Table 2.   Descriptive statistics for major variables used for analyzing the impact of groundwater markets on farmer income.

	Mean	Standard deviation	Minimum	Maximum
Dummy of buying water from private tubewells	0.25	0.43	0	1
Dummy of using water from collective tubewell	0.46	0.50	0	1
Dummy of fiscal subsidies for tubewell investment	0.19	0.39	0	1
Dummy of bank loans for tubewell investment	0.10	0.30	0	1
Share of village irrigated area serviced by groundwater	0.93	0.20	0	1
Village water scarcity indicator variable	0.32	0.47	0	1
Age of household head (year)	46	10	8	65
Education of household head (year)	7	3	0	15
Arable area per capita of household (hectare)	0.13	0.08	0.01	0.50
Household cropping income per capita (yuan)	1195	933	36	4951
Household total income per capita (yuan)	2403	1669	241	8723

Data source: Authors' survey.

Appendix Table 3.   First stage regression analysis of impact of groundwater markets on crop water use, crop yield and farmer income.

	If buy water from private tubewell (Plot level)	If use water from collective tubewell (Plot level)	If buy water from private tubewell (Household level)	If use water from collective tubewell (Household level)
Instrument variable				
Dummy of fiscal subsidies for tubewell investment	0.152 (3.42)***	−0.115 (2.49)**	0.179 (2.14)**	−0.050 (0.52)
Dummy of bank loans for tubewell investment	−0.111 (2.04)**	0.253 (4.46)***	−0.013 (0.13)	0.216 (1.83)*
Production environment				
Share of village irrigated area serviced by groundwater	0.165 (1.86)*	−0.251 (2.72)***	−0.221 (1.44)	−0.057 (0.32)
Village water scarcity indicator variable	−0.008 (0.32)	−0.009 (0.33)	0.113 (1.65)*	−0.088 (1.10)
Household characteristics				
Age of household head	−0.013 (0.82)	0.018 (1.09)	0.019 (0.91)	−0.008 (0.34)
Age of household head, squared	0.000 (1.39)	−0.000 (1.75)*	−0.000 (0.60)	0.000 (0.13)
Education of household head	0.003 (0.19)	−0.002 (0.09)	−0.007 (0.61)	0.026 (2.08)**
Area of plot	−0.375 (0.36)	−0.971 (0.90)		
Arable area per capita of household			−0.676 (1.77)*	−1.020 (2.30)**
Plot characteristics				
Loam soil	−0.059 (1.36)	0.133 (2.95)***		
Clay soil	0.010 (0.22)	0.056 (1.19)		
Distance to home	0.017 (0.29)	0.107 (1.81)*		
Constant	0.071 (0.20)	0.756 (2.01)**	−0.019 (0.04)	0.781 (1.56)
Observations	148	148	200	200
Chi	41	69	22	20

Appendix Table 4.   The relationship between IV and other control variables in equation (1).

	Dummy of fiscal subsidies for tubewell investment		Dummy of bank loans for tubewell investment	
	No	Yes	No	Yes
Share of village irrigated area serviced by groundwater	0.92	0.98	0.93	1.00
Village water scarcity indicator variable	0.26	0.67	0.34	0.55
Age of household head	44.97	46.48	45.62	41.36
Education of household head	7.06	5.82	6.69	8.00
Area of plot	0.18	0.15	0.18	0.16
Loam soil	0.22	0.45	0.27	0.27
Clay soil	0.53	0.21	0.47	0.27
Distance to home	0.75	0.79	0.76	0.80
Yield reduction due to production shocks	5.04	11.38	6.58	4.91

Data source: Authors' survey in 74 randomly selected wheat plots in 2 provinces (Hebei and Henan) of CWIM.

Subject index

Author index

SERIES IAH-Selected Papers

Volume 1–4 Out of Print

5. Nitrates in Groundwater
   Edited by: Lidia Razowska-Jaworek and Andrzej Sadurski
   2005, ISBN Hb: 90-5809-664-5

6. Groundwater and Human Development
   Edited by: Emilia Bocanegra, Mario Hérnandez and Eduardo Usunoff
   2005, ISBN Hb: 0-415-36443-4

7. Groundwater Intensive Use
   Edited by: A. Sahuquillo, J. Capilla, L. Martinez-Cortina and X. Sánchez-Vila
   2005, ISBN Hb: 0-415-36444-2

8. Urban Groundwater – Meeting the Challenge
   Edited by: Ken F.W. Howard
   2007, ISBN Hb: 978-0-415-40745-8

9. Groundwater in Fractured Rocks
   Edited by: J. Krásný and John M. Sharp
   2007, ISBN Hb: 978-0-415-41442-5

10. Aquifer Systems Management: Darcy's Legacy in a World of
    impending Water Shortage
    Edited by: Laurence Chery and Ghislaine de Marsily
    2007, ISBN Hb: 978-0-415-44355-5

11. Groundwater Vulnerability Assessment and Mapping
    Edited by: Andrzej J. Witkowski, Andrzej Kowalczyk and Jaroslav Vrba
    2007, ISBN Hb: 978-0-415-44561-0

12. Groundwater Flow Understanding – From Local to Regional Scale
    Edited by: J. Joel Carrillo R. and M. Adrian Ortega G.
    2008, ISBN Hb: 978-0-415-43678-6

13. Applied Groundwater Studies in Africa
    Edited by: Segun Adelana, Alan MacDonald, Tamiru Alemayehu
    and Callist Tindimugaya
    2008, ISBN Hb: 978-0-415-45273-1

14. Advances in Subsurface Pollution of Porous Media: Indicators,
    Processes and Modelling
    Edited by: Lucila Candela, Iñaki Vadillo and Francisco Javier Elorza
    2008, ISBN Hb: 978-0-415-47690-4

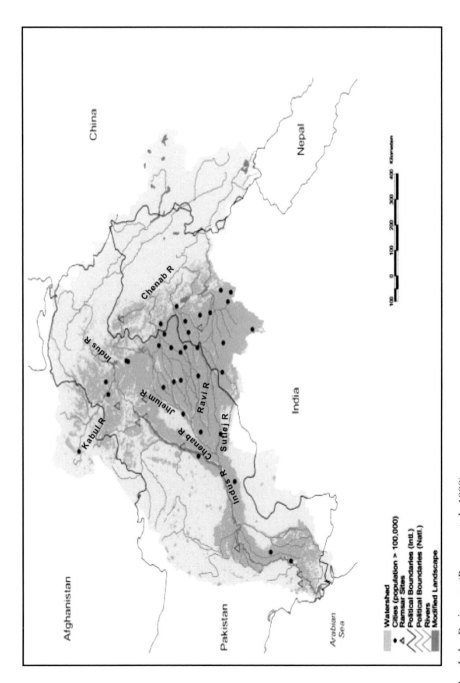

Figure 1. Indus Basin map (Revenga et al., 1998).

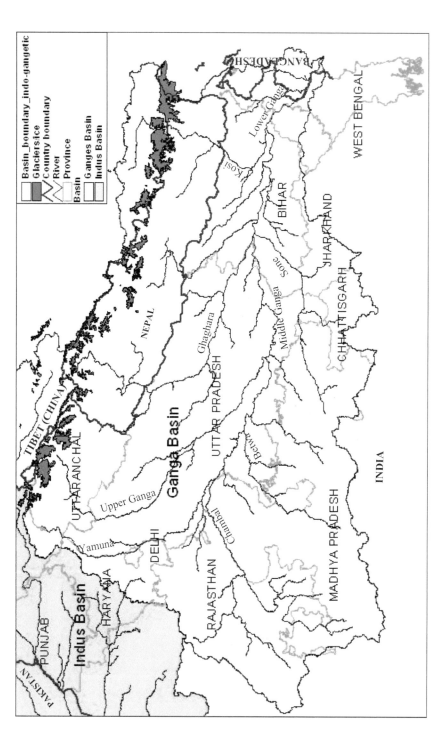

Figure 2.   Ganga Basin map (IWMI 2008, available at http://dw.iwmi.org/).

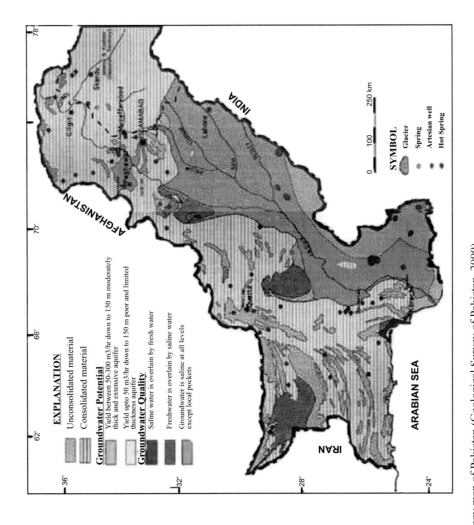

Figure 4. Hydrogeology map of Pakistan (Geological Survey of Pakistan, 2000).

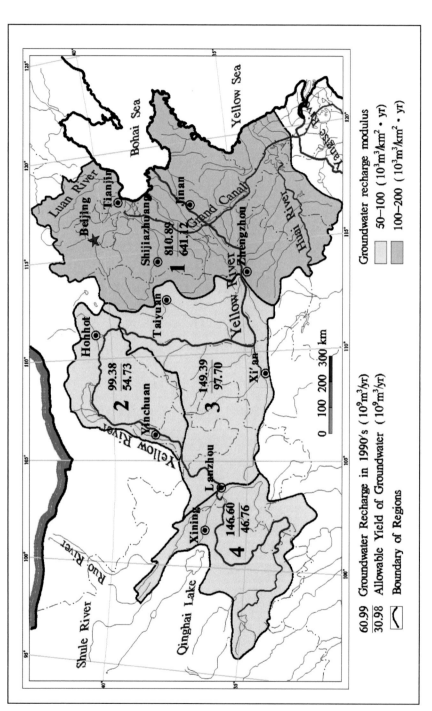

Figure 7.   Groundwater resources in the Yellow River basin and the Huang-Huai-Hai plain.

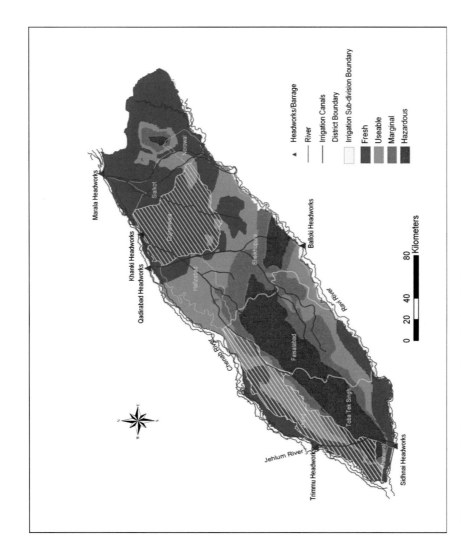

Figure 3.   Selected Zones/Districts overlaid with water quality in Rechna Doab.

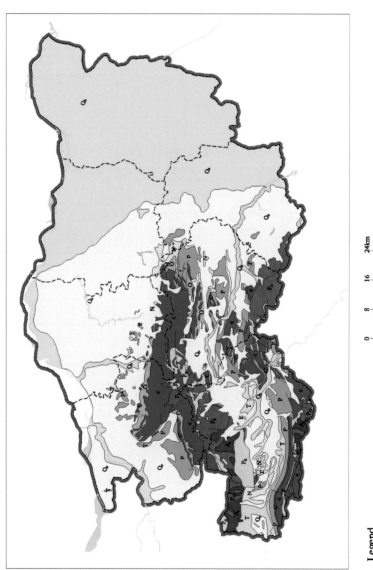

Legend

- Q$_h$ Holocene
- Q$_p$ Pleistocene
- N Neogene
- E Paleogene
- T Triassic
- P Permian
- C Carboniferous

- O Ordovician
- ∈ Cambrian
- Z Sinian
- P$_t$ Middle proterozoic
- P$_t$ Late proterozoic
- A$_r$ Archean
- Geological borderline

0   8   16   24km

Figure 4.   Geologic map of Zhengzhou.

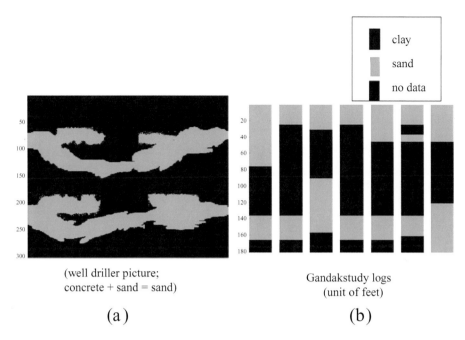

(well driller picture;
concrete + sand = sand)

Gandakstudy logs
(unit of feet)

(a)

(b)

Figure 5. Comparison of conceptual image (a) with logs provided in the Gandak report
(b) (NIH, 2000).

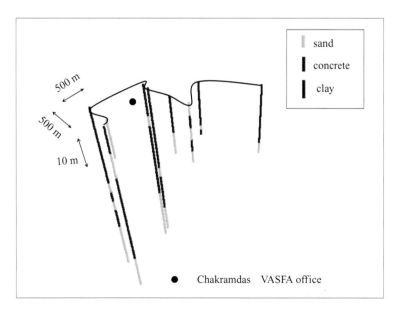

Figure 6. Collected well logs in and around Chakramdas village.

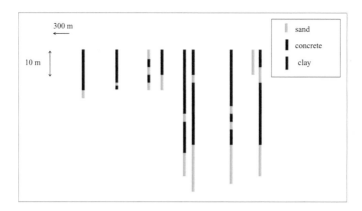

Figure 7.   2-D planar representation of the well logs in Chakramdas village.

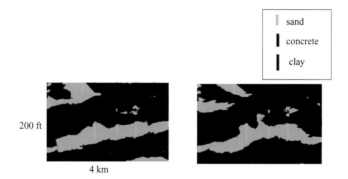

Figure 9.   Two equally likely depictions of the aquifer, anchoring the collected well logs to the training image.

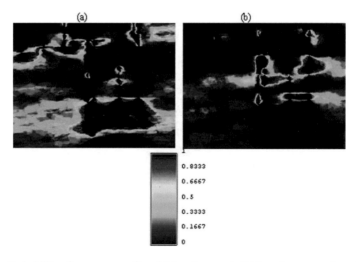

Figure 10.   Probability of occurrence of sand (a) and concrete (b) based on averaging of 15 equally likely depictions.